EAI/Springer Innovations in Communication and Computing

Series Editor

Imrich Chlamtac, European Alliance for Innovation, Ghent, Belgium

The impact of information technologies is creating a new world yet not fully understood. The extent and speed of economic, life style and social changes already perceived in everyday life is hard to estimate without understanding the technological driving forces behind it. This series presents contributed volumes featuring the latest research and development in the various information engineering technologies that play a key role in this process. The range of topics, focusing primarily on communications and computing engineering include, but are not limited to, wireless networks; mobile communication; design and learning; gaming; interaction; e-health and pervasive healthcare; energy management; smart grids; internet of things; cognitive radio networks; computation; cloud computing; ubiquitous connectivity, and in mode general smart living, smart cities, Internet of Things and more. The series publishes a combination of expanded papers selected from hosted and sponsored European Alliance for Innovation (EAI) conferences that present cutting edge, global research as well as provide new perspectives on traditional related engineering fields. This content, complemented with open calls for contribution of book titles and individual chapters, together maintain Springer's and EAI's high standards of academic excellence. The audience for the books consists of researchers, industry professionals, advanced level students as well as practitioners in related fields of activity include information and communication specialists, security experts, economists, urban planners, doctors, and in general representatives in all those walks of life affected ad contributing to the information revolution.

Indexing: This series is indexed in Scopus, Ei Compendex, and zbMATH.

About EAI - EAI is a grassroots member organization initiated through cooperation between businesses, public, private and government organizations to address the global challenges of Europe's future competitiveness and link the European Research community with its counterparts around the globe. EAI reaches out to hundreds of thousands of individual subscribers on all continents and collaborates with an institutional member base including Fortune 500 companies, government organizations, and educational institutions, provide a free research and innovation platform. Through its open free membership model EAI promotes a new research and innovation culture based on collaboration, connectivity and recognition of excellence by community.

Luis Carlos Méndez-González •
Luis Alberto Rodríguez-Picón •
Iván Juan Carlos Pérez Olguín
Editors

Innovation and Competitiveness in Industry 4.0 Based on Intelligent Systems

Springer

EAI
RESEARCH MEETS INNOVATION

Editors
Luis Carlos Méndez-González
Industrial Engineering and Manufacturing
Universidad Autónoma de Ciudad Juárez
Ciudad Juárez, Chihuahua, México

Luis Alberto Rodríguez-Picón
Industrial Engineering and Manufacturing
Universidad Autónoma de Ciudad Juárez
Ciudad Juárez, Chihuahua, México

Iván Juan Carlos Pérez Olguín
Industrial Engineering and Manufacturing
Universidad Autónoma de Ciudad Juárez
Ciudad Juárez, Chihuahua, México

ISSN 2522-8595 ISSN 2522-8609 (electronic)
EAI/Springer Innovations in Communication and Computing
ISBN 978-3-031-29774-8 ISBN 978-3-031-29775-5 (eBook)
https://doi.org/10.1007/978-3-031-29775-5

This Springer imprint is published by the registered company Springer Nature Switzerland AG
The registered company address is: Gewerbestrasse 11, 6330 Cham, Switzerland

To our beloved families

"For being a source of inspiration,
the reason to face challenges."

Preface

Industry 4.0 has established a watershed for methodologies and techniques in different areas of science and engineering to establish a new way of producing products through intelligent factories, which use a fundamental tool for operation, the Internet of Things (IoT). Over 10 years have passed since the first proposal for an intelligent factory was made at the Hannover Messe Industrial Technology Hall. Technological advances from that day until today have involved new actors that have allowed Industry 4.0 to spread and develop with greater penetration in different product sectors. This book aims to provide the reader with new technological advances related to Industry 4.0 and intelligent systems such as Artificial Intelligence, Machine Learning, Fuzzy Techniques, and soft Computing applied to different fields of industry.

Innovation and Competitiveness in Industry 4.0 Based on Intelligent Systems are divided into three sections. In Part I, Artificial Intelligence and Fuzzy Techniques Applications in the Industry 4.0, readers will be able to find how, through Machine Learning, Augmented Reality, Mixed Reality, and diffuse technique techniques, solutions are given to problems that are not only focused on the industrial field but also on social and health aspects. The chapters presented in this section, readers will be able to learn how the use of industry 4.0 methodologies and intelligent systems can provide solutions to an infinite number of situations, and that can be taken as a basis to extend knowledge in new areas. In Part II, Analytical Strategies for Productive Processes Based on Industry 4.0, readers can find methodologies that are based on different mathematical and statistical models focused on productive processes within the industry 4.0. The chapters present case studies and analytical approaches to provide solutions to specific situations through modifications to classical techniques. The case studies presented by the authors aim to optimize resources, reduce costs, and improve product designs. Finally, in Part III, Soft Computing Application in the Industry 4.0, readers will be able to find the application of fuzzy logic techniques, neural networks, and evolutionary algorithms that, together with industry 4.0 techniques, offer a solution to problems that may be from the point of view of non-linear systems.

The editors of this book hope that this work will be well received by researchers, students, and practitioners of Industry 4.0 and intelligent systems. May this book help the readers find inspiration to improve designs or processes in any branch or even derive new research related to the topics covered in this book.

Finally, the editors of this work appreciate the efforts and dedication of each of the authors of the chapters that purchase this work. Without the results of the investigations carried out by the authors, this work's integration into the scientific community would not have been possible. However, we want to thank each of the chapter reviewers since, through knowledge in expertise, the chapters were enriched and of high quality. For all of you, thank you and our sincere appreciation.

Ciudad Juárez, Chihuahua, México — Luis Carlos Méndez-González
Ciudad Juárez, Chihuahua, México — Luis Alberto Rodríguez-Picón
Ciudad Juárez, Chihuahua, México — Iván Juan Carlos Pérez Olguín

Contents

Part I Artificial Intelligence and Fuzzy Techniques Applications in the Industry 4.0

Machine Learning and Edge Computing for Industry 4.0 Applications: Concepts and Extensive Review ... 3
Leonardo Barboni

Failure Detection System Controlled by a Mixed Reality Interface ... 21
Alan Yamir Rodríguez Gallegos, Luis Carlos Méndez-González, Alan Iván Hernández Holguín, and Luis Alberto Rodríguez-Picón

Industry 4.0 in the Health Sector: System for Melanoma Detection ... 43
Verónica Angelica Villalobos Romo, Soledad Vianey Torres Arguelles, Jose David Diaz Roman, Jesus Martin Silva Aceves, Salvador Noriega Morales, and Claudia Georgina Nava Dino

Assistive Device for the Visually Impaired Based on Computer Vision ... 71
Alan Iván Hernández Holguín, Luis Carlos Méndez-González, Luis Alberto Rodríguez-Picón, Iván Juan Carlos Pérez Olguin, Abel Euardo Quezada Carreón, and Luis Gonzalo Guillén Anaya

Part II Analytical Strategies for Productive Processes Based on Industry 4.0

Development and Evaluation of a Machine Learning Model for the Prediction of Failures in an Injection Moulding Process ... 101
A. Rojas-Rodríguez, F. S. Chiwo, H. Arcos-Gutiérrez, C. Ovando-Vázquez, and I. E. Garduño

An Approach to Select an Open Source ERP for SMEs Based on Industry 4.0 and Digitization Considering the SHERPA and WASPAS Methods ... 123
Juan Vicente Barraza de la Paz, Luis Alberto Rodríguez-Picón, Iván Juan Carlos Pérez-Olguín, and Luis Carlos Méndez-González

The Technological Role of Steepest Ascent Optimization in Industry 4.0 Modeling 145
Paulo Eduardo García-Nava, Luis Alberto Rodríguez-Picón, Luis Carlos Méndez-González, Iván Juan Carlos Pérez-Olguín, and Roberto Romero-López

The Role of Industry 4.0 Technologies in the Energy Transition: Conceptual Design of Intelligent Battery Management System Based on Electrochemical Impedance Spectroscopy Analysis 175
W. J. Pech-Rodríguez, Enrique Rocha-Rangel, Eddie N. Armendáriz-Mireles, Gladis G. Suarez-Velázquez, and L. C. Ordóñez

Performance Analysis of Eight-Channel WDM Optical Network with Different Optical Amplifiers for Industry 4.0 197
Anuj Kumar Gupta, Raju Sharma, Digvijay Pandey, Vinay Kumar Nassa, Binay Kumar Pandey, A. Shaji George, and Pankaj Dadheech

Part III Soft Computing Application in the Industry 4.0

Traffic Signs Configuration with a Geo-simulation Approach 215
Ariadna C. Moreno Román and Mailyn Moreno Espino

Emotional Diagnosis for Employees Within the Framework of Industry 4.0: A Case Study in Ciudad Juarez 243
Florencio Abraham Roldan-Castellanos, Ivan Juan Carlos Pérez Olguín, Luis Carlos Méndez-González, and Luis Ricardo Vidal-Portilla

Architecture for Initial States Algorithm for Blockchain Scalability in Local OnPrem IIoT Environments 275
Alfonso José Barroso-Barajas, Jesús Andrés Hernández-Gómez, Roberto Antonio Contreras-Masse, and Salvador A. Noriega-Morales

Distribution Route Optimization Using Floyd-Warshall Weighted Graph Analysis Algorithm with Google Maps Integration in Industry 4.0 Context 287
Uriel Ángel Gómez Rivera, Iván Juan Carlos Pérez Olguín, Luis Asunción Pérez Domínguez, Luis Alberto Rodríguez-Picón, and Luis Carlos Méndez-González

Feature Selection in Electroencephalographic Signals Using a Multicriteria Decision Analysis Method 307
Alexis Edmundo Gallegos Acosta, María Dolores Torres Soto, Aurora Torres Soto, Eunice Esther Ponce de León Sentí, and Carlos Alberto Ochoa Ortiz Zezzatti

Index 337

Part I
Artificial Intelligence and Fuzzy Techniques Applications in the Industry 4.0

Machine Learning and Edge Computing for Industry 4.0 Applications: Concepts and Extensive Review

Leonardo Barboni

1 Introduction

This article aims to introduce the academic state of the art in Industry 4.0 and to introduce several concepts as the edge computing and machine learning, as well as in which way they contribute to the implementation of Industry 4.0 and how they solve key challenges.

There is a lot of literature in journals and conferences, as well as in websites, but no official government's documentation on this topic, such as environmental standards or regulations, has been found. Neither were documents found on policies that motivate the implementation of Industry 4.0. This study of the literature shows positive and negative aspects. First, as a negative aspect, there is a lack of information from official sources, other than journals or conferences, that is, from the academic world. It has been found that the implementation of Industry 4.0 is being carried out at the initiative of businessmen/women and researchers rather than by government policies. As a positive impact (significance and contributions), the reader will finish with clear concepts and an overview of what the academy is doing toward Industry 4.0 implementation. It also has a selected database of representative articles on the subject.

The absence of documentation on the web regarding government policies makes this study based mainly on academic articles, and there is not found in literature a survey like this. In addition, the articles found are fundamentally from the year 2020 onward and are more frequently found in fast-publishing peer-reviewed conferences than in journals. For this reason, it is not possible to make a detailed comparison

L. Barboni (✉)
Dpto de Electronica, Facultad de Ingenieria, Udelar, Montevideo, Uruguay
e-mail: lbarboni@fing.edu.uy

L. C. Méndez-González et al. (eds.), *Innovation and Competitiveness in Industry 4.0 Based on Intelligent Systems*, EAI/Springer Innovations in Communication and Computing, https://doi.org/10.1007/978-3-031-29775-5_1

between works, because there are many cases of incipient or early work studies presented recently, all on different aspects of Industry 4.0.

The information search methodology is based on searching keywords in search engines such as GOOGLE, EBSCO host, IEEE, Scopus, and JSTOR. The keywords were Industry 4.0, machine learning for smart factory, smart sensors for Industry 4.0, image processing for Industry 4.0, policies for Industry 4.0, Industry 4.0 and environment, edge computing in Industry 4.0. It has been performed a systematic browsing of abstracts, conclusions, and full text in terms of technologies used, production areas, and solved problems. What is obtained from said search is reported below.

The manuscript is organized as follows: Sect. 2 introduces the concept of the Industry 4.0 or better said, what exactly it means. Section 3 refers to the idea of edge computing in Industry 4.0. Section 4 describes concepts of machine learning that are used in order to implement Industry 4.0. In Sect. 5, we discuss how image processing and computer vision are the main reported techniques that are used in Industry 4.0. Section 6 presents an overview of characteristics of sensors that Industry 4.0 needs. In Sect. 7, edge computing in Industry 4.0 is discussed again, taken into account case studies reported in literature.

Finally, the conclusion offers a final opinion and future perspectives from the author's point of view.

Next we present Sect. 2, where we establish the concept of Industry 4.0.

2 Concepts of Industry 4.0

Before discussing and comparing the literature, it must be noted that the Fourth Industrial Revolution or Industry 4.0 is an industrial revolution driven by information systems and automation technologies. It refers to the construction of digitized, no data-centric, interconnected factories that continuously monitor and control production and work as autonomously as possible.

In this sense, Industry 4.0 is moving toward automation and data exchange through information technologies (industrial Internet of things or IIoT), robots, cyber-physical systems, and processes within manufacturing lines. Industry 4.0 enables factories to monitor the entire production process and make autonomous decisions by deploying wirelessly connected machines and intelligent sensors on the production line. These technologies focus primarily on vehicle-assisted driving systems, outdoor video surveillance, sensing sensors, data fusion, and failure tracking. Additionally, all processes are moving toward event-driven processes and real-time automated plants that deliver higher productivity, less waste, higher product quality, more flexibility, and lower operating costs than not automated plants.

In Industry 4.0, machines will be connected to make manufacturing processes faster and smarter than not automated plants, allowing them to adapt to changing factory conditions and production processes. As a result, new technologies will extend the life of machines while improving product quality and reducing manufacturing costs.

This is accomplished by incorporating sensors into the production lines. Many sensors are intelligent and have a microcontroller that executes data processing algorithms based on machine learning (ML). Such sensors or systems are called edge computing or simply "fog computing" because they take place at the edge of the network, not in the cloud.

Conversely, as data volumes increase, the costs of bandwidth, data storage, computing, and data science increase. Therefore, it is important to manage a balance and minimize costs. To do so, we must create new business models and produce changes in industrial processes. Therefore, deploying edge computing is essential to the realization of Industry 4.0 (below, we explain what edge computing is). The topic is cutting edge, with dedicated conferences such as the IEEE International Workshop on Metrology for Industry 4.0 and IIoT and the IEEE Emerging Trends in Industry 4.0.

Several papers have defined Industry 4.0. All mention the placement of intelligent sensors with connectivity on the production line and production control for preventive maintenance, product quality control, handling large data volumes, and new business models. These involve minimizing the infrastructure costs of a smart factory and maximizing profits [1]. deeply discussed the concept of Industry 4.0, where the key idea is smart manufacturing used to rectify faults and monitor machines and production lines during manufacturing. It also includes smart manufacturing technologies in the industry. This paper will discuss the following key technologies: the Internet of things (IoT), cybersecurity, big data, cloud computing, automation, augmented reality, ML, and virtual reality. They have been enabled in Industry 4.0 because they are used to make the manufacturing process more intelligent. Industry 4.0 is a concept that can be applied to any production process, not limited to any factory or product. For instance, [2] discusses adopting Industry 4.0 technologies in the construction industry. Sensor connectivity is thought to be the key to ensuring that all facilities are interconnected and fully effective.

According to [3], there is an urgent need to implement Industry 4.0 in most developed countries. The authors refer to Industry 4.0 as a reconfigurable hybrid manufacturing system and mention the challenges that its implementation may face.

Finally, we recognize that this selection is somewhat subjective and that studies not included in this chapter have also made contributions. Before diving deeper, it should be noted that the selection of papers in this chapter aims to provide examples and samples of the state-of-the-art aspects related to edge computing and ML in Industry 4.0. Although the review is extensive, it does not cover everything published until now. Hence, we have tried to select the most representative articles in the field to share the ideas developed in this chapter.

3 Concepts of Edge Computing in Industry 4.0

Edge computing (also called fog computing or soft computing) refers to the installation of microcontrollers, minimal data centers (nodes or sensor nodes), and systems at the edge of the factory network (even in a corrupt environment). This allows data processing locally, close to the source, in the machine and production process, or line. For example, programmable logic controllers usually have network connection buses (Ethernet, Wi-Fi, RS485). This is not just a calculation or sensing center but also an actuator. Edge computing is a framework poised to address some new assembly and challenges in factories outside of data centers and cloud data processing (centralized data processing). It is useful in processes that require real-time action, such as quality control of generated products. It is important to consider that the traditional ones were sensing and sending data to a centralized system ("cloud computing"). The system required time-consuming data transfer and operator intervention for decision-making, and actions on the production line were sometimes not taken in real time. Edge computing can connect sensors, devices, machines, and processes without sending data to the cloud. In this way, data can be stored, processed, and analyzed locally and on machinery in real time. This has the advantage of reducing the number of processes in the system network, facilitating data management and automated control and not necessarily requiring a cloud IIoT platform. There are several advantages to using edge computing. In addition to real-time sensing and action, data are processed at the edge instead of being sent to the cloud, greatly reducing the risk of interception and manipulation (as data volumes increase, both time delay and security issues can arise). This allows factory systems to keep technology operational and information secure on the network. Edge computing also enables predictive maintenance by collecting and analyzing data from sensors attached to the equipment and making real-time decisions. By processing most of these data at the edge, information can be focused on what is most important, dramatically reducing data costs. Preventive maintenance is a critical factor in the industry. Shutting down a production line due to a machine failure or unexpected event can be costly to a factory. Therefore, preventive maintenance is a major application area of edge computing, with the most widely reported literature on data processing algorithms for prevention, fault detection, and abnormal processes in production lines.

Conversely, sending data to the cloud for analysis and decision-making results in long and unpredictable latency. Edge computing increases the opportunity to act on data. It is a desirable feature if the operation is critical or the process is running. Another critical aspect of the industry is that in the event of a failure, it is often necessary to replace parts or the entire machine for quick repair. Edge computing allows machines from different vendors to connect using a resilient hardware abstraction layer that can perform rapid local actions. This means that performance can be optimized with improved worker's safety without worrying about data analysis in the cloud. Additionally, this provides instant data analysis with greater speed (ready to use).

Industry 4.0 can be successfully implemented by adopting edge computing and the IIoT. These technologies help to adopt real-time processes and intelligent use of data by triggering local actions with reduced latency. Furthermore, because edge computing is an extension of the cloud, it is the basis for Industry 4.0.

4 Concepts of ML in Industry 4.0

ML is a branch of artificial intelligence (AI) that works with algorithms and the data machines constantly received from sensors. It is, therefore, able to learn without the need for reprogramming. In this way, it can recognize patterns and more structured and organized forms of information, independently concluding and resolving classification problems.

ML is a resource for the entire supply and production chain. Hence, applying these algorithms will be essential for Industry 4.0. The main benefits of using ML include improved planning and predictive capabilities. These solutions will give companies a clearer vision of what will happen and act accordingly in real time. Additionally, by facilitating process automation, human resources can be used for value-added tasks rather than being dedicated to repetitive tasks. Supervised ML can predict future failure and improve equipment operation in a plant to avoid unexpected production stoppages and processes in general. Predictive maintenance enables the establishment of prediction and repair protocols to prevent machine breakdown (downtime = financial loss). It also helps produce higher-quality products, maintain customer satisfaction, and reduce shrinkage and return-saving logistics costs by making the manufacturing process intelligent or unsupervised.

Moreover, computer vision is a technology that uses supervised ML algorithms on images received by cameras. It is the most widely used supervised ML technology in Industry 4.0. It considerably increases the speed and reliability of visual inspections and eliminates human factors such as inherent fatigue and biases in operating conditions. It is now easier to determine the quality of resources on the production line, identify defects that may not be visible to the human eye, and take action accordingly in real time.

Supervised ML is a remarkable concept that refers to an automatic algorithm that uses past similar cases to learn parameters and programming structures. The algorithm can automatically infer new information from new data. This is not a magical process, and as discussed below, large datasets are required to achieve useful supervised ML algorithms. Moreover, a new generation of programmers and technical operators with technical knowledge of ML is required. ML algorithms are required to install, commission, and maintain a given system. Returning to the use of algorithms, ML and quality control involve computer vision and variables measured by sensors, such as temperature, pressure, vibration, and level, according to the literature and case studies.

Furthermore, ML can handle data related to materials and human resources. Therefore, the information interpreted from the ML algorithms provides a realistic and up-to-date view of everything happening in a smart factory. This makes it possible to define effective strategies that correspond to real situations. For instance, it is possible to adjust the workload, skills, experience, and authorization of employees, as well as their position in the factory or production line, to optimize task organization.

As mentioned earlier, computer vision is the most widely used ML technique, and the remainder of this section will focus on this aspect. A point related to previous ones is that sensors and cameras can identify the aspects to classify the products based on the measured parameters. Depending on the presence or absence of defects recognized by computer vision in a product or machine, some products will be directly marketed, whereas others with more defects will be discarded. Therefore, supervised ML is not a lucrative technology for controlling industrial operations. Instead, this is a way to implement smart factories.

Creating a database is one of the main problems in supervised ML techniques. For example, for the automatic visual detection of faults in a product or failure in the production line or machine, an extensive database of faults is needed to train the supervised ML model to achieve online inference and rule out defective products. Therefore, inference through supervised ML models requires large databases. However, studies have been conducted to reduce and apply them based on the needs of Industry 4.0. For example, large-scale image databases are unavailable for defective products or misbehaving machines.

The industry's predominant supervised ML technique for image analysis is called deep learning. It is based on deep neural networks (multilayers) and convolutional neural networks (CNNs). In this sense, [4] provides the basis for various use cases of industrial deep learning. The authors discuss how deep learning plays an essential role in Industry 4.0 manufacturing lines, with multiple ML-based image processing products currently on the market. Furthermore, the authors propose generative adversarial networks, generating a handful of images to train other CNNs and making them ideal candidates for industrial applications where data are scarce and confidential.

Neural networks are the most widely used ML technology, not only for visual inspection but also for other tasks. The remainder of this chapter will provide examples from the literature on the use of artificial neural networks (ANNs) in Industry 4.0 and explain the benefits according to the authors of the works.

For example, [5] presents the diagnostic trouble code (DTC) as a fundamental diagnostic indicator for machines. It is used to understand the mechanics between components and associated failure. A novel technique called sequential DTC vector embedding is proposed on the basis of recurrent neural networks and CNNs. Empirically proven results are presented in vehicle diagnostic systems.

Furthermore, [6] explains the roles of employees, which may change significantly, and how to perform control and decision-making tasks. In this work, neural networks are used to identify human factors as intelligent systems to model skills for human resource management in a production line.

The article [7] presents a methodological approach for the cutting process in a kitchen manufacturing line. The implemented ANN can predict and control defects.

Thus, the implemented ANN is suitable for predictive maintenance of kitchen production lines.

Additionally, [8] summarized the theory of CNNs and showed how they could be applied to Industry 4.0 challenges. But defect and anomaly detection for maintenance appears to be the killer application.

A question may arise when reading the literature. Can ML techniques be applied to industries of any size? However, its implementation is costly. What, then, are the cost-benefits? In this sense, [9] explains the problem of small- and medium-sized factories that are unable to computerize and automate their factories because of inadequate capital and factory scale. With this motivation, this paper presents an approach that uses triaxial sensors to assist in machine monitoring. The data obtained are analyzed for abnormalities using neural networks, achieving low costs and easy implementation.

As another application, [10] presents the detection of tiny visual defects in die-cast aluminum mechanical components. The analytical tools were based on ANN on images. It described the dataset creation, memory, computational requirements, and an implemented version of the technique that achieved 75% accuracy.

The article [11] explains how a combination of macronutrients and micronutrients results in optimal soil health and how a deficiency of any of these nutrients affects plant's health and longevity. In paper [11], a back-propagation neural network with trapezoidal fuzzy numbers is applied to optimize the exact nutrient requirements in the soil. The optimal amount of macronutrients and micronutrients applied is calculated until the maximum yield of the cultivar is obtained.

The cost-benefit is largely determined by the dedicated computer that must be placed on the production line to run the ML algorithm. In this case, we deal with neural networks. However, reading the application examples in the literature may raise another question. Is it necessary to deploy such dedicated computers, or can cost be reduced by implementing neural networks on smaller, lower-cost hardware platforms?

In this way, [12] introduces efforts by open communities such as tiny ML (TinyML) implementations. They are currently focusing on a deep learning framework for ultralow power neural networks and related applications targeting microcontrollers rather than computers. The work [12] aims to compare two deep learning frameworks. One of them is TensorFlow, which is cataloged as one of the most widely used deep learning tools by the industry community aimed at deploying ML in the field. The strengths and weaknesses of the framework are discussed when using a hybrid binary neural network in Industry 4.0 anomaly detection, achieving up to 98.6% accuracy. Training a neural network is complex and requires a huge amount of case datasets for training, which is expensive. However, the engineering and scientific community is working on such problems (fast and low-cost training). In this sense, [13] introduces how neural networks can be used to solve a discrete multi-nomenclature production management problem that can be solved by numerical minimization. Each new task of optimal selection of parameter production requires building on previously obtained solutions in the cumulative training of neural networks throughout their life cycle (knowledge accumulation).

Finally, [14] discusses the great interest in developing intelligent sensors in the context of Industry 4.0. The article [14] shows how AI can process vibration data to determine the current state of a machine. It simplifies the development of such special-purpose processing and simultaneously brings it as close as possible to the sensor (Edge-AI). This paper uses a MEMS capacity accelerometer to sense the vibration of a system. It also demonstrates using the tools and IDE STM32Cube to build and train deep neural networks using an open-source deep learning framework and code library for the microcontrollers.

5 Image Processing and Computer Vision

As this status survey showed, computer vision image processing and supervised ML technologies have an important state in Industry 4.0 and are receiving particular attention. Several representative case studies can be found in the literature. The work [15] applied a supervised ML technique to the mining industry to analyze geoscientific data using cloud computing technology. The authors in [16] show an image vision methodology suitable for the quality monitoring of a tire assembly factory. The proposed model is suitable for Industry 4.0. The system integrates a camera system with a laser function to control the exact tire positions. Furthermore, [17] investigates the application of a deep learning algorithm in weather detection problems, particularly foggy weather conditions. The system is used as a vehicle-assisted driving system in an outdoor factory. The authors in [18] focus on production traceability, image vision quality inspection, and predictive maintenance in Industry 4.0 implementing IoT devices (edge computing). The application is a facility in the pasta industry that integrates information from silo capacity, pollution aspiration systems, production traceability, in-line pasta quality analysis, and energy monitoring to improve pasta production as opposed to traditional machines or line production. Elsewhere, [19] introduces glasses that integrate cloud datasets and technology lines to speed up production and assist process operators it presents. The algorithm used is an ANN algorithm belonging to the deep learning family, capable of recognizing the current occupancy of storage trays in a production line. The authors in [20] also describe the advantages of using image processing for the presence or absence of inspections on a factory production line. This technique uses Faster R-CNN, TensorFlow's object detection API, and computer vision to detect the presence or absence of an industrial part. It has been claimed that such a solution offers a cost-effective method. In [21], an image segmentation technique is used to estimate the defect extent of manufactured parts (defected holes and tolerance gaps in fabricated parts). Image segmentation sensitivity was provided by varying the camera distance from the sample. The work [22] also presents an optical inspection system using standard hardware suitable for Industry 4.0 and the implementation of IoT devices to control the manufacturing processes. The images must be preprocessed to enable inspection of weld seams of different sizes so that a deep neural network can be used. The authors of [22] highlight a fast decision process that can be used directly on the manufacturing line. In [23], as another example, the authors proposed an approach that uses computer vision technologies such as image processing, analysis, and human–machine interactions with a camera embedded in a robot. Finally, [24] argues that in the Industry 4.0 vision, humans in smart factories should be equipped with remarkable communication capabilities. Despite advances in deep neural network algorithms, there remains a need to develop high-performance deep learning models that achieve greater accuracy. Here, we compare three models: long short-term memory (LSTM), CNNs, and a hybrid CNN-LSTM. Experiments show that the hybrid model CNN-LSTM performs better than LSTM or CNN in classifying human activities.

6 Sensors in Industry 4.0

The realization of smart factories requires complex algorithms and hardware platforms (computers) for execution and the development of intelligent, low-cost sensors suitable for Industry 4.0. Additionally, sensors provide data to be processed and are the core of Industry 4.0. For this reason, we will briefly review the current status of how the development of sensors will drive Industry 4.0.

In this sense, [25] describes an Industry 4.0 scenario for predictive maintenance based on triaxial accelerometer sensors and AI platforms for monitoring machines. The goal is to evaluate the lifetime of the equipment.

Specialized hardware design is often desired. The article [26] presents a lossless compression algorithm for signals and images applied to the IoT and its chip design to respond to this problem. Sensing thermal, photographic, gas, and image signals are important in IoT and Industry 4.0. Therefore, lossless signal compression techniques have been proposed on the basis of adaptive prediction and hybrid entropy coding. To meet the miniaturization and low power consumption requirements of IoT (edge computing), the proposed algorithm was realized using very large-scale integration (VLSI) technology and was designed and synthesized using the TSMC 0.18-μm CMOS process. The proposed algorithm offers advantages such as higher compression rates and lower memory requirements than JPEG-LS-based designs.

In [27], an infrared thermography approach is applied to monitor fresh foods (meat) in a refrigerator. The K-means algorithm is applied to the local processing of radiometric images to highlight oxidation effects and infer the behavior that defines food quality.

In [25] also presents an Industry 4.0 scenario for predictive maintenance based on a triaxial accelerometer sensor and an AI platform for machine monitoring. The scenario aims to assess the lifetime of equipment by monitoring the machine's operation over time.

Sensors for Industry 4.0 are usually not classical but have some levels of intelligence (integrated circuits) or are built for a specific purpose in a production line. The question then becomes how to proceed with calibration. In [28], the authors assume that Industry 4.0 requires ISO calibration standards for electronic equipment. This study [28] proposes a new calibration method that uses a worldwide network for efficient services and communications with clients.

To meet the needs of smart factories, [29] proposed a novel printing method: aerosol jet printing to produce piezoelectric force sensors that meet the requirements of various custom applications. This technology is inexpensive and can be quickly integrated into machinery and factory processes.

As for the devices used in daily life, [30] introduced an optical monitoring system that uses Bluetooth wireless transmission and smartphone technology to instantly monitor production status.

The article [31] also analyzed printed electronics (PE – aerosol printing) technology to fabricate customized sensors. This can capture data on production lines, health equipment, tools, and environmental conditions. The authors emphasize that to realize a smart factory, sensors must be placed everywhere in industrial facilities. They also provided examples of sensors fabricated using PE.

Moreover, [32] describes the integration of thin film SAW resonators directly into industrial metals in the field of structural health monitoring.

The article [33] refers to optical sensing technologies in the new paradigm, Industry 4.0. The emphasis is on fiber optic sensing and, where feasible, fiber Bragg grating sensing technology.

The work [34] argues the problem of machine communication for cooperation and product quality monitoring in an Industry 4.0 scenario. An optical access network is then presented, providing a networking framework for smart factories.

7 Edge Computing in Industry 4.0

First, the concept of edge computing (also called fog computing) must be defined. In Industry 4.0, the data processing chain begins with the sensors on the production line or machinery and continues toward the nodes at the edge of the network. These nodes communicate with each other and provide communication between the machine or production line and the network in the cloud. Data are often sent directly to the cloud for processing, but latency is generally high. To avoid this and facilitate real-time and local actions on production lines and machines, the algorithms are intended to work on devices located at the edge of the network immediately after the sensors that provide the data. This does not mean there will be no communication or processing in the cloud, but these will be more structured forms of information that are not associated with real-time actions on production lines, processes, and machinery. Hence, the concept of edge computing is about implementing algorithms on these node devices at the edge of the network. This is because these nodes are devices with limited computing resources (often based on microcontrollers rather than dedicated computers) and cannot run any ML algorithm. Instead, they can run ML algorithms tailored to the hardware resources and processes controlled by the production line nodes. However, compared with the latency of communicating with more powerful computers in the cloud, the advantage of being located on a production line or machinery and taking action in real time is worth the network security issues within a company, factory, or industry. ML algorithms, also known as edge computing or fog computing, are called TinyML and are the subject of research and a source of cutting-edge technology.

There is extensive literature on edge computing, with a reduced but representative overview of the subject.

For example, [35] addresses the communication problem between nodes at the edge. Multilayer, multi-access edge computing (part of 5G technology) has shown advantages for data handling within Industry 4.0 requirements. The authors of [35] investigate its role in smart factories to identify research scenarios.

Furthermore, [36] investigated Deep Q-Network to solve the job-shop scheduling problem in smart factory processes. The authors of [36] introduced an edge computing system that replaces cloud computing to avoid communication delay time and security issues. The proposed framework is compared and analyzed.

The work [37] discussed privacy and security issues in IoT scenarios. The authors of [37] proposed a fingerprinting authentication that ensures secure and private communication between edge devices in IoT and Industry 4.0. The proposed method was compared with other and found to outperform other methods, such as the AES-128-bit key encryption method. Incorporating conventional ML algorithms into edge computing has been intensively studied.

In this sense, [38] argues that employing cumbersome CNNs for IoT applications is inappropriate and that model compression studies are needed. Furthermore, the article [38] introduces multi-scale representation for knowledge transfer.

In [39], a residue number system was designed as an arithmetic building block for digital signal processing, image processing, and cryptography. This was based on using parallel prefix tree adders that speed up multiplications. Such blocks are very important to speed up the multiplication in CNNs. Implementing hardware for ML inference is a very active research area.

The article [40] presented a novel spectroscopic inspection approach using principal component analysis as a distinctive example of IoT food inspection in the industry. They used this technique to assess food quality.

Moreover, [41] discusses how the intelligent edge and intelligent cloud enable factory autonomy. It mainly uses two modules: (i) stacked recurrent neural networks for continuous process control and (ii) cloud CNNs. The proposed platform was tested on a semiconductor chip production line to provide AI-based defect detection and preventive maintenance.

Additionally, [42] describes the need to process large amounts of data at high speeds and how cloud computing enables centralized data processing. However, it requires much time for data transfer. This makes it unsuitable for certain real-time applications. For this reason, sensing and computation are decentralized at the edge of the network. The authors of [42] introduces cutting-edge technology. The authors of [43] present a genetic algorithm-based resource management integrated with ML for predictive maintenance of fog (or edge) computing. Predictive maintenance models are built by logistic regression using real-time datasets.

The authors of [44] adopt a novel concept of software-defined cloud manufacturing and describe the advantages of real-time response, reconfiguration, and operational manufacturing systems.

How and where to incorporate edge computing has been studied as an optimization topic. To this end, [45] presents the problem that cloud computing systems in factories are centrally managed and may not be able to perform heavy computational tasks from thousands of IoT devices deployed on machines in a factory. Fog and edge devices placed in strategic sites to solve this problem are the main challenges for Industry 4.0. This paper [45] presents a case study and solves this NP-hard device placement problem with a meta-heuristic algorithm. In [46], mobile edge computing is used, and the authors plan to develop strategies for the optimal allocation of computational and wireless resources.

The work [47] proposed a hybrid computing framework and designed an intelligent resource scheduling strategy. With edge computing support and resource scheduling strategy, a prototype was evaluated to fulfill the real-time requirements in smart manufacturing. As a result, the excellent real-time operation was achieved in a case study.

The manuscript [48] points out that security and scalability are becoming major concerns as IoT and IIoT connected to the Internet continue to grow and the amount of data being generated increases. This article [48] introduces the critical infrastructure for IoT and IIoT in Industry 4.0 and presents the blockchain and edge computing paradigms to address it. The convergence of these two paradigms will result in a secure and scalable critical infrastructure.

In [49], the edge-cloud collaborative intelligent platform is presented for smart industries. This platform effectively reduces the response latency of industrial applications, and the authors of [49] evaluate the effectiveness of IndustEdge in two key case studies.

The paper [50] investigates the problem of connectivity between machines, in terms of perspectives of market, engineering system, product, and innovation, and introduces the concept of Software-Defined Cloud Manufacturing (SDCM). The SDCM is then the technology for connecting versatile and reconfigurable manufacturing machines/hardware, where it represents an interface between the physical world and the cyberspace and can be programmed to provide different functions, fulfilling the manufacturing needs. It is an attempt to standardize the different models and implementations of edge computing.

The article [51] shows a case study in which way edge computing solves key challenges of the Industry 4.0 implementation. The research of a new spectroscopic inspection approach as an attempt to standardizations of methods is addressed. It is applied in real-time monitoring process in food facilities. Moreover, [52] shows that the quality of service (QoS) of smart factories can be enhanced by means of edge computing, reducing costs by exploiting real-time data processing for monitoring and controlling the smart machinery that dynamically change according to the requirements of the production line.

The work [53] discusses the intelligent automation of manufacturing processes. It reports a case study of neural networks' usage on video taken from the closed circuit of cameras used for accident prevention. As edge device is used, the platform Raspberry Pi with TensorFlow Lite is used for machine learning applications.

Innovative aspects are always present and solutions are often ad hoc. For instance, [54] proposed a cost-effective, embedded edge computing system for high-precision injection molding process. It is based on temperature and pressure sensor interface with Arduino Mega and ESP 32D, by using also TensorFlow Lite.

Finally, we leave two surveys for the end. In [55], an interesting discussion about the industrial needs in the field of edge computing is offered, where some cases of industry needs are explained and exemplified. In addition, it is discussed how these problems are solved with technological solutions available today. But the most important thing is that it is pointed out or highlighted that universities are beginning to educate for future jobs in Industry 4.0.

It is an interesting work because it is based on a question-survey system.

Finally, in [56], a systematic literature review about edge computing in Industry 4.0 is offered, where it is claimed that it is needed for flexible production systems and autonomous systems and that the implementation requires investments in infrastructure, and in the development of employee's knowledge.

8 Conclusions

This manuscript outlines the concept and current state (from the academy's point of view) of ML applied to Industry 4.0 and edge computing. Issues of scalability, communication, computing, and security are raised, which is why ML and edge computing technologies are intensively studied for their applications and benefits in Industry 4.0. What can be inferred from the current state is that these technologies are not yet mature, and further research is required. However, they have a promising future to contribute to the realization of smart factories.

The problem of communication between machines, handling huge amounts of data, and standardization of the use of sensors are marked in the literature as the most challenging for Industry 4.0. These problems are addressed by the academy instead of establishing governments' policies or actions of specialized institutions such as standardization organizations.

The review of the state of the art shows that there is not clear established trend. It depends on the kind of smart factory we analyze because they are very different among them. Perhaps intelligent sensors are more used than computer vision techniques, which is deduced from the number of publications and the ease of finding them, but it is impossible to quantify at the industry level the tendencies of the field. Nevertheless, from the point of view of research, the points where more efforts seem to be being made are in aspects related to communication between sensors and cybersecurity.

Moreover, we have not found any information on how to solve problems or implement an upgrade system in a smart factory and what operating costs it has. There are not reports on this, and at the academic level, there is not study or example of how smart factories' upgrades are implemented, or how often they are necessary, or information about costs and maintenance times. In addition to this, another weak aspect of the implementation of Industry 4.0 is that there are no government policies available, and there is still nothing reported regarding whether Industry 4.0 can be friendly from the environmental point of view. The absence of documentation in the Web regarding government policies makes this study based on academic articles of case studies. A similar conclusion is achieved in [55, 56].

References

1. M. Maheswari, N.C. Brintha, Smart manufacturing technologies in Industry-4.0, in *2021 Sixth International Conference on Image Information Processing (ICIIP)*, (IEEE, 2021). https://doi.org/10.1109/ICIIP53038.2021.9702613
2. C.J. Turner, J. Oyekan, L. Stergioulas, D. Griffin, Utilizing Industry 4.0 on the construction site: Challenges and opportunities. IEEE Trans. Industr. Inform. **17**(2) (2021). https://doi.org/10.1109/TII.2020.3002197
3. I. Garbie, A. Garbie, Outlook of requirements of manufacturing systems for Industry 4.0, in *2020 Advances in Science and Engineering Technology International Conferences (ASET)*, (IEEE, 2020). https://doi.org/10.1109/ASET48392.2020.9118244
4. E. Ntavelis, I. Kastanis, L. Van Gool, R. Timofte, Same but different: Augmentation of tiny industrial datasets using generative adversarial networks, in *2020 7th Swiss Conference on Data Science (SDS)*, (IEEE, 2020). https://doi.org/10.1109/SDS49233.2020.00011
5. K.R. Thoorpu, N. Prafulla, Sequential DTC vector embedding using deep neural networks for Industry 4.0, in *2020 IEEE 7th International Conference on Industrial Engineering and Applications (ICIEA)*, (IEEE, 2020). https://doi.org/10.1109/ICIEA49774.2020.9102090
6. R.B. Khalifa, K. Tliba, M.L. Thierno Diallo, O. Penas, N.B. Yahia, J.-Y. Choley, Modeling and management of human resources in the reconfiguration of production system in Industry 4.0 by neural networks, in *2019 International Conference on Signal, Control and Communication (SCC)*, (IEEE, 2019). https://doi.org/10.1109/SCC47175.2019.9116104
7. A. Massaro, I. Manfredonia, A. Galiano, B. Xhahysa, Advanced process defect monitoring model and prediction improvement by artificial neural network in kitchen manufacturing industry: A case of study, in *2019 II Workshop on Metrology for Industry 4.0 and IoT (MetroInd4.0&IoT)*, (IEEE, 2019). https://doi.org/10.1109/METROI4.2019.8792872
8. C. Monsone, Á.B. Csapó, Charting the state-of-the-art in the application of convolutional neural networks to quality control in Industry 4.0 and smart manufacturing, in *2019 10th IEEE International Conference on Cognitive Infocommunications (CogInfoCom)*, (IEEE, 2019). https://doi.org/10.1109/CogInfoCom47531.2019.9089932
9. T.-Y. Lin, Y.-M. Chen, D.-L. Yang, Y.-C. Chen, New method for Industry 4.0 machine status prediction – A case study with the machine of a spring factory, in *2016 International Computer Symposium (ICS)*, (IEEE, 2016). https://doi.org/10.1109/ICS.2016.0071
10. D. Pau, F. Previdi, E. Rota, Tiny defects identification of mechanical components in die-cast aluminum using artificial neural networks for micro-controllers, in *2021 IEEE International Conference on Consumer Electronics (ICCE)*, (IEEE, 2021). https://doi.org/10.1109/ICCE50685.2021.9427592
11. S. Anita Shanthi, G. Sathiyapriya, L.D.C. Henry, Evaluating the impact of abiotic factors on wheat crop production using back propagation fuzzy neural network, in *2021 Emerging Trends in Industry 4.0 (ETI 4.0)*, (IEEE, 2021). https://doi.org/10.1109/ETI4.051663.2021.9619255
12. D. Pau, M. Lattuada, F. Loro, A. De Vita, G.D. Licciardo, Comparing industry frameworks with deeply quantized neural networks on microcontrollers, in *2021 IEEE Interna-*

tional Conference on Consumer Electronics (ICCE), (IEEE, 2021). https://doi.org/10.1109/ICCE50685.2021.9427638
13. A.I. Chernoded, V.A. Vasiliev, A.V. Tsyrkov, Application of neural networks in production control tasks, in *2021 International Conference on Quality Management, Transport and Information Security, Information Technologies (IT&QM&IS)*, (IEEE, 2021). https://doi.org/10.1109/ITQMIS53292.2021.9642816
14. S. Akhtari, F. Pickhardt, D. Pau, A. Di Pietro, G. Tomarchio, Intelligent embedded load detection at the edge on Industry 4.0 powertrains applications, in *2019 IEEE 5th International Forum on Research and Technology for Society and Industry (RTSI)*, (IEEE, 2019). https://doi.org/10.1109/RTSI.2019.8895598
15. H. Ouanan, E.H. Abdelwahed, Image processing and machine learning applications in mining industry: Mine 4.0, in *2019 International Conference on Intelligent Systems and Advanced Computing Sciences (ISACS)*, (IEEE, 2019). https://doi.org/10.1109/ISACS48493.2019.9068884
16. A. Massaro, I. Manfredonia, A. Galiano, N. Contuzzi, Inline image vision technique for tires Industry 4.0: Quality and defect monitoring in tires assembly, in *2019 II Workshop on Metrology for Industry 4.0 and IoT (MetroInd4.0&IoT)*, (IEEE, 2019). https://doi.org/10.1109/METROI4.2019.8792911
17. S. Goswami, Towards effective categorization of weather images using deep convolutional architecture, in *2020 International Conference on Industry 4.0 Technology (I4Tech)*, (IEEE, 2020). https://doi.org/10.1109/I4Tech48345.2020.9102678
18. A. Massaro, I. Manfredonia, A. Galiano, L. Pellicani, V. Birardi, Sensing and quality monitoring facilities designed for pasta industry including traceability, image vision and predictive maintenance, in *2019 II Workshop on Metrology for Industry 4.0 and IoT (MetroInd4.0&IoT)*, (IEEE, 2019). https://doi.org/10.1109/METROI4.2019.8792912
19. M. Kozek, Transfer learning algorithm in image analysis with augmented reality headset for Industry 4.0 technology, in *2020 International Conference Mechatronic Systems and Materials (MSM)*, (IEEE, 2020). https://doi.org/10.1109/MSM49833.2020.9201739
20. G. Dighvijay, D.S. Vaishnav, R. Mohan, A faster R-CNN implementation of presence inspection for parts on industrial produce, in *2021 Emerging Trends in Industry 4.0 (ETI 4.0)*, (IEEE, 2021). https://doi.org/10.1109/ETI4.051663.2021.9619228
21. A. Massaro, A. Panarese, G. Dipierro, E. Cannella, A. Galiano, V. Vitti, Image processing segmentation applied on defect estimation in production processes, in *2020 IEEE International Workshop on Metrology for Industry 4.0 & IoT*, (IEEE, 2020). https://doi.org/10.1109/MetroInd4.0IoT48571.2020.9138278
22. A. Spruck, J. Seiler, M. Roll, T. Dudziak, J. Eckstein, A. Kaup, Quality assurance of weld seams using laser triangulation imaging and deep neural networks, in *2020 IEEE International Workshop on Metrology for Industry 4.0 & IoT*, (IEEE, 2020). https://doi.org/10.1109/MetroInd4.0IoT48571.2020.9138205
23. Z. Zouhal, K. Benfriha, M. El Helou, C. El Zant, Q. Charrier, O. El Assal, G. Najmi, Approach for industrial inspection in the context of Industry 4.0, in *2021 International Conference on Electrical, Computer, Communications and Mechatronics Engineering (ICECCME)*, (IEEE, 2021). https://doi.org/10.1109/ICECCME52200.2021.9591119
24. S. Mohsen, A. Elkaseer, S.G. Scholz, Industry 4.0-oriented deep learning models for human activity recognition. IEEE Access **9** (2021). IEEE. https://doi.org/10.1109/ACCESS.2021.3125733
25. N. Gligoric, S. Krco, D. Drajic, Digital transformation in Industry 4.0 using vibration sensors and machine learning, in *2021 International Balkan Conference on Communications and Networking (BalkanCom)*, (IEEE, 2021). https://doi.org/10.1109/BalkanCom53780.2021.9593121
26. S.-L. Chen, C.-H. Liao, T.-K. Chi, T.-L. Lin, C.-A. Chen, Flexible signals and images lossless compression chip design for IoT and Industry 4.0, in *2018 14th IEEE/ASME International Conference on Mechatronic and Embedded Systems and Applications (MESA)*, (IEEE, 2018). https://doi.org/10.1109/MESA.2018.8449205

27. A. Massaro, A. Panarese, A. Galiano, Infrared thermography applied on fresh food monitoring in automated alerting systems, in *2020 IEEE International Workshop on Metrology for Industry 4.0 & IoT*, (IEEE, 2020). https://doi.org/10.1109/MetroInd4.0IoT48571.2020.9138207
28. R. Benitez, R. Benitez, C. Ramirez, J.A. Vazquez, Sensors calibration for metrology 4.0, in *2019 II Workshop on Metrology for Industry 4.0 and IoT (MetroInd4.0&IoT)*, (IEEE, 2019). https://doi.org/10.1109/METROI4.2019.8792886
29. T. Fapanni, M. Borghetti, E. Sardini, M. Serpelloni, Novel piezoelectric sensor by aerosol jet printing in Industry 4.0, in *2020 IEEE International Workshop on Metrology for Industry 4.0 & IoT*, (IEEE, 2020). https://doi.org/10.1109/MetroInd4.0IoT48571.2020.9138219
30. S.-T. Shih, I.C. Li, A light monitoring system with smartphone control based on Industry 4.0, in *2021 International Symposium on Intelligent Signal Processing and Communication Systems (ISPACS)*, (IEEE, 2021). https://doi.org/10.1109/ISPACS51563.2021.9651055
31. M. Borghetti, E. Cantù, E. Sardini, M. Serpelloni, Printed sensors for smart objects in Industry 4.0, in *2021 IEEE 6th International Forum on Research and Technology for Society and Industry (RTSI)*, (IEEE, 2021). https://doi.org/10.1109/RTSI50628.2021.9597209
32. P. Mengue, S. Hage-Ali, O. Elmazria, S. Zhgoon, SAW sensors directly integrated onto industrial metallic parts for manufacturing 4.0, in *2020 IEEE International Workshop on Metrology for Industry 4.0 & IoT*, (IEEE, 2020). https://doi.org/10.1109/MetroInd4.0IoT48571.2020.9138176
33. J.L. Santos, Optical sensors for Industry 4.0. IEEE J. Sel. Top. Quantum Electron. **27**(6) (2021). https://doi.org/10.1109/JSTQE.2021.3078126
34. Z. Guo, K. Zhang, H. Xin, M. Bi, H. He, W. Hu, An optical access network framework for smart factory in the Industry 4.0 era supporting massive machine connections, in *2017 16th International Conference on Optical Communications and Networks (ICOCN)*, (IEEE, 2017). https://doi.org/10.1109/ICOCN.2017.8121481
35. N.-N. Dao, Y. Lee, S. Cho, E. Kim, K.-S. Chung, C. Keum, Multi-tier multi-access edge computing: The role for the fourth industrial revolution, in *2017 International Conference on Information and Communication Technology Convergence (ICTC)*, (IEEE, 2017). https://doi.org/10.1109/ICTC.2017.8190921
36. J. Moon, J. Jeong, Smart manufacturing scheduling system: DQN based on cooperative edge computing, in *2021 15th International Conference on Ubiquitous Information Management and Communication (IMCOM)*, (IEEE, 2021). https://doi.org/10.1109/IMCOM51814.2021.9377434
37. M. Golec, S.S. Gill, R. Bahsoon, O. Rana, BioSec: A biometric authentication framework for secure and private communication among edge devices in IoT and Industry 4.0. IEEE Consum. Electron. Mag. **11**(2) (2022). https://doi.org/10.1109/MCE.2020.3038040
38. F. Shipeng, Z. Li, K. Liu, S. Din, M. Imran, X. Yang, Model compression for IoT applications in Industry 4.0 via multiscale knowledge transfer. IEEE Trans. Industr. Inform. **16**(9) (2020). https://doi.org/10.1109/TII.2019.2953106
39. B.K. Patel, J. Kanungo, Efficient tree multiplier design by using modulo $2^n + 1$ adder, in *2021 Emerging Trends in Industry 4.0 (ETI 4.0)*, (IEEE, 2021). https://doi.org/10.1109/ETI4.051663.2021.9619220
40. T. Konishi, T. Nakamichi, R. Kamikawa, Y. Yamasaki, Spectroscopic inspection optimization for edge computing in Industry 4.0, in *2020 22nd International Conference on Transparent Optical Networks (ICTON)*, (IEEE, 2020). https://doi.org/10.1109/ICTON51198.2020.9203553
41. J. Ying, J. Hsieh, D. Hou, J. Hou, T. Liu, X. Zhang, Y. Wang, Y.-T. Pan, Edge-enabled cloud computing management platform for smart manufacturing, in *2021 IEEE International Workshop on Metrology for Industry 4.0 & IoT (MetroInd4.0&IoT)*, (IEEE, 2021). https://doi.org/10.1109/MetroInd4.0IoT51437.2021.9488441
42. S. Trinks, C. Felden, Edge computing architecture to support real time analytic applications: A state-of-the-art within the application area of smart factory and Industry 4.0, in *2018 IEEE International Conference on Big Data (Big Data)*, (IEEE, 2018). https://doi.org/10.1109/BigData.2018.8622649

43. Y.K. Teoh, S.S. Gill, A.K. Parlikad, IoT and fog computing based predictive maintenance model for effective asset management in Industry 4.0 using machine learning. IEEE Internet of Things J. (Early Access) (2023). IEEE. https://doi.org/10.1109/JIOT.2021.3050441
44. C. Yang, S. Lan, W. Shen, L. Wang, G.Q. Huang, Software-defined cloud manufacturing with edge computing for Industry 4.0, in *2020 International Wireless Communications and Mobile Computing (IWCMC)*, (IEEE, 2020). https://doi.org/10.1109/IWCMC48107.2020.9148467
45. C.-C. Lin, J.-W. Yang, Cost-efficient deployment of fog computing systems at logistics centers in Industry 4.0. IEEE Trans. Industr. Inform. **14**(10) (2018). IEEE. https://doi.org/10.1109/TII.2018.2827920
46. N. Unnisa, M. Tatineni, Intelligent allocation strategy of mobile users for multi-access edge computing resources, in *2021 Emerging Trends in Industry 4.0 (ETI 4.0)*, (IEEE, 2021). https://doi.org/10.1109/ETI4.051663.2021.9619420
47. X. Li, J. Wan, H.-N. Dai, M. Imran, M. Xia, A. Celesti, A hybrid computing solution and resource scheduling strategy for edge computing in smart manufacturing. IEEE Trans. Industr. Inform. **15**(7) (2019). IEEE. https://doi.org/10.1109/TII.2019.2899679
48. Y. Wu, H.-N. Dai, H. Wang, Convergence of blockchain and edge computing for secure and scalable IIoT critical infrastructures in Industry 4.0. IEEE Internet Things J. **8**(4) (2021). IEEE. https://doi.org/10.1109/JIOT.2020.3025916
49. Y. Wang, S. Yang, X. Ren, P. Zhao, C. Zhao, X. Yang, IndustEdge: A time-sensitive networking enabled edge-cloud collaborative intelligent platform for smart industry. IEEE Trans. Industr. Inform. **18**(4) (2022). https://doi.org/10.1109/TII.2021.3104003
50. C. Yang, L. Shulin, S. Weiming, W. Lihui, Q. Huang George, Software-defined cloud manufacturing with edge computing for Industry 4.0, in *Proceedings of 2020 International Wireless Communications and Mobile Computing (IWCMC)*, (IEEE, 2020), pp. 1618–1623. https://doi.org/10.1109/IWCMC48107.2020.9148467
51. T. Konishi, T. Nakamichi, R. Kamikawa, Y. Yamasaki, Spectroscopic inspection optimization for edge computing in Industry 4.0, in *Proceedings of 2020 22nd International Conference on Transparent Optical Networks (ICTON) Transparent Optical Networks (ICTON)*, (IEEE, 2020). https://doi.org/10.1109/ICTON51198.2020.9203553
52. B. Armir, C. Antonio, F. Luca, P. Lorenzo, S. Andrea, Enhancing the performance of Industry 4.0 scenarios via serverless processing at the edge, in *Proceedings of ICC 2021-IEEE International Conference on Communications, ICC*, (IEEE, 2021). https://doi.org/10.1109/ICC42927.2021.9500286
53. M. Martin, S. Tomas, Edge computing implementation of safety monitoring system in frame of IIoT, in *Proceedings of 2022 23rd International Carpathian Control Conference (ICCC)*, (IEEE, 2022), pp. 125–129. https://doi.org/10.1109/ICCC54292.2022.9805918
54. C. Shia-Chung, J.M. Jibin, F. Ching-Te, H. Tzu-Jeng, An innovative method to monitor and control an injection molding process condition using artificial intelligence based edge computing system, in *Proceedings of 2022 8th International Conference on Applied System Innovation (ICASI)*, (IEEE, 2022), pp. 41–45. https://doi.org/10.1109/ICASI55125.2022.9774445
55. D. Stadnicka et al., Industrial needs in the fields of artificial intelligence, internet of things and edge computing. Sensors (14248220) **22**(12) (2022). https://doi.org/10.3390/s22124501
56. K. Kacper, D. Grzegorz, S. Dorota, Possible applications of edge computing in the manufacturing industry – Systematic literature review. Sensors (14248220) **22**(7), 2445, 24p (2022). https://doi.org/10.3390/s22072445

Failure Detection System Controlled by a Mixed Reality Interface

Alan Yamir Rodríguez Gallegos, Luis Carlos Méndez-González, Alan Iván Hernández Holguín, and Luis Alberto Rodríguez-Picón

1 Introduction

Over the years, technologies have become more and more a part of our lives, with the promise of improving and solving increasingly complex tasks and avoiding errors everywhere. One of the main reasons for this is productivity and failure reduction. Humans are prone to make many mistakes or errors in their daily lives, including in their work life. Mistakes tend to be very costly within jobs, as these regularly mean operating longer, replacing some material, or purchasing survival.

Specifically, in the manufacturing industry, the quality area verifies that the products created within the facilities meet the company's standards. For this purpose, inspection stations are used, which put an operator in charge of reviewing the product and determining with his judgment whether it is discarded or moves on to the next station. This has been a problem for some industries, such as the medical industry, since it requires intense and strict inspections [1]. It is common for those responsible for performing these operations to make mistakes in determining whether there is a fault. With the emergence of industry 4.0, several fields were combined to generate new technologies in this framework, such as augmented and virtual reality, artificial intelligence and machine learning, cloud computing, and other engineering fields. Some of the technologies derived from these fields are intelligent robots, mixed reality interfaces, the Internet of Things, cloud services, object detection tools, and other intelligent systems with great adaptability to industrial tasks.

A. Y. R. Gallegos · L. C. Méndez-González (✉) · A. I. H. Holguín · L. A. Rodríguez-Picón
Department of Industrial Engineering and Manufacturing, Institute of Engineering and Technology, Universidad Autónoma de Ciudad Juárez, Ciudad Juárez, Chihuahua, Mexico
e-mail: a150191@alumnos.uacj.mx; luis.mendez@uacj.mx; al216737@alumnos.uacj.mx; luis.picon@uacj.mx

L. C. Méndez-González et al. (eds.), *Innovation and Competitiveness in Industry 4.0 Based on Intelligent Systems*, EAI/Springer Innovations in Communication and Computing, https://doi.org/10.1007/978-3-031-29775-5_2

The current section introduces key concepts to understand the theoretical framework of the proposed innovation and presents the most relevant approaches found in the literature. Section 2, Problem Analysis, describes the purpose and scope of the proposed innovation and shows how machine vision and artificial intelligence systems can be key in developing automatic inspection systems that have strict precision, reducing failures due to human error. Section 3, Methodology, explains the steps followed to create a mixed reality interface and link it to a vision algorithm from the cloud to create a robust system that detects instrument failures. Section 4, Results, presents the data obtained after numerous tests of different aspects of the instruments to identify how accurate and effective the failure identification could be. Section 5, Conclusions, summarizes what was achieved during the research and specifies the improvements within the project for future work. The appendix contains the specs of the HoloLens hardware.

1.1 State of the Art

The fields of interest for this paper are augmented and virtual reality, artificial intelligence, and computer vision. These areas are some of the pillars of the industry 4.0 philosophy [2].

Augmented and virtual reality are frequently used to develop systems focused on product quality control [3] and process inspection [4] or to provide an integrated platform that allows the users to interact with an industrial environment [5]. Virtual reality (VR) consists of a computer-generated environment in which a user can interact in a tridimensional space with virtual objects [6]. Augmented reality (AR) is a variant of VR that allows the interaction between real-world and virtual objects in real time, so the environment is composed of real and computer-generated objects [7].

Some of the most relevant VR and AR implementations are a prototype of an error detection system for subtractive manufacturing processes such as machining metal parts [8]: a sorting system with a robotic arm that uses augmented reality to analyze the state of several trays and an algorithm that determines the occupied spaces and thus helps the robotic arm not to make a mistake by placing a part in a space that is already occupied [9]. Intending to boost the use of augmented reality in the learning environment, Woll et al. [10] designed an application that teaches users how to assemble a car power generator using augmented reality. Cardoso et al. [11] created a platform where Hardware and Software (VRCEMIG) are integrated to control electrical substations through a virtual reality environment. Puigvert et al. [12] proposed an augmented reality location system based on AR Core, an AR platform designed for cell phones, which consists of a point cloud created by AR core is used, and from there, an algorithm is calculated to determine the location. One of the biggest challenges in this type of implementation is to create a coherence between the virtual and the real world, so Rompapas et al. [13] designed and tested a display that measures the user's eye and adapts the virtual content to the real environment to provide a more immersive experience. Generally, augmented reality applications use anchors to display virtual information on real spaces or objects. For industrial use, this is not always possible. In these situations, the virtual information must be displayed in alternative positions. At the same time, the connections to the anchors must be recognizable, and Dauenhauer et al. [14] demonstrated that it is possible to prove that these anchors are better connections for superimposing information on objects or environments.

The artificial intelligence field is used in industrial applications to teach algorithms to perform specific tasks in variable conditions with a certain uniformity in the results obtained [15]. It can also be applied to robotic systems [16] and the administration of cyber-security [17] . Liu et al. [18] set the goal of improving mechanical manufacturing processes based on artificial intelligence. This above method is based on artificial intelligence technology and highlights the three characteristics of artificial intelligence technology: simulation, extensibility, and expandability. Ang et al. [19] proposed the smart manufacturing system with an artificial neural network (ANN) to predict the manufacturing line's healthiness: surface mount technology, manual welding, and functional testing. Yao et al. [20] performed a study about how symbolic AI (called AI 1.0), characterized by structured contents and centralized control structures, evolved to next-generation AI (called AI 2.0) characterized by unstructured contents, decentralized control structures, and machine learning (especially, deep learning). Lou et al. [21], to promote a cooperative relationship between manufacturing services and make manufacturing services obtain a high profit, proposed a behavioral learning mechanism to support the action selection of manufacturing services during the repeated interaction process.

Computer vision corresponds to one of the applications of artificial intelligence, focusing on extracting data from images and video using vision sensors, like cameras, which are used to produce machine learning models focused on the identification, classification, and detection tasks [22]. Some of its main applications are identifying failures and defects in the quality control process [23] and identifying certain types of materials or objects [24]. Li et al. [25] combined artificial intelligence technology to propose a new method for materials performance evaluation: Laser additive manufacturing is taken as the research basis, and three kinds of Ti6-Al-4V material microstructure images with different properties are used as datasets; according to the DenseNet model, a deep convolution neural network NDenseNet is trained to optimize the network model memory and improve the recognition accuracy. Khan et al. [26] proposed a system that a corporate office can use to log in its employees, which involves face detection using the Viola-Jones algorithm. Harikrishnan et al. [27] implemented a real-time surveillance and assistance system, which includes face-to-face attendance using a unique instant mode on smartphones for university classes and enhanced real-time facial recognition surveillance of laboratory facilities and workplaces. Zheng et al. [28] proposed a moving target tracking method based on computer vision and introduced a framework of computer vision theory through a bottom-up visual tracking processing method.

Several works with a similar approach to the innovation proposed in this paper were found in the literature. Muñoz et al. [29] developed an industrial mixed reality interface to inspect car bodies for quality control, allowing the user to mark defects over the inspected parts and generate a 3D model of the part with all the marks and annotations. Wang et al. [30] proposed a mixed reality interface deployed in HoloLens hardware to detect structural failures in plastic and metal parts by using a convolutional neural network connected to a cloud server. Arévalo et al. [31] also designed a mixed reality interface for the HoloLens hardware but based on the Unity engine and focused on the inspections of holograms and 3D CADs. Avalle et al. [32] created an augmented reality interface through HoloLens to a fault detection system customized for the analysis of manipulator robots by using an adaptive method that dynamically highlights the sections near the faults. De Pace et al. [33] also created a fault detection system for manipulator robots. However, they included a connection to Android devices in their system architecture to improve the data visualization and ease of usage. Hebenstreit et al. [34] proposed the implementation of a HoloLens mixed reality interface in an assembly line to reduce human failures by displaying data, status, and instructions about the manufacturing process to the workers. Alves et al. [35] combined an Internet of Things network with a mixed reality interface to improve the analysis of machine failures and the preventive maintenance process by using sensors to collect data in remote databases and HoloLens hardware to allow the interaction of the user with this data to make decisions.

2 Problem Analysis

Approximately 26% of each production batch from the company leader in the medical instruments field is trashed due to failures detected in the assembly process, from which 10% corresponds to missing components, 9% to misplaced components, and 7% to poorly assembled pieces. These products are used in delicate medical operations, so their failures would affect irreversible damage to patients. There have been attempts to reduce the prevalence of these problems by implementing more quality filters and assigning more staff to inspect the products; however, the human eye limits the quality control environment.

Mixed reality systems can amplify visual range and object details when implemented with users in specific environments. This represents an augmentation in the data obtained from the visual medium, which improves the analysis and response of the user to specific events. With a failure detection system like the proposed implementation, the quality inspection process could be improved by a control interface with visualization tools powered by a mixed reality interface and computer vision techniques, like object detection. Also, these systems can be upgraded with more sophisticated and powerful sensors to perform deeper analysis, determine the material composition, or check if they comply with the recommended specs. In health, applying failure detection systems to medical equipment is justified by events in the medical field, like failures in surgery, which can potentially harm patients.

The causes of these problems range from defects in the assembly process of the instruments to inadequate quality control mechanisms.

2.1 Assembly Process for Medical Instruments

The main area of application is the medical industry manufacturing field, particularly, the manufacturing of medical tools used to perform inspections and medical processes. The focus of the proposed approach is the identification of failures in a medical instrument composed of the components shown in Table 1.

The medical instrument above-mentioned is presented in Fig. 1, which was used to define the classes and the image dataset for the training process of the object detection algorithm. This tool needs to be inspected by the user of the mixed reality interface to detect failures.

2.2 Failures in the Assembly Process

The most common failures detected in the assembly process are related to the cosmetic details of the instrument, problems in manipulating the device due to bad

Table 1 Main components of the medical instrument

Component
Product label 1
Product label 2
Pipe
Air valve
Lube filter
Superior cap

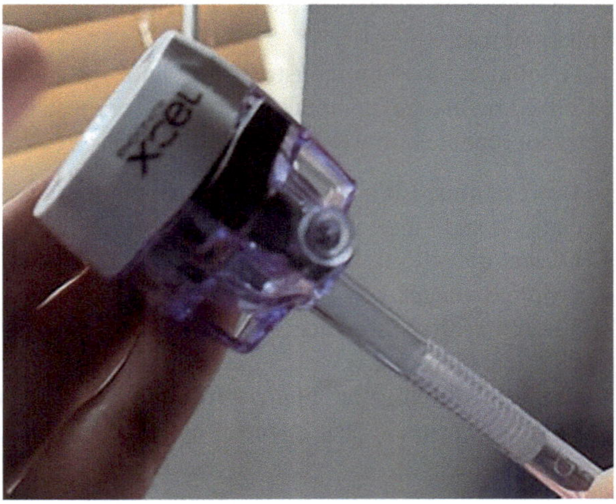

Fig. 1 Example of a medical instrument

Table 2 Classification of the failures detected in the medical instruments

Failure class	Description
Class 0	Patient's death
Class 1	Device malfunctioning
Class 2	Operation failures (difficulty of operation due to unexpected failures)
Class 3	Cosmetic details

manufacturing, or complete malfunctioning that prevents the user from using the tool's capabilities. Table 2 presents a classification of the most common failures.

2.3 Proposed Innovation

Mixed reality systems allow us to expand the range of visualization and detail an individual could have in an environment. By having a greater amount of data and information within our range of vision, the number of possibilities to act and respond to a specific situation increases. In this case, the inspection process would be favored by the control and visualization interface provided by the RM, with the help of AI algorithms that can provide data that would otherwise be difficult to obtain in this activity. These systems could be strengthened by adapting more powerful sensors that would allow us to analyze objects in greater depth, such as obtaining the composition of a material and knowing from the supplier if it has the necessary specifications for its use.

The reasons that lead to developing a solution that helps to increase the quality standards in this type of medical instrument are centered on the fact that there have been cases where patients suffer permanent damage during surgeries because of a failure in the instrument due to a wrongly placed part, in the wrong place, or because of the absence of such a part. These situations are caused by deficiencies in the inspection systems in the manufacturing process, where human error plays an important role.

A solution based on industry 4.0 technologies, a failure detection system based on mixed reality and computer vision, was proposed. The system can be segmented into four modules (illustrated in Fig. 2):

- **Data acquisition:** Capture of data through system sensors (the HoloLens' camera) and image processing
- **Object detection:** Detection of bounding boxes in object of interest detected in captured images
- **Virtual reality interface:** Mixed reality interface to allow the user to interact with real objects in a virtual interface
- **Artificial intelligence algorithm:** Processing of the data generated to detect failures in medical instruments. Implementation of computer vision

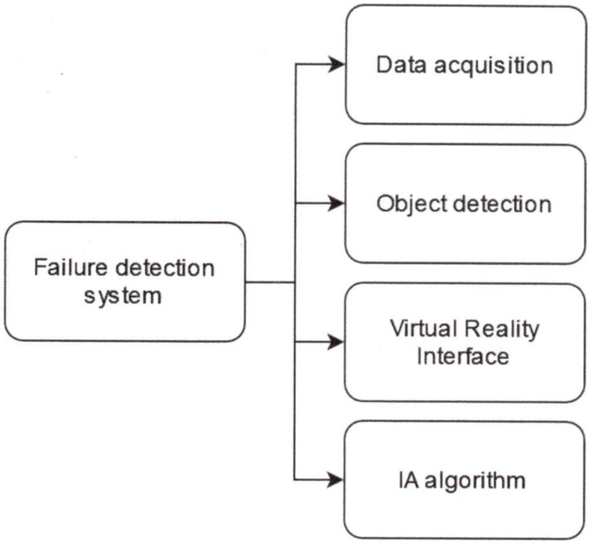

Fig. 2 Composition of the failure detection system

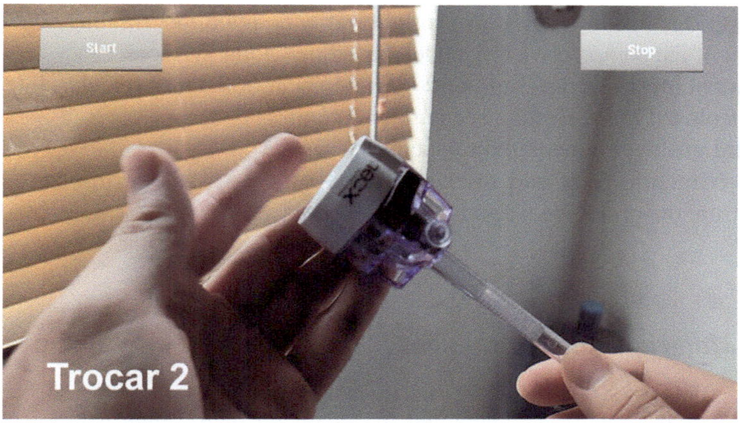

Fig. 3 Proposed implementation of the failure detection system

This implementation allows the user to detect failures in a mixed reality environment, so he can interact with a graphic interface and obtain data from the object analyzed in real time, as Fig. 3 depicts.

To focus the approach on a more practical implementation, the proposed system was limited by the following:

- The mixed reality interface is controlled by an interface through HoloLens 2.
- The control interface was designed in Unreal Engine (UE) 4.26.
- The computer vision algorithm was powered by Azure Cloud Services.
- The camera in the HoloLens 2 hardware was used to capture images of suspected faulty medical instruments.

- The device implementation is aimed at the manufacturing field. The training dataset is a custom image dataset. The project focuses solely on detecting faults in the instrument assembly, so the technical analysis of the characteristics of the faults was not covered. The main features of the interface require a stable Internet connection.

The HoloLens 2 was the hardware selected to deploy the software, due to its great adaptability to industrial environments [36].

3 Methodology

This section describes the equipment and methods used to develop the proposed system. The flow of work (depicted in Fig. 4) includes the following stages:

- **System characterization:** Definition of the core aspects of the system: frameworks, software, and hardware utilized; delimitation of the computer vision algorithm; and the draft of the control interface
- **Computer vision algorithm production:** Configuration, training, and deployment of the computer vision model via Azure Cloud Services
- **Control interface development:** Development of the mixed reality interface to be deployed in the HoloLens environment
- **HoloLens deployment:** Prototype of the system
- **Experimental tests and metrics:** Detection, usage, and performance tests to measure the prototype's capabilities

3.1 Materials and Equipment

The proposed system was integrated into a mixed reality interface composed of HoloLens glasses, a user interface developed in Unreal Engine, and a computer vision algorithm powered by Azure Cloud. The system's components are described in more detail in Table 3.

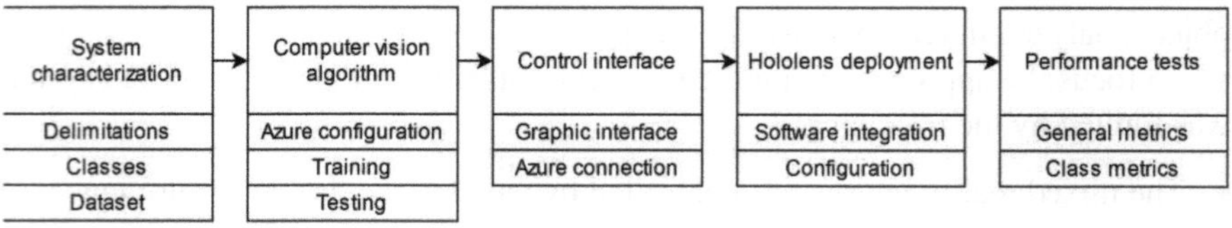

Fig. 4 Stages of development

Table 3 Components of the system

Component	Type	Description
HoloLens	Hardware	Augmented reality lens with support for cloud environments
Mixed reality user interface	Software	Interface implemented on the HoloLens to visualize data and control the systems functions
Computer vision algorithm	Software	Object detection algorithm trained to detect failures in several parts of the medical instrument assembly

3.2 System Characterization

The system was designed to be implemented in the quality control process in the manufacturing field, according to the following criteria:

- Real-time execution
- Object detection tool
- Failure detection model trained with an image dataset composed of photos of medical instruments
- Mixed reality interface deployed in HoloLens 2 hardware

3.3 Computer Vision Algorithm

The model for the computer vision algorithm was generated using Azure Cloud Services. An object detection algorithm was selected to be trained in the cloud environment due to the requirement of a multi-detection approach to identify several objects in a single frame. In comparison, an object classification technique would not be enough to perform well in the proposed implementation because it would require multiple classifiers to identify more than one object in a single frame. The model was trained using a custom dataset with images labeled according to the failures contained in each one.

The production of the computer vision algorithm consists of five phases:

1. **Configuration:** Selection of model and parameters for training in Azure Cloud.
2. **Dataset:** Load the dataset to prepare the image to be processed in the training process.
3. **Training:** Iterative processing of the images to train the model. The final product is a detection model, which can identify failures in selected components of the medical instrument.
4. **Validation:** Regulates the model's performance by comparing the results obtained between training and a validation set. Provides metrics for general and per-class performance.

5. **Deployment:** The validate model is deployed and connected to the control interface in the HoloLens.

The overall process is resumed in Fig. 5.

3.3.1 Dataset

A custom image dataset was created to train the computer vision algorithm. It contains a collection of images segmented into the following classes (Table 4, with their respective labels. The dataset was created from scratch using photos of the parts that compose the medical instrument. This approach was selected because an image dataset focused on the classes selected for the algorithm was not found. Also, a custom dataset can provide better detection performance for tasks like the proposed failure detection. All the images were re-scaled to a standard resolution to normalize the dataset using the Azure environment tools. An image augmentation technique was performed on the dataset using the training environment features. The augmentation allowed the generation of more examples for the training process, which resulted in a better overall performance for the detection model.

Fig. 5 Production process for the computer vision model

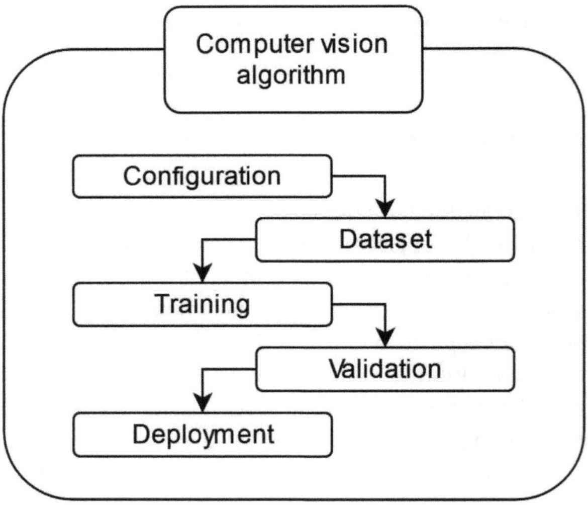

Table 4 Dataset classes

Tag	No. images
Size_label	46
Air_valve	80
Product_label	20
Produc_label2	18
Component_1	21
Component_2	22

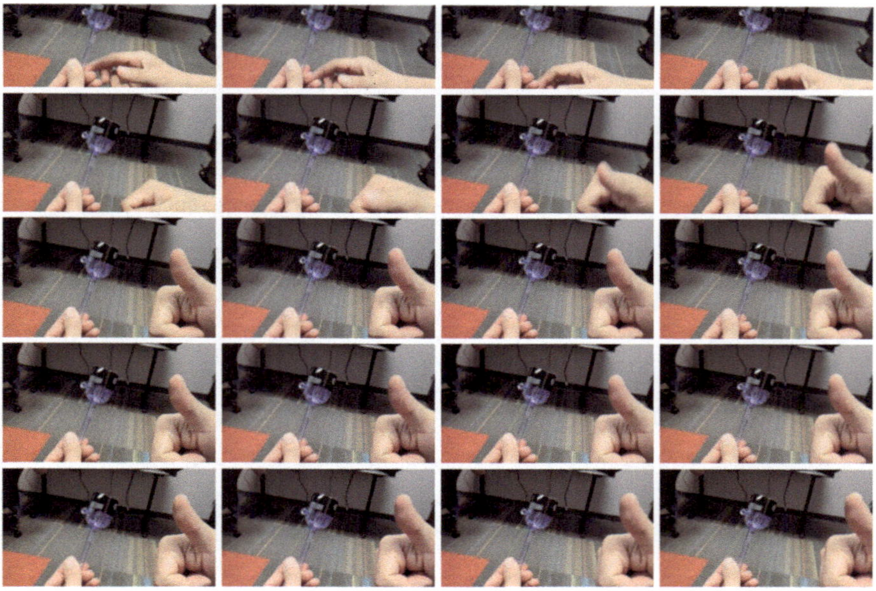

Fig. 6 Example of images found in the dataset

Each class contains the number mentioned above of images, which were taken from different angles (as seen in Fig. 6 to reduce the like-hood of an over-fitting model [37].

3.3.2 Training

The training process was executed in a session of Azure Cloud Services using the custom dataset and a detection algorithm to generate a model with the capabilities to detect failures in images of medical instruments. In Fig. 7, a flow chart describes this process.

The convergence point was reached during training after 9 hours, with a precision threshold defined as 50%. As seen in Fig. 8, the precision of the training model is 55.9%, which is considered sufficient for implementation purposes. The mAP of the model is 82.7%. Thus, the classes adjust adequately to the data provided in training. In comparison, the algorithm trained by Wang et al. [30] achieved a mAP score of 85% with a model configured with fewer classes than the current approach, so the metric obtained is good enough. The model trained in Azure was composed of more classes than those mentioned above, so it can be considered more complex.

The performance per class is slightly irregular. The class *product_label2* achieved the highest precision with a score of 80%, but the class with the lowest

Fig. 7 Training and validation process

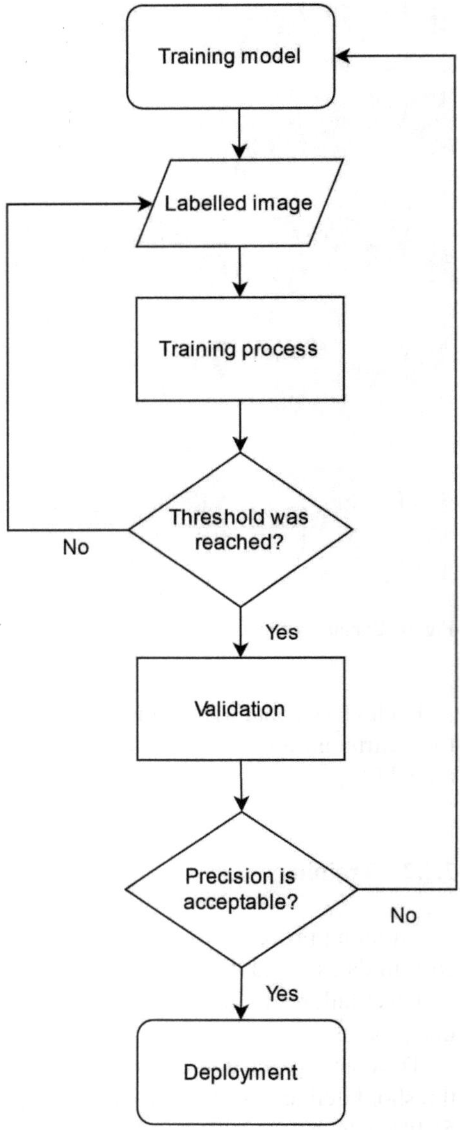

precision score is *component_1* with 14.3%. Nevertheless, according to Table 5, all the classes surpass the precision threshold. The irregular performance could be produced by bad data batches (low-quality images with low variability), class complexity, or a lack of more examples per class in the training dataset.

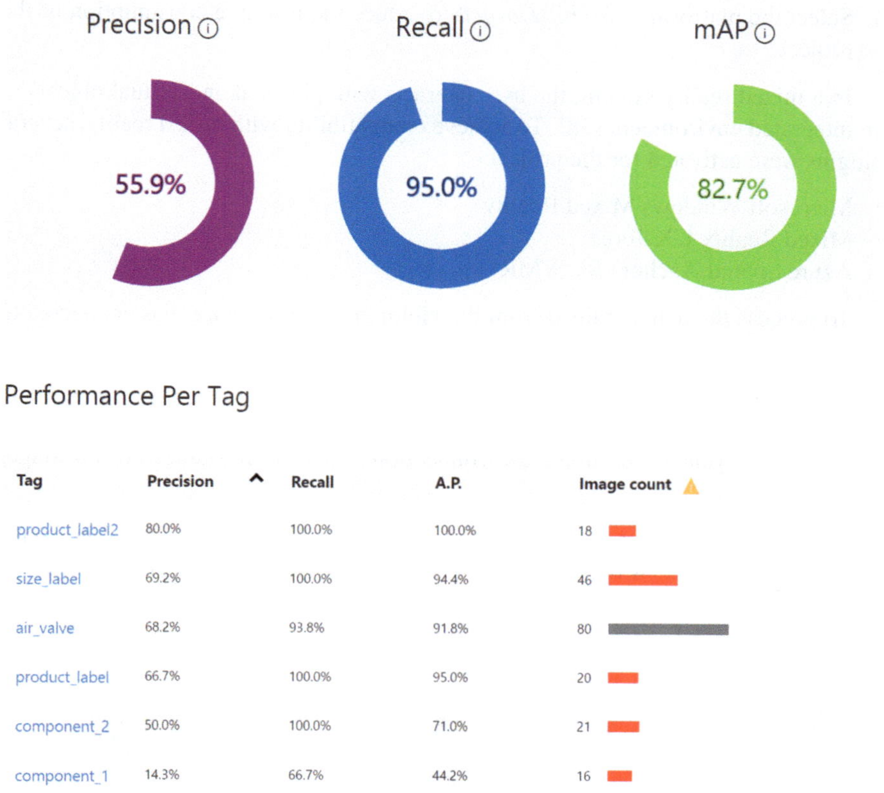

Fig. 8 Metrics obtained after training the model

Table 5 Performance metrics per class obtained from the training process

Class	Precision	A.P	No. images
Size_label	69.2%	94.4%	46
Air_valve	68.2%	91.8%	80
Product_label	66.7%	95.0%	20
Product_label2	80.0%	100%	18
Component_1	14.3%	44.2%	16
Component_2	50%	71.0%	21

3.4 Control Interface

The control interface was designed in Unreal Engine 4.26, using a blank template with the following parameters:

1. Create a blank project in UE 4.26.
2. Select the option *Scalable 3D* or *2D*, to reduce the object's graphic quality and improve performance.

3. Deactivate Raytracing.
4. Select the platform as *Mobile/Target*, to reduce the resource consumption of the project.

In a mixed reality system, the user interacts with physical and virtual objects in an integrated environment [38]. To achieve compatibility with mixed reality, several plugins were activated for the project:

- Microsoft Windows Mixed Reality
- Mixed Reality UX Tools
- Azure Spatial Anchors for WMR

To process the data obtained from the HoloLens, the interface was connected to the Azure environment (Fig. 9), using an API key method:

1. Obtain an Azure API key.
2. Create an event in UE to login to Azure, using Azure ID and API key.
3. Create an event to terminate the connection when the resources are no longer being used.

The graphic interface was developed using Blueprint, a block-based language used in UE. The code was debugged and deployed in the HoloLens. The interface is shown in Fig. 10.

Fig. 9 Connection between HoloLens, mixed reality interface, and Azure services

Fig. 10 Schematic view of the interface developed in UE

Fig. 11 App deployed in the HoloLens interface

3.5 HoloLens Configuration

The HoloLens was configurated to test the failure detection system. Due to the control interface was created from scratch, several features were activated and adjusted [39]:

- Developer mode was enabled to allow the execution of external apps obtained from other means than the Windows Store.
- The app *Microsoft HoloLens* was installed on an external PC to connect it to the lens.
- The HoloLens was initialized through the app mentioned above.
- A connection was established between the HoloLens interface and Azure Cloud using an Azure API key. This API allows the execution of the computer vision algorithm using the HoloLens camera's data. A stable Internet connection is required to use the failure detection system in the proposed implementation.

The configured *Microsoft HoloLens* app was launched to load the interface and initialize the modules of the HoloLens implementation. Figure 11 shows the app deployed in the lens.

4 Results

The main products obtained are the mixed reality interface, developed in UE and deployed using a graphic interface in HoloLens, and the classification model, trained in Azure Cloud with a dataset of medical instruments.

Fig. 12 Home screen after launching the app

4.1 HoloLens App

The final product of the project consists of an HoloLens application which allows fault detection in medical instruments through a mixed reality interface assisted by a computer vision algorithm from Azure. The essential operation of the interface is described below: The fault detection system screen (named Robust Process MR) has the function of the main (home) screen (Fig. 12).

The control buttons are displayed on each side of the app interface and are composed by:

- **Start:** Starts the process and authenticates the Azure credentials to activate the failure detection system
- **Stop:** Stops the main process and deactivates the recognition system
- **Piece:** Provides information about the scanned piece according to the database stored in Azure

4.2 Failure Detection System Powered by Azure

Ten tests were performed for each instrument component to verify the accuracy of fault detection. These tests were performed using instruments in good condition and instruments with faults in the assembly of their parts. As can be seen in Fig. 13, the part identified as Component1 was detected as positive due to the correct assembly of that part.

In Table 6, the data obtained from Azure is shown, including the precision scores obtained in the test performed to evaluate the system. This data represents the object detection algorithm performance in the identification of the classes contained in the dataset.

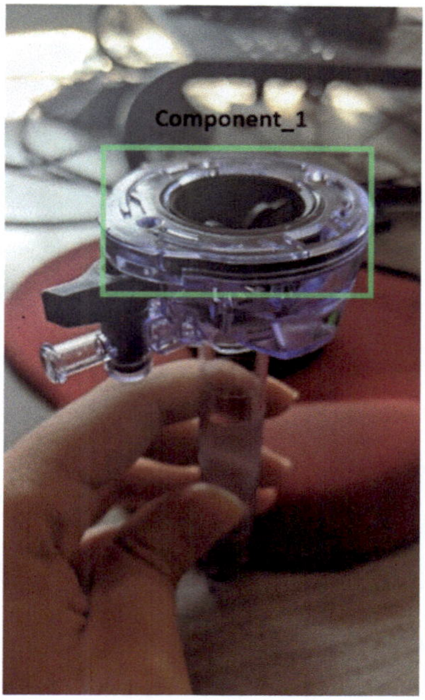

Fig. 13 The system detects a failure in the instrument label

Table 6 Results obtained from the failure detection tests

Tag	Precision	Confidence	# of tests
Product_label2	80.0%	100%	10
Size_label	91.7%	94.4%	10
Air_valve	88.2%	91.8%	10
Product_label	97.3%	95.0%	10
Component_2	84.2%	71.0%	10
Component_1	94.2%	88.6%	10

Figure 14 shows how the component is poorly positioned within the assembly, and the system can detect it and enclose it within a red-colored figure. The accuracy level in this component's failure detection according to the results obtained by Azure is 94.2, which is a more than acceptable result.

The label is a crucial aspect of the company's quality policies; the wrong engraving on the instrument represents a quality failure. It is a reason for return during human eye inspections. Identifying when it should be taken as a failure is often challenging. However, the system can identify when the instrument's engraving is in poor condition. Figure 15 shows the result of inspecting a labeling failure.

Based on the data obtained from the Azure platform, the percentage accuracy of the tag detection is 97.3%. The system was able to detect, for the most part, the faults presented during the tests.

Fig. 14 The systems detect a failure in component 1

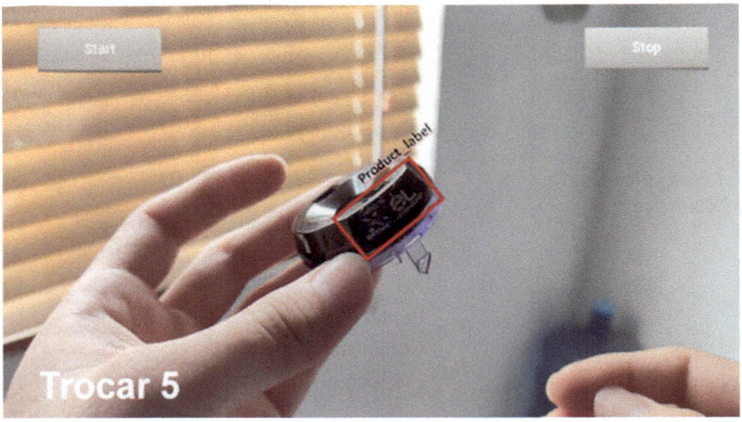

Fig. 15 The systems detects a failure in the product label

5 Conclusion and Future Work

The system developed in this research project allowed combining computer vision technologies, cloud computing with the Microsoft Azure platform, and mixed reality through the HoloLens viewer to generate a fault detection system for medical instruments in the industry. Compared to the other works reviewed in the state-of-the-art section, the proposed HoloLens interface adds the novelty of an object detection tool deployed in a cloud environment, which allows to configure and deploy other types of algorithms on the fly, such as facial recognition and image segmentation models. One of the main benefits of the system is that it can be used without requiring much technical knowledge, which is an improvement in failure detection within the manufacturing industry, which in turn allows greater flexibility and dynamism, allowing other types of functions to make inspections faster and without relying on the human eye. In addition, since the detection algorithm runs in a cloud environment, it is easy to update and improve it for future new functions that are planned to be added. Because the mixed reality interface is already prepared, and only keys need to be entered to access the resource in Azure, it is easy to customize. Within the scope of accessibility, the system is simple to use and designed so that any work within the company can operate this system without requiring extensive training or certification. So, in the context of industry 4.0, the main benefits of the proposed innovation are easiness of implementation in the quality control process and improved quality of the final product. Another technical benefit is the possibility of documentation of quality control inspections performed using the HoloLens interface in a remote database. Also, the proposed interface could be integrated into an IoT network to improve its data representation capabilities.

Regarding the economic scenario, the mixed reality interface proposes a quality control technique that could reduce the incidence of faulty batches of medical instruments and produce better products with improved quality, representing a significant cost reduction in the quality control process.

The system developed in this research can lay the foundation for future research on developing fault detection systems in other types of instruments or products outside the medical area (such as automotive). There are several areas of opportunity in the area to make detection systems much more robust, such as implementing voice commands that would allow switching between the different options simply by voice. A virtual assistant could be incorporated to guide the user during detections and give real-time instructions on the tasks to be performed after inspecting the object.

The following areas have room for improvement:

- Precision of the system and complexity of the failure detection
- Integration of voice commands to provide a free-hand interface
- Integration of virtual assistance to provide support to the user

As recommendations, the object detection algorithm could be improved using a state-of-the-art technique, like the current version of the YOLO algorithm or Mobilenet SSD. Also, the algorithm parameters could be tuned using a more specialized technique to improve performance after training. The control interface could be enhanced using Unreal Engine 5 (a more recent version than the one used in the project), and software optimization could be applied to achieve a better performance in the mixed reality interface.

Appendix

HoloLens 2 specs	
Optics	Clear holographic lens
Head tracking	4 visible-light cameras
Eye tracking	2 infrared cameras
Camera	Image resolution, 8 Mpx; video resolution, 1080p 30 fps
Microphones	5 channels matrix
Degrees of freedom	6DoF
SoC	Qualcomm Snapdragon 850
Memory	DRAM LPDDR4x 4 GB
Storage	64 GB
Wi-Fi	Wi-Fi 5 (802.11ac 2x2)
USB	Type C
IMU	Accelerometer, gyroscope y magnometer
Depth	Time-of-flight depth sensor 1 Mpx

References

1. T.K. Lepasepp, W. Hurst, A systematic literature review of industry 4.0 technologies within medical device manufacturing. Future Int. **13**(10), 264 (2021)
2. E. Oztemel, S. Gursev, Literature review of industry 4.0 and related technologies. J. Intell. Manuf. **31**(1), 127–182 (2020)
3. G.M. Santi, A. Ceruti, A. Liverani, F. Osti, Augmented reality in industry 4.0 and future innovation programs. Technologies **9**(2), 33 (2021)
4. T. Masood, J. Egger, Augmented reality in support of industry 4.0—implementation challenges and success factors. Robot. Comput.-Integr. Manuf. **58**, 181–195 (2019)
5. V. Reljić, I. Milenković, S. Dudić, J. Šulc, B. Bajči, Augmented reality applications in industry 4.0 environment. Appl. Sci. **11**(12), 5592 (2021)
6. D. Kamińska, T. Sapiński, S. Wiak, T. Tikk, R.E. Haamer, E. Avots, A. Helmi, C. Ozcinar, G. Anbarjafari, Virtual reality and its applications in education: survey. Information **10**(10), 318 (2019)
7. L.F. de Souza Cardoso, F.C.M.Q. Mariano, E.R. Zorzal, A survey of industrial augmented reality. Comput. Ind. Eng. **139**, 106159 (2020)
8. M.H. Sreekanta, A. Sarode, K. George, Error detection using augmented reality in the subtractive manufacturing process, in *2020 10th Annual Computing and Communication Workshop and Conference (CCWC)* (2020), pp. 0592–0597
9. M. Kozek, Transfer learning algorithm in image analysis with augmented reality headset for industry 4.0 technology, in *2020 International Conference Mechatronic Systems and Materials (MSM)* (2020), pp. 1–5
10. R. Woll, T. Damerau, K. Wrasse, R. Stark, Augmented reality in a serious game for manual assembly processes, in *2011 IEEE International Symposium on Mixed and Augmented Reality - Arts, Media, and Humanities* (2011), pp. 37–39
11. A. Cardoso, E. Lamounier, G. Lima, L. Oliveira, L. Mattioli, G. Júnior, A. Silva, K. Nogueira, P. do Prado, J. Newton, Vrcemig: A virtual reality system for real time control of electric substations, in *2013 IEEE Virtual Reality (VR)* (2013), pp. 165–166

12. J.R. Puigvert, T. Krempel, A. Fuhrmann, Localization service using sparse visual information based on recent augmented reality platforms, in *2018 IEEE International Symposium on Mixed and Augmented Reality Adjunct (ISMAR-Adjunct)* (2018), pp. 415–416
13. D.C. Rompapas, A. Rovira, S. Ikeda, A. Plopski, T. Taketomi, C. Sandor, H. Kato, Eyear: Refocusable augmented reality content through eye measurements, in *2016 IEEE International Symposium on Mixed and Augmented Reality (ISMAR-Adjunct)* (2016), pp. 334–335
14. R. Dauenhauer, T. Müller, An evaluation of information connection in augmented reality for 3d scenes with occlusion, in *2016 IEEE International Symposium on Mixed and Augmented Reality (ISMAR-Adjunct)* (2016), pp. 235–237
15. D. Mhlanga, Artificial intelligence in the industry 4.0, and its impact on poverty, innovation, infrastructure development, and the sustainable development goals: lessons from emerging economies? Sustainability **13**(11), 5788 (2021)
16. J. Lee, H. Davari, J. Singh, V. Pandhare, Industrial artificial intelligence for industry 4.0-based manufacturing systems. Manuf. Lett. **18**, 20–23 (2018)
17. A. Bécue, I. Praça, Artificial intelligence, cyber-threats and industry 4.0: challenges and opportunities. Artif. Intell. Rev. **54**(5), 3849–3886 (2021)
18. Y.-J. Liu, Research on the construction method of mechanical manufacturing system based on artificial intelligence technology, in *2020 5th International Conference on Smart Grid and Electrical Automation (ICSGEA)* (2020), pp. 149–152
19. E.C. Ang, S.A. Suandi, Smart manufacturing with an artificial neural network to predict manufacturing healthiness, in *2019 IEEE 15th International Colloquium on Signal Processing Its Applications (CSPA)* (2019), pp. 120–123
20. X. Yao, J. Zhou, J. Zhang, C.R. Boër, From intelligent manufacturing to smart manufacturing for industry 4.0 driven by next generation artificial intelligence and further on, in *2017 5th International Conference on Enterprise Systems (ES)* (2017), pp. 311–318
21. P. Lou, J. Guo, J. Yan, X. Jiang, J. Hu, Behavior simulation of manufacturing services in a cloud manufacturing environment, in *2018 3rd International Conference on Information Systems Engineering (ICISE)*, (Los Alamitos, CA, USA) (IEEE Computer Society, Washington, 2018), pp. 137–141
22. R. Rai, M.K. Tiwari, D. Ivanov, A. Dolgui, Machine learning in manufacturing and industry 4.0 applications. Int. J. Prod. Res. **59**(16), 4773–4778 (2021)
23. J. Villalba-Diez, D. Schmidt, R. Gevers, J. Ordieres-Meré, M. Buchwitz, W. Wellbrock, Deep learning for industrial computer vision quality control in the printing industry 4.0. Sensors **19**(18), 3987 (2019)
24. A. Choudhury, The role of machine learning algorithms in materials science: A state of art review on industry 4.0. Arch. Comput. Methods Eng. **28**(5), 3361–3381 (2021)
25. Y. Li, H. Yan, Y. Zhang, A deep learning method for material performance recognition in laser additive manufacturing, in *2019 IEEE 17th International Conference on Industrial Informatics (INDIN)*, vol. 1 (2019), pp. 1735–1740
26. S.A. Khan, M.N.I. Shuzan, M.H. Chowdhury, M.M. Alam, Smart entrance system using computer vision at corporate environment, in *2019 IEEE International Conference on Electrical, Computer and Communication Technologies (ICECCT)* (2019), pp. 1–5
27. J. Harikrishnan, A. Sudarsan, A. Sadashiv, R.A. Ajai, Vision-face recognition attendance monitoring system for surveillance using deep learning technology and computer vision, in *2019 International Conference on Vision Towards Emerging Trends in Communication and Networking (ViTECoN)* (2019), pp. 1–5
28. Y. Zheng, S. Xiao, Performance analysis of a moving target tracking method based on computer vision, in *2016 Eighth International Conference on Measuring Technology and Mechatronics Automation (ICMTMA)* (2016), pp. 467–470
29. A. Muñoz, X. Mahiques, J.E. Solanes, A. Martí, L. Gracia, J. Tornero, Mixed reality-based user interface for quality control inspection of car body surfaces. J. Manuf. Syst. **53**, 75–92 (2019)
30. S. Wang, R. Guo, H. Wang, Y. Ma, Z. Zong, Manufacture assembly fault detection method based on deep learning and mixed reality, in *2018 IEEE International Conference on Information and Automation (ICIA)* (2018), pp. 808–813

31. F. Arévalo, D. Sunaringtyas, C. Tito, C. Piolo, A. Schwung, Interactive visual procedure using an extended fmea and mixed-reality, in *2020 IEEE International Conference on Industrial Technology (ICIT)* (2020), pp. 286–291
32. G. Avalle, F. De Pace, C. Fornaro, F. Manuri, A. Sanna, An augmented reality system to support fault visualization in industrial robotic tasks. IEEE Access **7**, 132343–132359 (2019)
33. F. De Pace, F. Manuri, A. Sanna, D. Zappia, An augmented interface to display industrial robot faults, in *Augmented Reality, Virtual Reality, and Computer Graphics*, ed. by L.T. De Paolis, P. Bourdot (Springer International Publishing, Cham, 2018), pp. 403–421
34. M. Hebenstreit, M. Spitzer, M. Eder, C. Ramsauer, An industry 4.0 production workplace enhanced by using mixed reality assembly instructions with microsoft HoloLens, in *Mensch und Computer 2020 - Workshopband*, ed. by C. Hansen, A. Nürnberger, B. Preim (Gesellschaft für Informatik e.V., Bonn, 2020)
35. F. Alves, H. Badikyan, H. António Moreira, J. Azevedo, P.M. Moreira, L. Romero, P. Leitão, Deployment of a smart and predictive maintenance system in an industrial case study, in *2020 IEEE 29th International Symposium on Industrial Electronics (ISIE)* (2020), pp. 493–498
36. G. Evans, J. Miller, M.I. Pena, A. MacAllister, E. Winer, Evaluating the microsoft HoloLens through an augmented reality assembly application, in *Degraded environments: sensing, processing, and display 2017*, vol. 10197 (International Society for Optics and Photonics, Bellingham, 2017), p. 101970V
37. L. Perez, J. Wang, The effectiveness of data augmentation in image classification using deep learning (2017). Preprint arXiv:1712.04621
38. X. Shi, X. Xiang, L. Ye, Design and implementation of virtual-real interactive system for mixed reality, in *2019 IEEE Conference on Multimedia Information Processing and Retrieval (MIPR)* (IEEE, Piscataway, 2019), pp. 475–479
39. A.G. Taylor, HoloLens hardware, in *Develop Microsoft HoloLens Apps Now* (Springer, Berlin, 2016), pp. 153–159

Industry 4.0 in the Health Sector: System for Melanoma Detection

Verónica Angelica Villalobos Romo, Soledad Vianey Torres Arguelles, Jose David Diaz Roman, Jesus Martin Silva Aceves, Salvador Noriega Morales, and Claudia Georgina Nava Dino

1 Introduction

One of the most common cancers in humans is skin cancer [1], classified into two large groups: non-melanoma and melanoma [2, 3]. The latter is the most lethal type of cancer, as it ranks third in mortality in Mexico with 7.9% and represents 75% of the causes of death from skin cancer in the country. A study conducted between 2014 and 2018 in Mexico yielded a total of 3973 patients who died from melanoma [2]. Also, according to the American Society of Clinical Oncology ASCO and [4], early diagnosis is essential to combat this type of cancer. There are techniques that are used to make this diagnosis, such as visual inspection which is a non-invasive technique and invasive techniques such as biopsy, which help determine whether a skin lesion is benign or malignant [5].

Visual inspection is based on identifying certain defined clinical features to determine if a mole is progressing to melanoma. There are visual examination techniques to identify such characteristics, among which the criteria [6] for clinical diagnosis stand out: the ABCDE rule, the 7-point list, and the Menzies rule. The ABCDE rule represents asymmetry, irregular or round edges, different colors, diameter greater than 6 mm, and evolution. For its part, "the 7-point list identifies three major signs, (1) change in size, (2) shape, and (3) color, and four minor signs, (4) inflammation, (5) crusting or bleeding, (6) sensory change, and (7)

V. A. Villalobos Romo (✉) · S. V. T. Arguelles · J. D. D. Roman · J. M. S. Aceves · S. N. Morales
Universidad Autonoma de Ciudad Juarez, Chihuahua, Mexico
e-mail: veronica.villalobos@uacj.mx; vianey.torres@uacj.mx; david.roman@uacj.mx; jesilva@uacj.mx; snoriega@uacj.mx

C. G. Nava Dino
Universidad Autonoma de Chihuahua, Chihuahua, Mexico
e-mail: ndino@uach.mx

L. C. Méndez-González et al. (eds.), *Innovation and Competitiveness in Industry 4.0 Based on Intelligent Systems*, EAI/Springer Innovations in Communication and Computing, https://doi.org/10.1007/978-3-031-29775-5_3

smaller diameter equal to 7 mm, which allows to identify the formation of malignant melanoma." Finally, the Menzies rule is a method that evaluates 11 dermoscopic criteria which are divided into two negative criteria that must not be present for the diagnosis of melanoma and nine positive criteria, i.e., one of them must be met for the diagnosis of melanoma [7].

In recent years, the use of computer-aided diagnosis (CAD) tools for the early detection of melanoma has become common [8–10] and [11]. One of the advantages of using CAD is to obtain a reliable diagnosis, by means of acquisition, preprocessing, segmentation, feature extraction, and selection, to finally classify dermoscopic images and thus overcome obstacles to improve the automatic diagnosis of dangerous lesions such as melanoma [12].

Studies show similar results to those obtained by certified dermatologists have been achieved by applying convolutional neural networks (CNNs) [13, 14] and [15]. The use of filters within the architecture of a convolutional neural network that examines the different structures of the input images manages to distinguish melanoma from different benign lesions, thus improving accuracy. Deep learning has proven to be very useful, and with the use of deep neural networks [16], for segmentation [17], object detection [14, 18] and [19], in addition to classification, has demonstrated its performance in improving the performance of neural networks for recognizing and classifying medical images [20]. Thus, in the architecture performed in studies [16, 21], they show the use of multiple layers to process and classify images to extract the desired data.

Melanoma accounts for 75% of deaths from skin conditions [2]. It is worth mentioning that the classification of melanoma reaches 65–80% of effectiveness when depending on the clinical examination by a medical expert [13, 22].

Although the cause of skin cancer or melanoma is not known, there are several factors that influence the occurrence of cancer. One of these factors is the continuous and prolonged exposure to solar radiation [23]. A mole (nevus) is a benign skin growth that arises from melanocytes (the majority of persons develop moles) [24]. Although almost all moles are inoffensive, some can increase the risk of skin cancer [25]. One type of mole that looks like a melanoma is sometimes called a Spitz nevus. This mole is most common in children and adolescents, although it occasionally appears in adults. In general, these tumors are benign and do not metastasize. However, physicians sometimes have difficulty distinguishing a Spitz nevus from a malignant melanoma, even when viewed under a microscope [26]. As mentioned, visual inspection of the areas where skin lesions are present (either directly or by means of dermatoscopic images) is essential for the proper diagnosis of early stage cancer, and this evaluation should be performed by an experienced medical specialist in order to recognize the descriptive characteristics of the lesion and thus be able to classify them correctly. This type of task is not exempt from human error; particularly the risk of misdiagnosis increases when operator visual fatigue is present. Additionally, a physician with little experience in the evaluation of dermoscopic images will have little precision in the identification of lesions and may miss a malignant case that, if not confirmed, may eventually develop into a melanoma with little chance of cure. It is in this sense that computational diagnostic tools become relevant, since they lack the subjectivity or associated problems of the medical professional in charge of evaluating the lesions. Another aspect to highlight of this type of tools is that they open the possibility of telediagnosis, allowing a greater coverage in medical care when diseases that are mainly detected with images, especially those associated with skin lesions, occur. In the present work, we propose the development of a dermatoscopic image classification model focused on melanoma detection based on deep learning. This artificial intelligence methodology is chosen as it has been shown to be robust and with high degrees of accuracy in image classification in any context and in particular in medical images. As will be observed below, the proposed model achieves an accuracy close to 90% with test images with an area under the receiver operating characteristic (ROC) curve of 0.95, which demonstrates a high performance in the classification task of the constructed model.

2 Machine Learning Methods for Melanoma Diagnoses

There are invasive and non-invasive diagnostic techniques: invasive techniques involve the removal of part of the skin for subsequent clinical analysis, while non-invasive techniques involve visual inspection by a dermatologist to diagnose whether the skin lesion is prone to become melanoma. However, the diagnosis depends on each expert's point of view and may lack accuracy. For these reasons, alternative methods based on artificial intelligence (AI) are sought for early melanoma diagnosis [6].

Artificial intelligence (AI) is at the center of one of the most profound scientific developments of recent years: the study of information and knowledge and how this information is collected, stored, understood, processed, used, and transmitted [27]. The research led to the development of a new branch of computer science that studies the principles: knowledge obtained and used, goals created and achieved, information transmitted, collaboration achieved, concept formation and language used, and on which the design of artificial intelligence computers has focused, including speech recognition, inference, problem-solving, perception, classification, and image recognition, among others.

2.1 Machine Learning

Machine learning (ML) is a branch of AI in which the system learns from information rather than explicit programming [28]. ML is a complex procedure; according to the algorithm it trained, it manages to build more accurate models from the extracted information. The result of a training information applied to a machine learning algorithm is called a machine learning model. Once the training has been applied, information is entered at the input to validate whether the model gives the correct output [29].

2.1.1 Supervised Learning

This technique usually originates with a fixed group of information and the main idea of how to classify that data. According to [30] *it "is defined by its use of labeled datasets to train algorithms to classify data or predict outcomes accurately. As input data is fed into the model, the model adjusts its weights until it has been fitted appropriately. This occurs as part of the cross validation process to ensure that the model avoids overfitting or underfitting. Supervised learning helps organizations solve a variety of real-world problems at scale, such as classifying spam in a separate folder from your inbox."*

2.1.2 Unsupervised Learning

"Unsupervised learning uses machine learning algorithms to analyze and cluster unlabeled datasets. These algorithms discover hidden patterns or data groupings without the need for human intervention. This method's ability to discover similarities and differences in information make it ideal for exploratory data analysis, cross-selling strategies, customer segmentation, and image and pattern recognition," as explained in [30]. This technique is applied iteratively with the goal of analyzing data without the need for human involvement. It can be used to classify spam in e-mail messages, without the need for an operator to check this process.

2.1.3 Reinforcement Learning

It is a model that, unlike the machine learning model, does not need to be trained with input information, since it learns by empirical methods. Consequently, some successful solutions lead to process improvement, as it is the process that solves problems more efficiently [30].

2.1.4 Supervised Learning Algorithms

This procedure is based on the introduction of data, previously labeled, in the training of the algorithm; this allows categorizing the data or predicting the results accurately at the moment the information is introduced. Since the data are previously labeled (hence the name supervised), the model is adjusted according to the data input. Linear regression and the ranking algorithm are examples of this type of process. Examples of this techniques are linear and logistic regression, decision trees, K-NN (k-nearest neighbor), and neural networks [30].

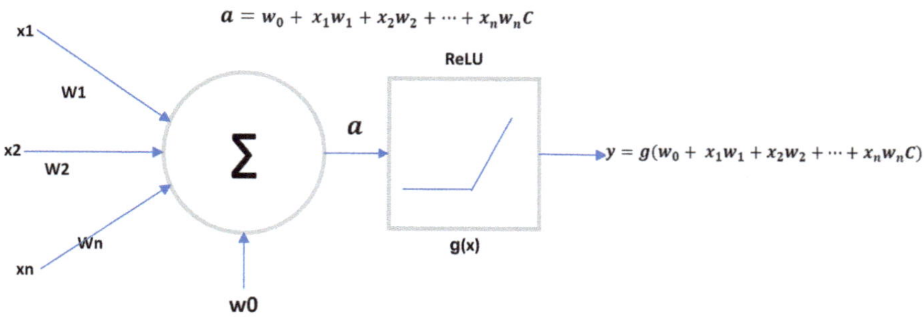

Fig. 1 Processing unit or neuron

2.1.5 Neural Networks

This type of system is named for its relationship to a human nervous system, which is composed of interconnected neurons or processing units (PU) [31]. Synchronization of the PUs is required, and they can work in layers, as shown in Fig. 1. The example figure shows an input layer $x_1, x_2, x_3, \ldots x_n$ with information representing the inputs; in addition in the block, there may be one or more hidden layers called $g(x)$; finally the output layer $y = g(w_0 + x_1w_1 + x_2w_2 + \cdots + x_nw_n)$ is shown. The neurons can be connected with different weights, and the information must pass through the neurons at different levels. Finally, the result is sent to the output layer [32].

Initially, the network starts training with the input records, makes predictions by adjusting the initial weights, and makes adjustments to them until the correct prediction is found (this requires repeating the process several times until a stopping criterion is met) [32].

2.1.6 Multilayer Neural Network

It is an artificial neural network consisting of many layers, being able to solve problems that cannot be separated linearly, which is the main limitation of perceptrons (called simple perceptron). It consists of an input layer (*Input*) which receives the signal, one or several hidden layers (*Hidden*), and finally the output layer (*Output*). In this way, the perception of the second layer makes decisions based on the results obtained from the first layer, allowing it to make decisions at a more complex level than the perception of the first layer. As it progresses through the hidden layers, more complex decisions can be made based on the number of hidden layers in the network. Compared to the single vs. multilayer perceptron, this one contains hidden layers and uses a backpropagation algorithm. The backpropagation algorithm is a process that involves constantly adjusting the weights to minimize the error between the obtained value and the expected value [33]. By adjusting these weights and biases until the desired value is obtained, the neural network can be trained. The structure of a multilayer perceptron can be seen in Fig. 2 [34].

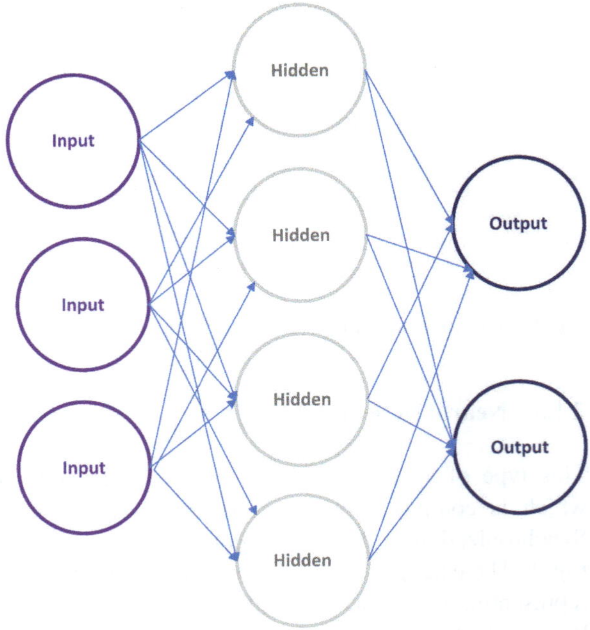

Fig. 2 Neural network with multilayer structure

2.1.7 Convolutional Neural Network

A convolutional neural network (CNN) is a deep learning architecture (this is a subfield of neural networks) which is trained directly with the input information, without requiring to obtain features by hand [35]. They can be used to recognize objects and find logical orders in images, faces, sceneries, etc. They can also be applied in the classification of information other than images, for example, audio signals and temporal signals in general [34, 36]. CNNs can be found in applications that use computer vision and object recognition, such as those used by autonomous vehicles and security systems where facial recognition is required.

2.1.8 CNN Architecture

CNN be made up of several layers, the input, output, and several hidden intermediate layers. Here the layers perform operations that modify data to learn specific characteristics of that information. According to [37] there are three most common layers: the convolution layer (convolutional filters are applied to the input data, in order to highlight certain features), the active layer or linear rectified unit ReLU (in this layer only zero to negative values are assigned, and positive values are kept, which allows you to speed up training time; they are also called activation, since only activated characteristics can pass to the layer that follows), and the reduction layer pooling (the amount of information to be introduced in the training of the network is reduced by decreasing the sampling frequency) [34].

2.1.9 Activation Functions

This function returns outputs in the range 0–1 for an input value. Simple derivatives are chosen to reduce computational cost [38].

2.1.10 Sigmoid Function

This sigmoid function changes the input values into a (0, 1) scale, with high values tending to an asymptote toward 1 and very low values tending to an asymptote toward 0 [38].

$$f(x) = \frac{1}{1 - e^{-x}} \tag{1}$$

2.1.11 Hyperbolic Tangent Function

The hyperbolic tangent function converts the values administered to a scale $(-1, 1)$, where the values obtained tend to an asymptote on that scale [38].

$$f(x) = \frac{2}{1 + e^{-2x}} - 1 \tag{2}$$

2.1.12 Función ReLU—Rectified Lineal Unit

This function adjusts the values enrolled and only allows positive values to pass through and cancels negative values as they entered [38].

$$f(x) = max(0, x) = \begin{cases} 0 & \text{for } x < 0 \\ x & \text{for } x \geq 0 \end{cases} \tag{3}$$

2.1.13 Función Leaky ReLU—Rectified Lineal Unit

The Leaky ReLU function converts the values registered by multiplying negative data by a correction coefficient and leaving the positive values as entered [38].

$$f(x) = \begin{cases} 0 & \text{for } x < 0 \\ a \cdot x & \text{for } x \geq 0 \end{cases} \tag{4}$$

2.1.14 Softmax Function

The Softmax function converts the outputs into a probabilistic representation, so that the result of summing the output probabilities is 1 [38].

$$f(z)_j = \frac{e^{z_j}}{\sum_{k=1}^{k} e^{z_k}} \tag{5}$$

2.1.15 Loss Function

When we train a neural network, the goal is to obtain an output response to the input of predetermined input values. The aim is to minimize the error obtained by adjusting the weights; therefore, efficient training requires constant adjustment of the weights to minimize the loss function. To choose a loss function for a specific difficulty, for instance, for the selected ML algorithm class. In general, loss functions can be divided into regression loss and classification loss. Equation 6 is an example of how to calculate the categorical cross entropy loss function [39]:

$$Loss = -\sum_{i=1}^{outputsize} y_i (log \hat{y}_i) \tag{6}$$

where $\hat{y}_i = n$th scalar value in the model output, y_i = target value, and the output size = scalar quantity in model output.

2.2 *Evaluation Parameters*

When it is required to evaluate classification performance, the metrics of precision, sensitivity, F1 score, accuracy, and confusion matrix are used. To evaluate performance in a classification model, a confusion matrix is normally used; an example is shown in Table 1, where the matrix represents combinations of known and predicted values and measures the performance of a classification model.

- TP: True positive indicates that the present value is positive and is predicted positive.

Table 1 Confusion matrix

	Predictive value	
Present value	Negative (0)	Positive (1)
Negative (0)	TP	FN
Positivo (1)	FP	TN

- TN: True negative indicates the present value is positive and is predicted to be negative.
- FP: False positive indicates that the current value is negative and is predicted to be negative.
- FN: False negative indicates that the current value is negative and is predicted to be positive.

2.2.1 Precision

Precision *"Refers to the dispersion of a set of values obtained from repeated measurements of a quantity"* and is used to measure the quality of the model. The lower the dispersion, the higher the accuracy can be obtained [40]. It is expressed as shown in Eq. 7.

$$precision = \frac{TP}{TP + FP} \tag{7}$$

2.2.2 Sensitivity or Completeness

This metric provides information about how much the model is capable of reporting [40].

$$Sensitivity = \frac{TP}{TP + FN} \tag{8}$$

2.2.3 Specificity

Specificity corresponds to the *"fraction of true negatives that are correctly identified"* [40] and is represented as follows:

$$Specificity = \frac{TN}{TN + FP} \tag{9}$$

2.2.4 Accuracy

It *"refers to how close a measurement is to the true value"*; in statistical terms, precision is interconnected with *"the bias of an estimate."* It is expressed in Eq. 10 [40].

$$Accuracy = \frac{TP + TN}{TP + TN + FP + FN} \tag{10}$$

2.2.5 F1-Score

"The F1 value is used to combine the precision and recall measurements into a single value." This is convenient since it facilitates the comparison of the combined accuracy and recalls performance between different solutions [40].

$$F1 = 2 \times \frac{Precision \times Sensitivity}{Precision + Sensitivity} \tag{11}$$

2.2.6 ROC Curve

The Receiver Operating Characteristic (ROC) curve is a statistical method that helps determine the diagnostic accuracy of test models. It is used for three specific purposes: to deduce the optimal values to reach the highest sensitivity and specificity and thus to be able to make good predictions in an evidential test, i.e., the capability to distinguish moles from melanoma and to compare the discriminatory power of two or more diagnostic examination by expressing their results as a continuous scale. For this purpose, the parameter to be estimated is the area under the ROC curve (AUC). The AUC reflects the ability of the test to distinguish images with and without melanoma. When constructing ROC curves, they are intended to aid in the accurate visualization of the results [41].

3 Proposed Architecture for Classification

Deep convolutional neural networks are used to solve real issues such as image classification. These typically contain a set of convolutional layers with each layer succeeded by clustering layers and finally totally coupled layers. The convolutional layers extract features, the clustering layers summarize them, and the fully connected layers are for classification [42].

The following is a description of the ResNet50 convolutional neural network architecture that has been widely used in pattern recognition to classify images of different nature, including medical images. This architecture has shown great effectiveness in image classification, and it is the one used to generate the skin cancer classification model proposed in this chapter.

3.1 ResNet50 Architecture

ResNet is a residual neural network, and the *"ResNet-50 model is a convolutional neural network (CNN) that stacks residual blocks on top of each other to form a network,"* which has 50 layers deep. This architecture consists of increasing the amount of layers by entering a residual connection [43].

4 Development of the Methodology for Injury Classification

This section describes the methodology used in the development of the algorithm for skin lesion classification, using artificial intelligence techniques and tools to evaluate the images. Also, the classifier that was trained to identify images containing skin cancer lesions is presented. Figure 3 shows a plan for the development of the methodology.

4.1 Materials Used

A personal computer can be used for the development of the project, and if it has a Graphics Processing Unit (GPU), the processing speed can be accelerated. In this case an Intel(R) Core(TM) i7-4810MQ CPU 2.80 GHz 16 GB of RAM was used as the first test computer; however, the processing times were up to 8 h for the first tests. There is another free tool with certain restrictions called *Google Colaboratory* or *Colab*, which also has paid versions; in this case we used the Colab Pro version,

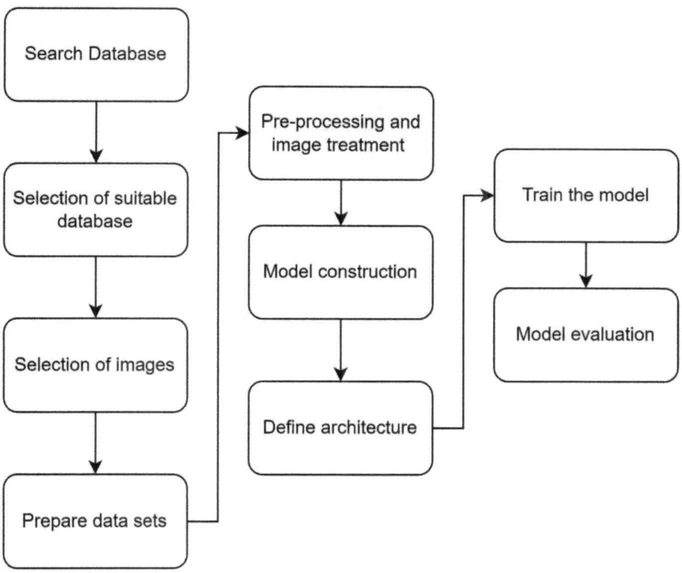

Fig. 3 Methodology planning diagram

with which we could speed up the training time, which allows to speed up the computation time and among others write and run *Python* code in HTML interpreter (usually known as browser), with the following features:

- No configuration required
- Access to Graphics Processing Unit (GPU)
- Ease of sharing

4.2 Database Selection

The algorithm was programmed in the *Python* language using Anaconda and the *Spyder* development environment.

Anaconda is an open source suite that includes a number of applications, libraries, and concepts designed for data science development with *Python*. It is a Python distribution that basically works as an environment manager and has a collection of over 720 open source packages. Anaconda Distribution is grouped into four technology sectors or solutions:

- *Anaconda Navigator*: Anaconda Python graphical interface
- *Anaconda Project*
- Data Science Libraries

An integrated development environment (IDE) called *Spyder* was used to develop the programs in Python language, since its interface is relatively simple. This IDE was used through the ANACONDA platform, since it is easy to install and update additional packages.

In order to find a suitable database for the project, a bibliographic review was carried out [14, 20], in which the set of images from the HAM10000 database was selected [44] (*Human Against Machine* it contains 10,000 training images) which is available at the following address https://dataverse.harvard.edu/dataset.xhtml?persistentId=doi:10.7910/DVN/DBW86T, since it has a wide range of dermoscopy images from multiple origins of common pigmented skin lesions which are mentioned in Table 2.

Table 2 Types of injuries

Injury	Percentage
Melanocytic nevus (nv)	67%
Melanoma (mel)	11%
Benign lesions (bkl)	11%
Carcinoma (bcc)	5%
Actinic keratosis (akiec)	3%
Vascular lesions (vasc)	2%
Dermatofibroma (df)	1%

On the website it is mentioned that *"more than 50% of lesions have been confirmed by pathology, while the ground truth for the rest of the cases was either follow-up, expert consensus or confirmation by in-vivo confocal microscopy."*

4.2.1 Image Selection and Classification

The HAM10000 database consists of seven skin cancer lesion classes within this dataset, which are listed below [45]:

- Melanocytic nevus
- Melanoma
- Benign keratosis-like lesions
- Basal cell carcinoma
- Actinic keratosis
- Vascular lesions
- Dermatofibroma

The reorganization of the images basically consisted of separating all the classes into folders in order to have the images sorted in their respective class. All this procedure is done by reading the metadata file that accompanies the HAM10000 database, which is read by means of the *Python* program as shown below to assign the image to the corresponding lesion type assigned in the HAM10000 metadata.csv file [46]. A sample of images that can be found within each injury classification folder can be seen in Fig. 4.

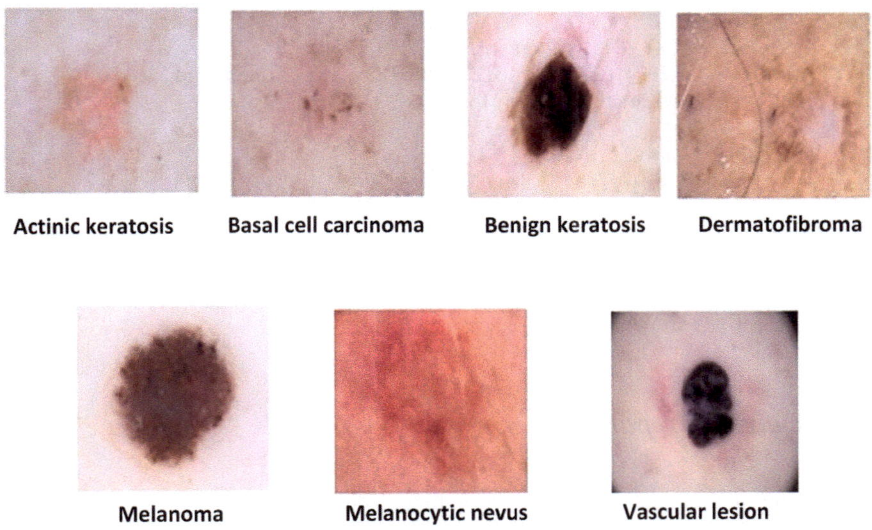

Fig. 4 Image samples by classification

The characteristics of the images are the following:

- Image size 600 × 450 pixels
- RGB color image
- .jpg format

Listing 1 Folder assignment code [46]

```
#Order images
import pandas as pd
import os
import shutil

# Empty all images into a folder and specify the path:
data_dir = os.getcwd() + "/data/all_images/"

# Path to the destination directory where we want to subfolders
dest_dir = os.getcwd() + "/data/reorganized/"

# Read the csv file containing the names of the
#images and the corresponding tags and the corresponding tags

skin_df2 = pd.read_csv('data/HAM10000/HAM10000_metadata.csv')
print(skin_df['dx'].value_counts())

label=skin_df2['dx'].unique().tolist()
#Extract labels into a list
label_images = []

# Copy images to new folders
for i in label:
    os.mkdir(dest_dir + str(i) + "/")
    sample = skin_df2[skin_df2['dx'] == i]['image_id']
    label_images.extend(sample)

    for id in label_images:
        shutil.copyfile((data_dir + "/"+ id +".jpg"),
        (dest_dir + i + "/"+id+".jpg"))
    label_images=[]
```

Thus, the folders were assigned the amount of pictures corresponding to each type of lesion, as shown in Table 3.

4.3 Dataset Preparation

In this step it was taken from the reorganized folder obtained in Sect. 4.2.1, where the images are divided in the folders of Table 3, to put them in three folders named train, valid, and test. Within these, there are folders corresponding to the label of the image itself. In turn, within these folders, we have the images. For

Table 3 Types of injuries [44]

Type of injury	Folder	Number of images
Melanocytic nevus	nv	6705
Melanoma	mel	1113
Benign lesions	bkl	1099
Carcinoma	bcc	514
Actinic keratosis	akiec	327
Vascular lesions	vasc	142
Dermatofibroma	df	115

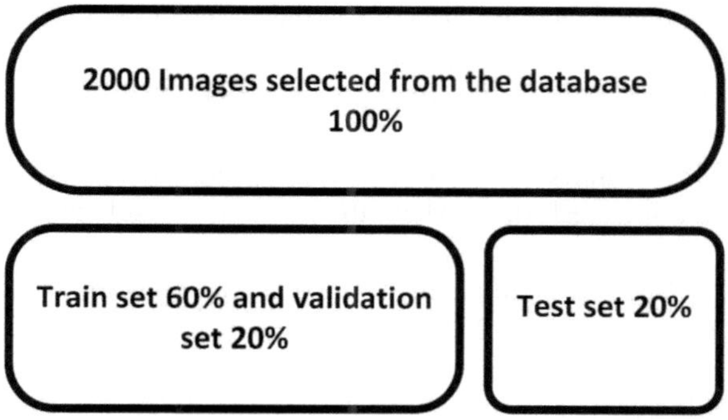

Fig. 5 Training and test dataset

training, a separation of the images was made: training set, which is 80%, from which 20% of this data was taken to form the validation set, and test set, which is 20% of the total selected images, as shown in Fig. 5. The training set was used to train from the network, which allowed the model to be built. These images allow the CNN to extract the necessary features to properly identify each of the classifications, melanoma and melanocytic nevus, while the test set serves to verify that the algorithm is able to properly classify images not known to the network.

4.3.1 Selected Injuries

Class imbalance is one of the most common problems when solving classification problems related to healthcare domain, as in this case classifying melanoma type skin cancer from among the lesion types found within the selected database is unbalanced; this means that we find more nv type cases than mel type with less number of samples. This produces a problem of unbalanced classification, since the classes of the dataset have an unequal number of samples. For this reason, an equal number of images of the randomly selected classes were taken to achieve a balance between classes. It is worth mentioning that this technique is known as random undersampling, since it consists of balancing the data with respect to the classes with the least amount, selecting the data from the class with the greatest amount randomly, as shown in Table 4.

Table 4 Number of images used

Selected images	
Class	Quantity
Melanoma (mel)	1000
Nevus (nv)	1000

4.3.2 Preprocessing and Image Processing

The images were treated by reducing them to 224 × 224 pixels. Since when an image is reduced there is a loss of sharpness, a bicubic interpolation was also applied since this method helps to preserve the definition of the edges and reduces possible losses in the details of the image. It is important to normalize the values of the pixels, before entering them into the network. Pixel colors have values ranging from 0 to 255, so images were converted to a numpy data matrix. Then a normalization of the data was performed where the operation (pixel value/255) was performed to obtain a value between 0 and 1.

When it is required to train a model with images of a large resolution, in this case 224 × 224, it is necessary to divide the training process into blocks or batches of smaller number of images, since trying to train the model requires a large memory capacity. Therefore, *Keras* has the class *ImageDataGenerator*, which allows generating such blocks, in addition to performing the technique called data *augmentation*. It is used for information analysis and is a *"technique used to increase the amount of data by adding slightly modified copies of already existing data,"* and this was achieved by artificially increasing the size of the dataset during training by transforming data [47]. In data enhancement, geometric transformations, rotation, stretching and shrinking of the image, and horizontal and vertical rotation were performed to achieve an artificially increased number of images.

4.4 Model Construction

To perform the classification of skin cancer lesions, which is a case for the use of deep neural networks, we chose to test four pre-trained neural network models to perform the classification of images of a large number of classes. Especially for these models, we have the classes melanoma and melanocytic nevus (moles). The model parameters are as follows:

- Image size 244 × 224
- Image batch size 16
- Number of classes 2
- Optimizer Adam
- Initial learning rate of 0.001
- Type of classification categorical

The training image generator was defined using a class for the training and validation images named *ImageDataGenerator*, where the transformations for the images were defined in the training where it was normalized and, apart from training the original image, it was trained with the same image but transformed by rotating 20°, zooming in 20% or zooming out 20%, and doing a horizontal and vertical rotation. To validate no transformation other than normalization was performed as shown in the code 2:

Listing 2 ImageDataGenerator code

```
train_datagen = ImageDataGenerator(
    rescale=1./255,
    rotation_range=20,
    zoom_range=0.2,
    width_shift_range=0.1,
    height_shift_range=0.1,
    horizontal_flip=True,
    vertical_flip=True)
valid_datagen = ImageDataGenerator(rescale=1./255)
```

Using the *flow_from_directory* statement, train_data and valid_data are constructed. This method is able to automatically identify the classes from the folder title and returns labels of categorical type, which supports multiple label output. In addition, the size of the images was defined, so, each image was resanpled, since the target size is different from the loaded image. The *batch size* parameter specifies the number of images to train in each block.

Listing 3 Dataset code

```
train_data = train_datagen.flow_from_directory(directory=
    train_dir, class_mode='categorical',
    batch_size=16,#16 images at a time
    interpolation='bicubic',
    target_size=(224,224), shuffle=True, seed=100)

valid_data = valid_datagen.flow_from_directory(directory=
    valid_dir, class_mode='categorical',
    batch_size=16,#16 images at a time
    interpolation='bicubic',
    target_size=(224,224), shuffle=False, seed=100)
```

4.5 Architecture Definition

In this step, it used *Keras* applications that are deep learning models, which are used for feature extraction, fine-tuning, and prediction. The aim of the library is to accelerate the creation of a neural network: for this purpose, *Keras* acts as a visual user interface API (short for *Application Programming Interfaces*) that allows access and development of various machine learning tools, where the *TensorFlow-Keras* package was applied. The following libraries were imported:

Listing 4 Code 4

```
import tensorflow as tf
from tensorflow import keras
from tensorflow.keras import layers
from tensorflow.keras.layers import Dense, Dropout, Flatten
from tensorflow.keras.layers import Conv2D,
from tensorflow.keras.callbacks import ModelCheckpoint,
EarlyStopping
from tensorflow.keras.applications import ResNet50
```

The inputs for the tested models were 224 × 224 × 3 as they are RGB images, and the arguments defined are as follows:

- Name of the model to be trained
- input_shape: tuple optionally, it must have exactly three input channels, and the width and height must not be less than 32. In this case, (224, 224, 3).
- include_top: True to include the three fully connected layers at the top of the network.
- input_tensor: optional Keras tensor (i.e., layer output. Input()) to use as image input for the model
- classes: 2 number of classes to classify the images, will only be specified if include_top is true and if no weight argument is specified.
- weights: None (random initialization).

Once the architecture was defined, the model was compiled with the following arguments:

- optimizer: Adam
- A loss function called with loss: categorical crossentropy.
- metrics: List of metrics that the model will evaluate during training and testing, in this case acc which is accuracy.

In training the model, it was helpful to reduce the learning rate as training proceeds with *LearningRateSchedule*, which uses an exponential decay schedule. This applies associate decay performed to the associate optimizer step, given a given initial learning rate. The program commands to decide the argument that produces a decayed learning rate once the present step of the optimizer changes. This is to change the value of the learning rate at different invocations of optimization functions. For the network learning schedule using the above-mentioned methodology, the code 5 was used:

Listing 5 Code network learning rate

```
Mel_modelRN50 = ResNet50(input_shape=(224,224,3),
include_top=True, input_tensor=None, classes=2,weights=None)
Mel_modelRN50 = tf.keras.models.Model(
```

```
inputs = Mel_modelRN50.input ,outputs=Mel_modelRN50.output)
initial_learning_rate = 0.001
lr_schedule = keras.optimizers.schedules.ExponentialDecay(
initial_learning_rate , decay_steps=100000, decay_rate=0.96,
staircase=True)
Mel_modelRN50.compile(optimizer=keras.optimizers.Adam(
learning_rate = lr_schedule), loss='categorical_crossentropy ',
metrics=['acc '])
Mel_modelRN50.summary()
```

In the case argument *staircase* is true, then *step / decay_steps* is a fraction of integers, and also the decay learning rate follows a stairway performance, so it can pass this program on to *tf.keras.optimizers.Optimizer* as the learning rate.

The *callback* or callback functions have the ability to do various processes at various stages during training, and were used for training are *EarlyStopping*, which stops the training if it sees that it does not improve the cost function after certain *epochs*; furthermore, if the training does not improve after certain specific *epochs*, it reduces the *learning rate* value of the model, which usually gets an improvement of the training. We also used *ModelCheckpoint* which saves the best model (in a checkpoint file), so that the model is loaded later to continue training from the saved state. As can be seen in the code 6:

Listing 6 Code callback functions and save best model

```
file_model = "/content/gdrive/MyDrive/Melanoma_1/Mel_500_RN50.h5"
checkpoint_cb = ModelCheckpoint(file_model , monitor='val_acc ',
verbose=1, save_best_only=True , mode='max')
early_stopping_cb = EarlyStopping(monitor="val_acc",patience=80)
```

4.6 Model Training

The next step was to train the models selected for classification mentioned in chapter "Failure Detection System Controlled by a Mixed Reality Interface" by applying the same parameters for their construction. For this, the image generators that were defined in Sect. 3.1 were passed to them. For this purpose, the function *Fit()* was used, which has the following parameters:

- Input data: train data training set
- Epochs: 500
- Verbose: 1
- Callbacks: checkpoint_cb,early_stopping_cb
- Validation_data: valid_data validation set
- Shuffle: True
- Workers: 1
- Use_multiprocessing: False

And the following code is used:

Listing 7 Training code

```
Melanoma_train_dropout = Mel_modelRN50.fit(train_data ,epochs=500,
callbacks=[checkpoint_cb ,early_stopping_cb], verbose=1, workers=1,
validation_data=valid_data)
```

The *fit* method takes the training data as arguments, to return a *history* object.

4.6.1 Visualization of the Training Model

The *fit* function goes back to a *History* object, while *History.history* registers the changes of the loss function and metrics for each epoch of training. It also contains the changes of the metrics in the validation set. With this, graphs were created from the collected historical data, where the following were obtained:

- A plot of accuracy in training and validation datasets during training epochs
- A plot of loss in training and validation datasets during the training epochs

Using the code 8 the graphics are obtained:

Listing 8 Code display graphics

```
loss = Melanoma_train_dropout.history['loss']
val_loss = Melanoma_train_dropout.history['val_loss']
epochs = range(1, len(loss) + 1)
plt.plot(epochs, loss, 'b', label='Lost in Train')
plt.plot(epochs, val_loss, 'r', label='Lost in Val')
plt.title('Loss in Training and Validation')
plt.xlabel('Epochs')
plt.ylabel('Lost')
plt.legend()
plt.show()


acc = Melanoma_train_dropout.history['acc']
val_acc = Melanoma_train_dropout.history['val_acc']
plt.plot(epochs, acc, 'b', label='Training acc')
plt.plot(epochs, val_acc, 'r', label='Validation acc')
plt.title('Precision training and validation')
plt.xlabel('Epochs')
plt.ylabel('Precision')
plt.legend()
plt.show()
```

4.7 Model Evaluation

The next step is to evaluate the model; at this point it is where the confusion matrix shows the number of hits and errors in the classification of the validation and test data of the trained model. Since the model was trained, it was tested by means of a classification with the test data to see the degree of successes that the network has had. With the code 9, the percentage of the function of *loss* and *precision* for the test set is obtained:

Listing 9 Code evaluate model

```
# loss and precision metrics
score = modelo.evaluate(test_data)

print ("%s: %.2f%%" % (modelo.metrics_names[0], score[0]*100))
print ("%s: %.2f%%" % (modelo.metrics_names[1], score[1]*100))
```

The confusion matrix is obtained using the code 10:

Listing 10 Confusion matrix code

```
from sklearn.metrics import classification_report,
confusion_matrix,ConfusionMatrixDisplay, accuracy_score
from keras.models import load_model

test_dir = '/content/gdrive/MyDrive/Melanoma_1/Test'
data_test = ImageDataGenerator(rescale=1. / 255)

datos_prueba = data_test.flow_from_directory(test_dir,
class_mode='categorical', batch_size=16, #16 images at a time
interpolation='bicubic', target_size=(224,224), shuffle=False)
#Resize images
modelo = load_model('/content/gdrive/MyDrive/
Melanoma_1/Mel_500_RN50.h5')

test_predicted = model.predict(test-data)
test_predicted_label = np.argmax(test_predicted,axis=1)
test_label = test_data.labels
target_names = test_data.class_indices.keys()
reporte2 = classification_report(test_label,
test_predicted_label, target_names=target_names)
print(reporte2)

cm2 = confusion_matrix(test_label, test_predicted_label)
print(cm2)

displ = ConfusionMatrixDisplay(confusion_matrix=cm2,
display_labels=target_names)
displ.plot()
plt.show()
```

To correctly interpret the predictions made by two-class classification models, the ROC curve was used, where the ROC curve is the graph that plots the ratio of true positives or sensitivity (y-axis) against the ratio of false positives or 1-specificity (x-axis). These ratios were obtained as a function of a series of thresholds defined between 0 and 1. It plots the "false alarm" vs. success rate, displayed by code 11.

Listing 11 Code ROC curve

```
from sklearn.metrics import roc_curve
from sklearn.metrics import roc_auc_score
from matplotlib import pyplot as plt

fpr, tpr, thresholds = roc_curve(test_label,
1-test_predicted[:,0])
plt.figure()
plt.plot(fpr, tpr, color='darkorange',
         label='ROC curve')
plt.plot([0, 1], [0, 1],'--')
plt.xlim([-0.05, 1.05])
plt.ylim([-0.05, 1.05])
plt.xlabel('False Positive Rate')
plt.ylabel('True Positive Rate')
plt.title('Receiver operating characteristic example')
plt.legend(loc="lower right")
plt.show()
```

5 Results

The results are presented below.

5.1 Results of the ResNet 50 Model

In this model, the accuracy and loss function were observed, and the following results were obtained:

- Accuracy in training with validation data 92%
- Loss in training 26%

In the validation it can be seen in Fig. 6a in blue that the graph follows an upward trend which shows that in the training it is increasing; in turn, it is observed that it is increasing along with the training, which suggests that it can continue to improve. The loss function in Fig. 6b can be observed as the difference between the actual values and the expected values of the network is decreasing significantly, which can be seen to have a higher efficiency for classification.

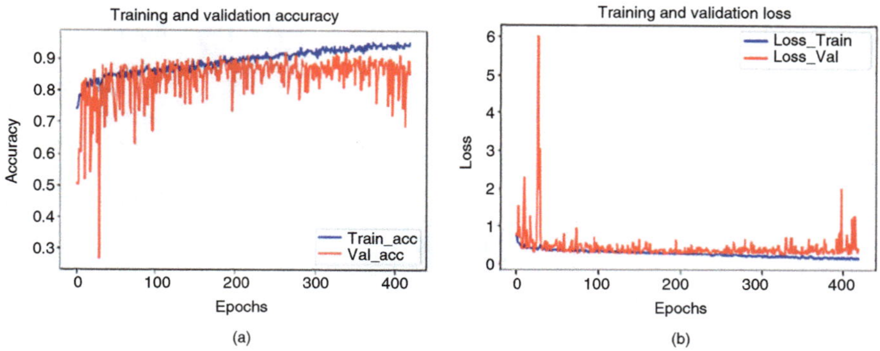

Fig. 6 ResNet50 training and validation historical charts. (**a**) Accuracy graph. (**b**) Loss graph

Fig. 7 Confusion matrix in validation ResNet50

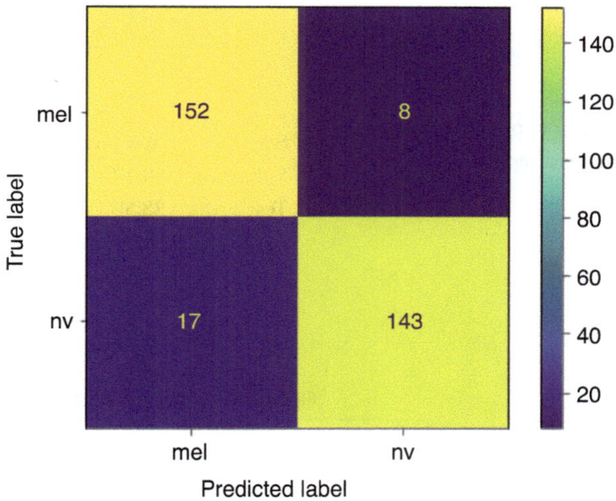

The confusion matrix for the validation set with a number of 320 images in Fig. 7 shows on the main diagonal 152 and 143 corresponding to the values correctly estimated by the model, both true positives TP=152 and true negatives TN=143. The other diagonal, therefore, represents the cases in which the model was wrong 17 false positives FP and 8 false negatives FN.

The confusion matrix for the test set, which contains images not previously seen by the trained model in Fig. 8, shows on the main diagonal 180 and 177 corresponding to the values correctly estimated by the model, both true positives TP=180 and true negatives TN=177. The other diagonal, therefore, represents the cases in which the model was wrong 23 false positives FP and 20 false negatives FN.

The data obtained in Table 5 can be defined as a good model to classify melanoma, since the accuracy in the test data is good, as can be seen in the ROC curve of the model in Fig. 9.

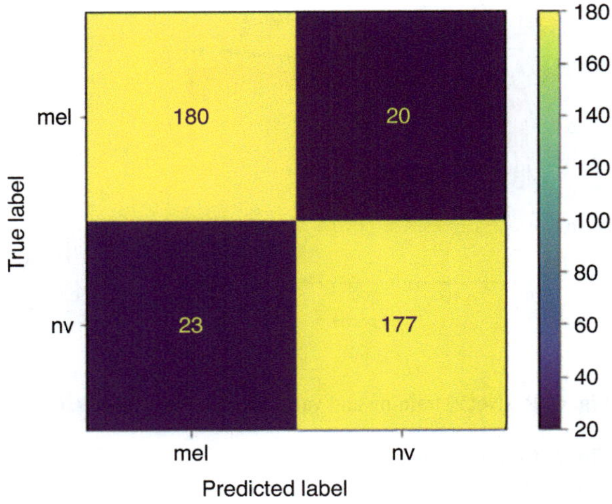

Fig. 8 Confusion matrix test set ResNet50

Table 5 ResNet50 training and test report

Set	Specificity	Accuracy	Sensitivity	Precision
Validation	89.37%	92.18%	95%	89.94%
Test	88.50%	89.25%	90%	88.66%

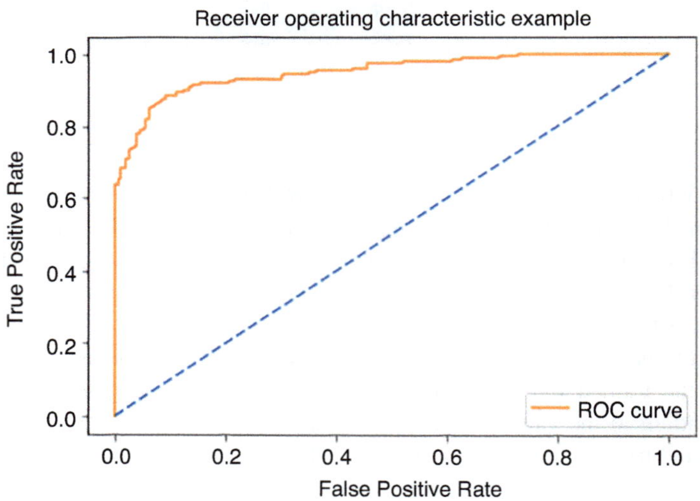

Fig. 9 ROC curve obtained from the ResNet50 model

Looking at the ROC curve of this model, which represents the sensitivity versus specificity for detecting melanoma cancer and benign moles, it can be seen that the network found a greater number of true positives than false positives. Where AUC = 0.95 is obtained, it means that there is 95% probability that the model distinguishes between the mel and nv classes. This means that the discriminative ability of a diagnostic test refers to its ability to succeed in classifying dermoscopic images of moles against images of melanoma.

6 Conclusions

In the present investigation, an algorithm based on deep learning was developed to classify melanoma lesions concerning melanocytic nevi or common moles.
Based on the specific objectives, the following results were obtained:

- For the selection of the database, a bibliographic review was carried out to find the appropriate dermoscopic images for the project. Thus, the HAM10000 database was chosen because it contains an extensive collection of dermoscopic images from multiple sources, with the most common pigmented skin lesions.
- The ResNet50 architecture, which specializes in solving the image classification problem, was analyzed, since it is widely used for pattern recognition in images of all kinds, including medical images.
- Regarding the design and adaptation of the architecture, the model was built by defining the parameters such as the size of the input image, the batch size, the number of classes, the optimizer to be used, the learning rate, and the type of classification.
- The training was carried out with 500 training epochs to analyze the behavior of the model when classifying two classes of images and the accuracy when validating the training, where the highest accuracy in the validation of 92% was obtained with the proposed algorithm.
- In the evaluation of the model carried out with the test images, an accuracy of 89% was obtained in the identification of melanoma against melanocytic nevus.

It can be concluded that the algorithm for training a model for classification based on deep learning works to identify melanoma from a common mole.

It is possible to develop an algorithm that can identify more than two classes of skin lesions, with the tools that are currently available. In this research it was possible to adapt an architecture such as ResNet50 to identify melanoma and melanocytic nevus. The algorithm can be improved with different techniques such as pattern recognition, segmentation, and data clustering. By applying them to image classification architectures to obtain better results, some models could be trained until the possibility of improvement is exhausted and a tool with greater capacity to identify more types of skin lesions could be found by combining the various known techniques to develop specialized algorithms for analyzing medical images to identify patterns of lesions and conditions, to provide a tool that is useful in medical diagnosis. Pre-trained network models can also be used to extract features and employ a classifier of any type, such as a support vector machine, a Bayesian classifier, random forests, or k-nearest neighbors. There are possibilities to develop, adapt, and/or improve on image detection.

References

1. M. Arnold, D. Singh, M. Laversanne, Global burden of cutaneous melanoma in 2020 and projections to 2040. JAMA Dermatol. 495–503 (2022). https://doi.org/10.1001/jamadermatol.2022.0160
2. A. Camara-Salazar, K. Santos-Zaldívar et al., Características individuales y por entidad federativa de la mortalidad por melanoma en México entre 2014 y 2018. Dermatología Revista Mexicana (Springer, Berlin, 2020). https://doi.org/10.1007/s10985-010-9161-1
3. M. V. Cuevas, M. E. Vega et al., Frecuencia de cáncer de piel; experiencia de 10 años en un centro de diagnóstico histopatológico en la ciudad de Durango, Durango, México. Dermatol Rev Mex **63**, 152–159 (2019)
4. H.G. Welch, B.L. Mazer, A.S. Adamson, The rapid rise in cutaneous melanoma diagnoses. N. Eng. J. Med. **384**(1), 72–79. https://doi.org/10.1056/NEJMSB2019760
5. M. Krishna Monika et al., Skin cancer detection and classification using machine learning, in *Materials Today: Proceedings*, vol. 33 (2020), pp. 4266–4270. https://doi.org/10.1016/j.matpr.2020.07.366
6. Secretaría de Salud México, Abordaje Diagnóstico del Melanoma Maligno. Catálogo Maestro de Guías de Práctica Clínica. IMSS-547-1 1–11 (2012)
7. P. Zaballos, C. Carrera et al., Criterios dermatoscópicos para el diagnóstico del melanoma. Med Cutan Iber Lat Am **32**, 3–17 (2004)
8. A. Khanna, B. Bhushan, An internet of health things-driven deep learning framework for detection and classification of skin cancer using transfer learning. Trans. Emer. Telecommun. Technol. **32**(7) (2021). https://doi.org/10.1002/ett.3963
9. A. Murugan, S.A.H. Nair, A.A.P. Preethi, K.P.S. Kumar, (2021) Diagnosis of skin cancer using machine learning techniques. Microprocess Microsyst. **81**. https://doi.org/10.1016/j.micpro.2020.103727
10. Z. Xu, F.R. Sheykhahmad, N. Ghadimi, N. Razmjooy, Computer-aided diagnosis of skin cancer based on soft computing techniques. Open Med. (Poland) **15**(1), 860–871 (2020). https://doi.org/10.1515/MED-2020-0131/MACHINEREADABLECITATION/RIS
11. N. Razmjooy et al., Computer-aided diagnosis of skin cancer: a review. Curr. Med. Imaging **16**(7), 781–793 (2020). https://doi.org/10.2174/1573405616666200129095242
12. E. Vocaturo, D. Perna et al., Machine learning techniques for automated melanoma detection, in *2019 IEEE International Conference on Bioinformatics and Biomedicine (BIBM)* (2017). https://doi.org/10.1109/BIBM47256.2019.8983165
13. T. Brinker, A. Hekler et al., Skin cancer classification using convolutional neural networks: systematic review. J. Med. Int. Res. 450–462 (2018). https://doi.org/110.2196/11936
14. T. Brinker, A. Hekler et al., A convolutional neural network trained with dermoscopic images performed on par with 145 dermatologists in a clinical melanoma image classification task. Eur. J. Cancer 148–154 (2019). https://doi.org/10.1016/j.ejca.2019.02.005
15. T.J. Brinker et al., Diagnostic performance of artificial intelligence for histologic melanoma recognition compared to 18 international expert pathologists. J. Am. Acad. Dermatol. **86**(3), 640–642 (2022). https://doi.org/10.1016/j.jaad.2021.02.009
16. T. Yan Tan, L. Zhang, C. Peng Lim, Adaptive melanoma diagnosis using evolving clustering, ensemble and deep neural networks, vol. 187 (2020), p. 104807. https://doi.org/10.1016/j.knosys

17. X. Tong, J. Wei, B. Sun, S. Su, Z. Zuo, P. Wu, Ascu-net: attention gate, spatial and channel attention u-net for skin lesion segmentation. Diagnostics **11**(3) (2021). https://doi.org/10.3390/diagnostics11030501
18. B. Albert, Deep learning from limited training data: novel segmentation and ensable algoritms applied to automatic melanoma diagnosis. https://ieeexplore.ieee.org/stamp/stamp.jsp?arnumber=8993822
19. H.P. Chan, L.M. Hadjiiski, R.K. Samala, Computer-aided diagnosis in the era of deep learning. Med. Phys. **47**(5), e218–e227 (2020). https://doi.org/10.1002/mp.13764
20. A. Namozov, Y.I. Cho, Convolutional neural network algorithm with parameterized activation function for melanoma classification. IEEE (2018). https://doi.org/10.1109/ICTC.2018.8539451
21. V. Pomponiu, H. Nejati et al., DEEPMOLE: deep neural networks for skin mole lesion classification. IEEE ICIP 2623–2627 (2016). https://doi.org/10.1109/ICTC.2018.8539451
22. R. García, R. Jiménez et al., Eficacia de la dermatoscopia en el diagnóstico de malignidad de lesiones circunscritas de la piel, mucosas y anexos cutáneos. Dermatol Peru 141–150 (2013)
23. C. Mayo, Cáncer de piel - Síntomas y causas. Mayo Clinic. A Available via DIALOG (2021). https://www.mayoclinic.org/es-es/diseases-conditions/skin-cancer/symptoms-causes/syc-20377605. Cited Sep 2022
24. L. Sainz-Gaspar et al., Nevo de Spitz y otros tumores spitzoides en la infancia. Parte 1: aspectos clínicos, histológicos e inmunohistoquímicos. Actas Dermosifiliogr **111**(1), 7–19 (2020). https://doi.org/10.1016/J.AD.2019.02.011
25. L. Sainz-Gaspar et al., Spitz nevus and other spitzoid tumors in children part 1: clinical, histopathologic, and immunohistochemical features. Actas Dermosifiliogr (2020). https://doi.org/10.1016/j.adengl.2019.12.006
26. American Society Cancer, Acerca del cáncer de piel tipo melanoma. American cancer Society. Available via DIALOG (2020). https://www.cancer.org/es/cancer/cancer-de-piel-tipo-melanoma/acerca/que-es-melanoma.html. Cited 20 Aug 2022
27. H. Banda, *Inteligencia Artificial: Principios y Aplicaciones* (Banda, España, 2014)
28. C. Tang, J. Ji, Y. Tang, S. Gao, Z. Tang, Y. Todo, A novel machine learning technique for computer-aided diagnosis. Eng. Appl. Artif. Intell. **92** (2020). https://doi.org/10.1016/j.engappai.2020.103627
29. CleverData, ¿Que es Machine Learning? Available via DIALOG (2019). https://www.ibm.com/mx-es/analytics/machine-learning. Cited 30 Sep 2022
30. IBM. Machine Learning. Available via DIALOG (2020). https://www.ibm.com/cloud/learn/machine-learning. Cited 2022
31. D. Popescu, M. El-Khatib, H. El-Khatib, L. Ichim, New trends in melanoma detection using neural networks: a systematic review. Sensors **22**(2) (2022). https://doi.org/10.3390/s22020496
32. IBM, El modelo de redes neuronales - Documentación de IBM. Available via DIALOG (2021). https://www.ibm.com/docs/es/spss-modeler/saas?topic=networks-neural-model. Cited 20 Sep 2022
33. J. Leonel, Multilayer Perceptron. A Available via DIALOG (2018). https://medium.com/@jorgesleonel/multilayer-perceptron-6c5db6a8dfa3. Cited 23 Jan 2022
34. S. Cooper, A Neural Networks: A Practical Guide for Understanding and Programming Neural Networks and Useful Insights for Inspiring Reinvention (2018)
35. L. Zhang, H.J. Gao, J. Zhang, B. Badami, Optimization of the convolutional neural networks for automatic detection of skin cancer. Open Med. (Poland) **15**(1), 27–37 (2020). https://doi.org/10.1515/MED-2020-0006/HTML
36. M.A. Kassem, K.M. Hosny, R. Damaševičius, M.M. Eltoukhy, Machine learning and deep learning methods for skin lesion classification and diagnosis: a systematic review. Diagnostics **11**(8) (2021). https://doi.org/10.3390/diagnostics11081390
37. Mathworks, Redes Neuronales Convolucionales. MATLAB & Simulink. Available via DIALOG (2020). https://la.mathworks.com/discovery/convolutional-neural-network-matlab.html?msclkid=d449a089d08a11ec9e6a91e0dc. Cited Sep 2022

38. D. Calvo, Función de activación-Redes neuronales. A Available via DIALOG (2018). https://www.diegocalvo.es/funcion-de-activacion-redes-neuronales/. Cited Sep 2022
39. Q. Wang, Y. Ma, K. Zhao et al., A comprehensive survey of loss functions in machine learning. Ann. Data. Sci. **9**, 187–212 (2022)
40. F. Rasul, N.K. Dey, M.M.A. Hashem, A comparative study of neural network architectures for lesion segmentation and melanoma detection, in *2020 IEEE Region 10 Symposium (TENSYMP)* (2020). pp. 1572–1575. https://doi.org/10.1109/TENSYMP50017.2020.9230969
41. J. Cerda, L. Cifuentes, Uso de curvas ROC en investigación clínica: Aspectos teórico-Prácticos. Revista chilena de infectología 138–141 (2012)
42. S. Mascarenhas, M. Agarwal, A comparison between VGG16, VGG19 and ResNet50 architecture frameworks for Image Classification, in *2021 International Conference on Disruptive Technologies for Multi-Disciplinary Research and Applications (CENTCON)* (2021). https://doi.org/10.1109/CENTCON52345.2021.9687944
43. K. He, X. Zhang, S. Ren J. Sun, Deep residual learning for image recognition, in *IEEE Conference on Computer Vision and Pattern Recognition (CVPR)* (2016), pp. 770–778. https://doi.org/10.1109/CVPR.2016.90
44. P. Tschandl, The HAM10000 dataset, a large collection of multi-source dermatoscopic images of common pigmented skin lesions (2018). https://doi.org/10.7910/DVN/DBW86T, Harvard Dataverse, V3, UNF:6:/APKSsDGVDhwPBWzsStU5A== [fileUNF]
45. P. Tschandl, C. Rosendahl, H. Kittler, The HAM10000 dataset, a large collection of multi-source dermatoscopic images of common pigmented skin lesions. Sci Data (2018). https://doi.org/10.1038/sdata.2018.161
46. B. Sreenivas, 202 - Two ways to read HAM10000 dataset into python for skin cancer lesion classification. (February 17, 2021). [Online video], Available at https://www.youtube.com/watch?v=qB6h5CohLbs
47. C. Shorten, T.M. Khoshgoftaar, A survey on image data augmentation for deep learning. J Big Data (2019). https://doi.org/10.1186/s40537-019-0197-0

Assistive Device for the Visually Impaired Based on Computer Vision

Alan Iván Hernández Holguín, Luis Carlos Méndez-González, Luis Alberto Rodríguez-Picón, Iván Juan Carlos Pérez Olguin, Abel Euardo Quezada Carreón, and Luis Gonzalo Guillén Anaya

1 Introduction

The artificial intelligence field is one of the focuses of industry 4.0. It has been used for varied applications like robotic implementations [1], object recognition [2], and maintenance tasks [3]. Classification and object recognition are subjects of interest for developing novelty devices to provide assistive technology for visual impairment.

Vision problems correspond to all those ailments and physical problems related to the gradual or sudden decrease in the sense of vision that prevails in a good part of the population and generates a certain degree of blindness. According to the WHO [4], in 2019, there were about 2.2 billion people with some vision problems, including cases of blindness, which corresponds to a state in which those affected have such impaired vision that they cannot distinguish essential characteristics of objects, such as colors, shapes, and spatial location.

Due to the variety of cases and the difficulty in comparing certain conditions and degrees of blindness, these belong to a spectrum of vision problems in which affected users may experience particular situations in each case. Many cases of blindness may be unique and may not match the traditional definition of blindness,

A. I. H. Holguín (✉) · L. C. Méndez-González · L. A. Rodríguez-Picón · I. J. C. Pérez Olguin
Department of Industrial Engineering and Manufacturing, Institute of Engineering and Technology, Universidad Autónoma de Ciudad Juárez, Chihuahua, México
e-mail: al216737@alumnos.uacj.mx; luis.mendez@uacj.mx; luis.picon@uacj.mx; ivan.perez@uacj.mx

A. E. Q. Carreón · L. G. G. Anaya
Department of Electrical and Computation, Institute of Engineering and Technology, Universidad Autónoma de Ciudad Juárez, Chihuahua, México
e-mail: abquezad@uacj.mx; lguillen@uacj.mx

L. C. Méndez-González et al. (eds.), *Innovation and Competitiveness in Industry 4.0 Based on Intelligent Systems*, EAI/Springer Innovations in Communication and Computing, https://doi.org/10.1007/978-3-031-29775-5_4

where a person does not have a sense of vision. There have been documented cases of people who are legally blind but can distinguish specific shapes or patterns with difficulty [5].

The proposed device focuses on implementing a light object classification model based on a Support Vector Machine (SVM) multi-class implementation and a detection model based on YOLO v4 to classify and detect objects in pictures and video inputs provided by the camera of a portable device. The user can interact with the recognition system through an Android or web app.

In literature, there are several approaches to image classification. In this paper, the SVM algorithm was selected because this approach produces lightweight models and can be easily adapted to high dimension problems with multiple features. It can be applied to image classification tasks.

1.1 Background and State of the Art

Some technologies implemented in assistive devices for visually impaired people are artificial intelligence, focused on computer vision, the Internet of Things, and cloud computing. Also, these techs integrate the core fields of industry 4.0.

Computer vision encompasses several techniques that involve the use of software tools for the design of algorithms and involves the use of software tools for the design of algorithms involving the application of machine learning techniques (such as neural networks) in order to train systems with information stored in databases to recognize patterns in certain databases so that they can recognize patterns in specific characteristics of a product, such as the color and shape of particular objects. These techniques, in conjunction with image processing and facial and object recognition, attempt to emulate the human sense of vision [6].

In computer vision, focusing on image processing, recognition systems have been designed for color identification tasks [7]. Accessibility tools have been adapted for blind people in systems that use audio feedback to inform the user of the shape or color of a particular object [8]. Object recognition and location systems have been developed in the environment of a house with support for the Internet of Things and cloud services [9]. Pattern analysis techniques have been implemented to feedback intelligent assistance systems to monitor and detect particularities in a user's daily routines [10].

Accessibility consists of providing the means to adjust an environment to the user's needs and preferences [11]. It is based on the implementation of systems, devices, and methods of assistance to facilitate people's daily activities, with a special focus on those with disabilities. In particular, blind people may have problems performing their daily activities, so alternatives such as the Braille system to allow writing and reading and pets and canes to facilitate navigation have been developed. Assistive devices based on technologies such as artificial intelligence, computer vision, and the Internet of Things have been implemented to provide alternative solutions [12].

In the field of accessibility systems applications, home automation systems focused on accessibility for blind people have been implemented, like implementations that allow the user to access information about their environment through a smartphone with an interface with descriptive audio [13]. A connectivity environment between smart devices has been developed that allows the integration of external systems such as canes and smartphones to facilitate the interaction between the user and their environment [14]. A wearable device robust to external disturbances, such as lighting and noise, has been designed to provide an accessibility interface to facilitate navigation, obstacle detection, and facial recognition [15]. Regarding rehabilitation, the learning curve for blind people to acquire the necessary skills to operate assistive devices for navigation, orientation, and Braille writing has been analyzed [16].

Cloud computing provides access to scalable and virtualized resources utilizing a service scheme through the Internet [17]. The Internet of Things (IoT) consists of an interconnected system of computing devices, machines, smart devices, and objects that can transfer information over a network.

In the development of the cloud computing and IoT field, these technologies have been implemented in home automation projects to propose improved systems focused on low power consumption and better connectivity between devices with cloud services [18]. Monitoring and control systems have been established for smart lamps [19]. An analysis of the implementation of the Internet of Things and computer vision has been developed to monitor the behavior of customers and users in given environments, such as businesses, to train intelligent systems to provide assistance and obtain information on their interests [20]. Due to the versatility of IoT, implementations have also focused on blind people, such as intelligent door systems with connectivity to cloud services and other smart devices [21]. The implementation of edge computing was used to reduce the bandwidth load and improve the response of IoT-based systems [22].

Some approaches were found in literature in the application of assistive devices focused on the visually impaired, which can vary according to the focus of the aid provided, portability, and software or hardware capabilities [23]. These assistive devices focus on three main tasks: navigation, detection, and classification.

Assistive systems that integrate navigation contain tools that allow greater ease of movement in outdoor or indoor environments; sensors can be incorporated into the navigation system, which can sometimes represent a high battery consumption. These systems also contain algorithms that allow predicting the position of objects concerning time and help to predict possible future obstacles in real time; in this case, in the literature, it is found that these systems may have a low accuracy [24]. It was also found in the literature that a dual system conformed by integrated lenses and shoes was developed to obtain better obstacle detection responses. However, this system is limited to detecting objects embedded in the ground [25].

Vision assistive systems have also been developed to detect various objects through a camera in real time. Patel et al. [26] proposed the inclusion of sensors through which it is possible to obtain other types of information, such as the distance and size of the detected objects; however, this system requires long execution times.

Other systems found in the literature focused on detection accuracy, for which they integrated algorithms based on neural networks, but they have the disadvantage of high consumption of computational resources, while to ensure object registration, GPS has been implemented to guarantee the detection of objects that have been detected with low accuracy [27].

On the other hand, classification-focused assistance systems are responsible for labeling images according to certain defined features, such as color and shape [7, 28].

The systems reported in the literature show a significant advance in detecting objects and their various characteristics, such as color, shape, and distance. In some cases, this detection is performed in real time. However, those that ensure accuracy require a long time for the identification or a high computational cost, and those that are fast reported low accuracy.

1.2 Proposed System

The system is a mobile device connected to an external camera to take pictures of objects, which will be classified using an SVM multi-class implementation executed through an Android app. A conceptual representation of the system is presented in Fig. 1.

The system process includes four stages:

1. **Voice recognition**: The users use voice commands to activate the app functions and can request the acquisition of an image through the camera.
2. **Image acquisition**: A photo is taken and loaded into the classification model. A simple image processing is applied to re-scale the image and represent its data.

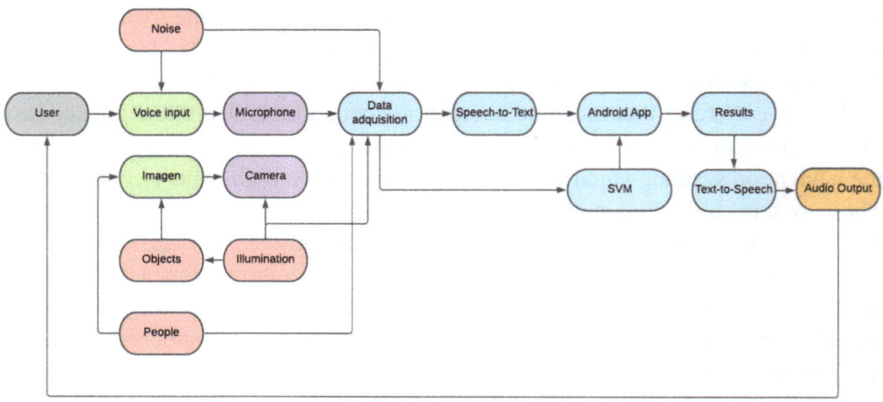

Fig. 1 Diagram of the proposed system

3. **Classification**: The users can choose which algorithm instance to be executed. If they choose the local version, the response will be faster but with reduced accuracy. If they choose the cloud version, the response will be more accurate and fast, but it requires a stable Internet connection.
4. **Results and feedback**: The model predicts the image's label, and the result is processed by a voice synthesis tool to provide the user with audio feedback on the detected class in the image.

2 Methodology

The SVM was developed in three stages: configuration, which includes the algorithm setup, dataset creation, training, and validation; classifications test, which uses new image data to verify the algorithm's performance; and implementation.

The configuration process (shown in Fig. 2 includes the following steps:

1. **Classes and parameters setup**: It consists of classes and images dataset selection, features of the Support Vector Machine, hyperparameters configuration, and cost function definitions.
2. **Training and validation set creation**: The image dataset is obtained from a repository, and the collection of images is segmented into two smaller subsets: the training set and the validation set. The former is the data the algorithm uses to train the model for predictions over new data, and the latter allows for validating the model's performance.

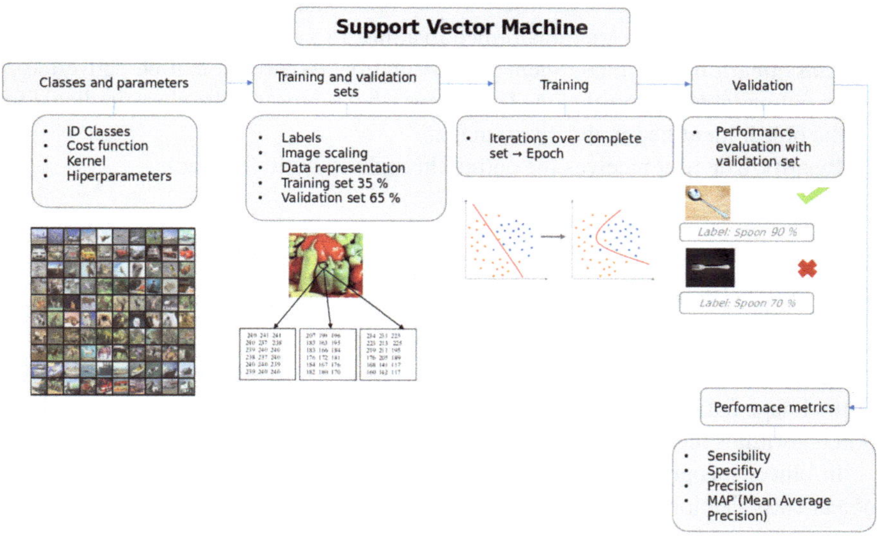

Fig. 2 SVM configuration process

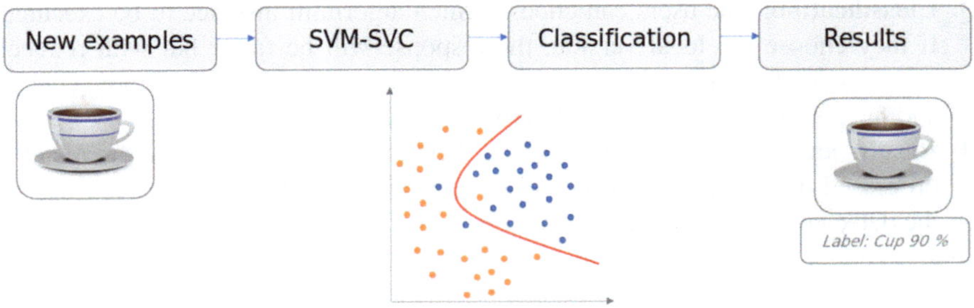

Fig. 3 SVM classification process

3. **Validation**: The model's performance is tested with the validation set (data not included in the training set) to know the model's reliability in making predictions over new data.
4. **Performance metrics and model correction**: The performance metrics are obtained after the model's validation, and, if required, the necessary adjustments or corrections to the model are made to improve his prediction ability.

The classification test and implementation stages require the deployment of the algorithm through an Android application (light model) and a cloud environment (robust model) to use the device's capabilities and make predictions over new images taken by the camera. The classification process (shown in Fig. 3 has four steps:

1. **Acquisition of new image data**: The device is used to take photos of objects. The images generated are loaded into the algorithm to start the classification.
2. **SVM-Support Vector Classifier**: The weight values are loaded into the model, and the input images are transformed to adjust to a default resolution.
3. **Classification**: The model generates a solution vector that can be converted in an output label by correlation. The output label represents the class predicted for the object identified in the input image.
4. **Results**: The user receives the output through an audio feedback.

2.1 Algorithms Comparison

Image classification tasks can be approached through different algorithms according to the image data used and the classes defined. The most common image classification algorithms identified in the literature are described in Table 1.

In object recognition, there are two approaches for algorithm implementation: object classification and object detection. In the former category, the most common algorithms are statistical, like Naive Bayes or random forest. Later, deep learning techniques are used through convolutional neural networks and other neural network

Table 1 Comparison between algorithms for image classification

Algorithm	Focus
Support Vector Machine	Classification and detection
k-nearest neighboor	Classification
Naive Bayes	Classification
Random forest neighbor	Classification
Perceptron	Classification and detection
Convolutional neural networks	Classification and detection

architectures. One of the most remarkable differences between these two categories is the computational power required to execute deep learning algorithms, which are more power-hungry than lighter algorithms like SVM.

2.2 *Objective Function*

The optimization objective is to maximize the margin value between the lines generated by the support vectors. This can also be represented by Eq. 1.

$$Min \frac{1}{2} \sum_{i=1}^{n} w_i^2 \tag{1}$$

The most common loss function in this type of algorithm is the hinge loss, described by Eq. 2, in which t is the value assigned to the class (which can be -1 or 1) and y is the raw output value of the prediction.

$$l(y) = Max(0, 1 - t \cdot y) \tag{2}$$

The cost function of the algorithm is obtained by combining the optimization and loss functions. For a binary classification problem, the cost function is described by 3

$$J(w) = Min \frac{1}{2} \sum_{i=1}^{n} w_i^2 + l(y) = Max(0, 1 - t \cdot y) \tag{3}$$

For a multi-class Support Vector Machine approach, Eq. 3 is extended to more than two classes and can be implemented with many features. The multi-class SVM can be defined using the One vs. One technique, which compares each class to the rest of the data and generates a classifier for each comparison. The number of classifiers is equal to Eq. 4.

$$Number\_of\_Classifiers = \frac{n(n-1)}{2} \tag{4}$$

The SVM implementation proposed consists of a Support Vector Classifier adapted to image classification of ten classes of different objects. The classifier was deployed using four different kernel functions to compare their performance. The SVM was adapted to multi-class using the "*One vs. One*" approach, which generates 45 instances of One vs. One.

3 Materials and Methods

3.1 Dataset

The classification and detection algorithms were trained using custom versions of the Open Images v4 dataset and the Imagenette dataset (a subset of the Imagenet dataset), respectively.

The custom YOLO v4 algorithm, implemented for object detection, was pre-trained using the OpenImages v4 dataset, an image dataset for object detection algorithms [29]. It is composed of a collection of images from 600 different classes, but only 9 were extracted to train and validate the proposed classification and detection system. The classes selected are described in Table 2. Each image has an associated text file that contains the label corresponding to the image's class and the coordinates of the bounding boxes for each object in an image. The image resolution varies between all the items in the dataset. Thus, they need to be re-scaled to adjust their resolution before being implemented in the training process.

Some selected images from the Open Images v4 dataset are shown in Fig. 4. A notable feature of this collection is that the illumination, object position, and size vary significantly between all the images, which helps to avoid the over-fitting problem of using too many similar images of each class.

The multi-class SVM was trained using a subset of the ImageNet dataset [30], called Imagenette [31]. The original dataset is composed by approximately 1000 object classes and is used for competitions and benchmarking purposes, focused on object detection and classification algorithms. The Imagenette subset is a

Table 2 Classes contained in the Open Images v4 subset

ID	Class	Description
0	Person	Varied age and genre
1	Scissors	Varied styles and colors
2	Coin	Varied values and sizes
3	Mug	Varied materials and colors
4	Knife	Varied styles and materials
5	Bottle	Different materials and colors
6	Socks	Varied patterns, colors, and sizes
7	Spoon	Varied materials and sizes
8	Fork	Varied materials and sizes

Fig. 4 Sample of the OpenImages v4 dataset

Table 3 Classes contained in the ImageNet subset

ID	Class	Description
0	Cassette_player	Electronic device
1	Chainsaw	Tool
2	Church	Building
3	English_springer	Dog breed
4	French_horn	Musical instrument
5	Gas_pump	Storage and dispenser of gasoline
6	Golf_ball	Sports equipment
7	Parachute	Device used for parachuting
8	Tench	Fish species
9	Trash_truck	Utilitary vehicle

classification-oriented collection of images integrated by ten classes, presented in Table 3. Each class contains a variable number of examples, which range between 50 and 100. The training and validation sets were defined with a ratio of 40 and 60% of the original sample, respectively.

Figure 5 shows a small sample of the images that can be found in the dataset utilized for the classification algorithm.

Fig. 5 Sample of images contained in the Imagenette subset

3.2 Hardware and Software

The algorithm was deployed in a mobile device. The system was constructed using the following hardware components:

- **Camera**: Integrated in the mobile device
- **Mobile device**: An Android smartphone device with an app powered by a light model of the detection algorithm

The software employed in the device is conformed by the following:

- **Android app**: Application deployed with a light detection model in the mobile device.
- **Light Support Vector Machine algorithm**: A compact version of the detection algorithm to provide fast results through the Android app.
- **Robust Support Vector Machine algorithm**: A more robust and complex version of the detection algorithm provides better prediction accuracy. This algorithm is deployed in a cloud environment connected to the Android app, so it requires a stable Internet connection.

3.3 Support Vector Machine

The SVM implementation proposed consists of a Support Vector Classifier adapted to image classification of ten classes of different objects. The classifier was deployed using four different kernel functions to compare their performance. The SVM was adapted to multi-class using the "*One vs. One*" approach, which generates 45 instances of One vs. One. The classifiers for the classes are described in Fig. 6.

For each classifier, the kernel functions used are linear kernel Eq. 5, RBF kernel (radius basis function) (Eq. 6), polynomial kernel (Eq. 7), and sigmoid kernel (Eq. 8).

$$k(x, y) = x \cdot y \tag{5}$$

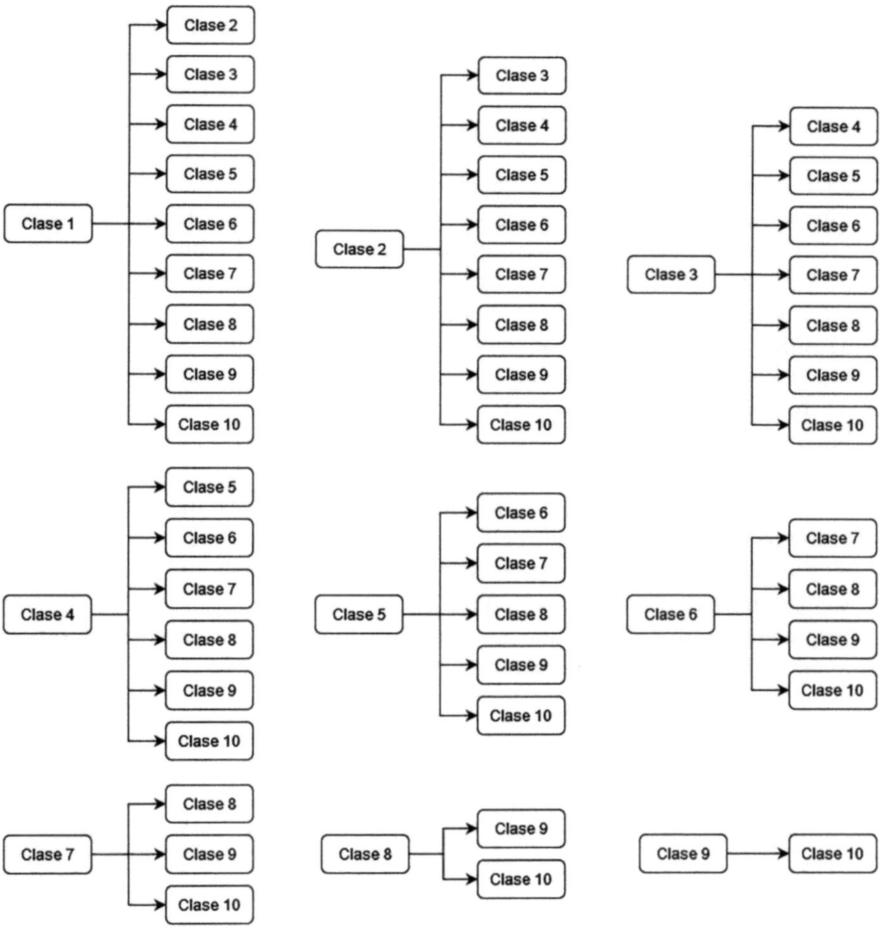

Fig. 6 Classifiers used for the model

$$k(x, y) = exp(-a||x - y||^2) \tag{6}$$

$$k(x, y) = (x \cdot y)^d \tag{7}$$

$$k(x, y) = tanh(ax \cdot y) \tag{8}$$

To create a model for the Support Vector Machine algorithm, the following tasks were completed:

- **Configuration of the algorithm**: The algorithm was characterized according to the multiple classifiers approach and using the loss function for Support Vector Machines. The environment for training, validation, and execution of the algorithm was prepared using a cloud service provided by Google through the Google Colab suite.
- **Preparation for the dataset**: The dataset was retrieved from the ImageNet repository, and ten classes were selected to create the dataset used for training and validation. This collection of images was split into two subsets to create the training set, which was used to train the algorithm, and the validation set, which was required to validate the performance of the trained model. As seen in Table 4, the training and validation sets were created with 35% and 65 & of the total images contained in the original dataset.
- **Parameter tuning**: The parameters that dictate the model's behavior were tuned. Several sets of parameters were obtained and compared with a short training performance test to obtain each set's performance metrics. The best-performing set was selected to configure the model.
- **Training**: The algorithm was trained using an iterative process that inputs the training set into the model to learn from the examples and achieve a good performance of object recognition. The trained model can receive an image as input and provide an output composed of the object class detected in the image and the model's confidence.
- **Validation**: After training, the validation set is used to test the performance in object recognition and provide performance metrics to compare the trained model with other similar implementations. A simplified overview of the training and validation process is presented in Fig. 7.
- **Implementation**: After obtaining a model with a good performance according to the metrics, it was deployed on the Android application.

Table 4 Sets for training and validation

Set	Total %	No. images
Training	35%	2800
Validation	65%	5200

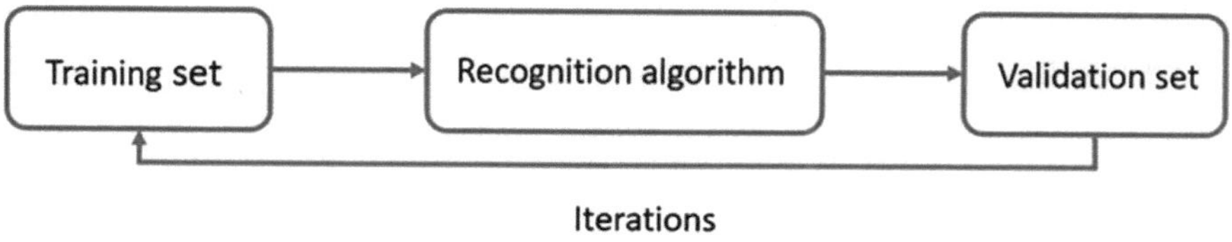

Fig. 7 Simplified training process

Fig. 8 Influence of several factors in the time complexity of the training process

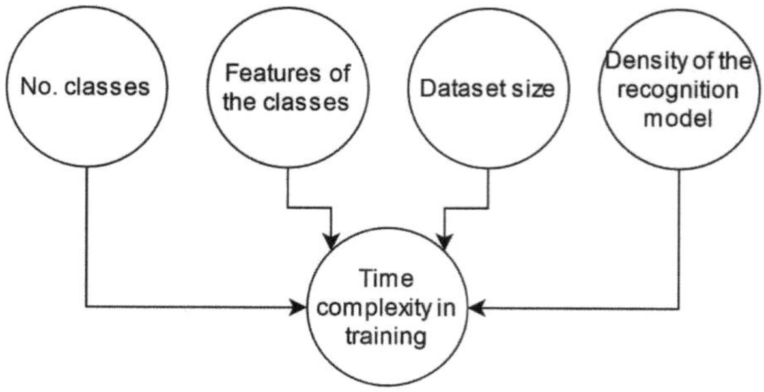

3.4 Problem Characterization

The structure of the SVM model is composed of 45 classifiers, and the object recognition implementation requires a high amount of features to process the data of interest. Hence, the model has a certain degree of complexity. This situation impacts several areas, such as training, validation, and performance (as seen in Fig. 8). In a complex model, the data required can be so big that the training time can exponentially grow. The parameters can be tuned and optimized to obtain the best values to achieve good performance without incurring limited training time to reduce training problems.

3.5 Algorithm Implementation

The classification algorithm was implemented in an Android app through a simplistic interface with voice recognition tools. The detection algorithm was based on a custom YOLO v4 [32] trained with the custom dataset labeled for object detection. The interaction model is represented in Fig. 9.

The app features Speech-to-Text and Text-to-Speech using Google built-in modules. The process of queries (Fig. 10 is composed of four stages:

1. **Stage 1**: The user starts the app, and all the assets are loaded. The audio feedback informs the user that the app is ready to start queries.
2. **Stage 2**: The user utilizes his voice to request a particular object. The system activates the camera, which the user can focus on to scan several scenes to search for the object of interest.

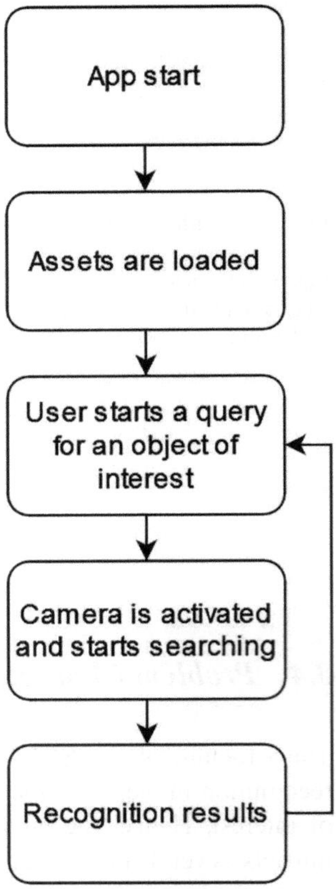

Fig. 9 Interaction process of the Android app

3. **Stage 3**: If the object was located, the audio feedback responded to the users indicating the status of the object. On the contrary, the audio feedback negatively responds to the user, informing that the object of interest was not found in the scenes scanned.
4. **Stage 4**: The query is finished, and the user can choose to start a new one or exit the app.

The command voice proposed for the voice command interface is enumerated in Table 5. The default languages supported by the app are English and Spanish, so any user with a specific proficiency in these languages can use the application with relative ease (as seen in Fig. 10.

3.6 Proposed Design of Experiments

The model parameters were optimized through a DOE (design of experiments) approach. First, the optimizing parameters were defined, and a DOE was formu-

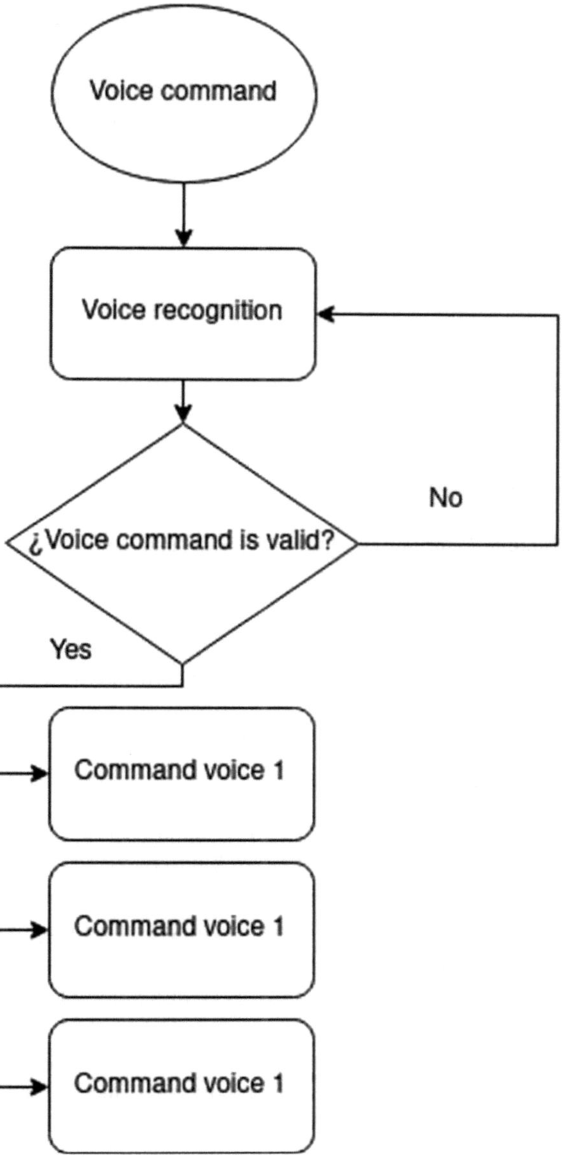

Fig. 10 Queries with voice commands

lated. Then, the parameters set were iterated using the DOE proposed, and several sets of values were obtained. The sets were compared through a performance test using the reduced-scale training process. The best-performing set was selected to train the recognition model using the images dataset. (The tuning process is shown in Fig. 11).

Table 5 Voice commands

English	Spanish	Description
Start	Inicio	Start the app and load all the assets
Search for + "object name"	Busca + "nombre de objeto"	The users select the object of interest
New query	Nueva consulta	The process returns to the initial state

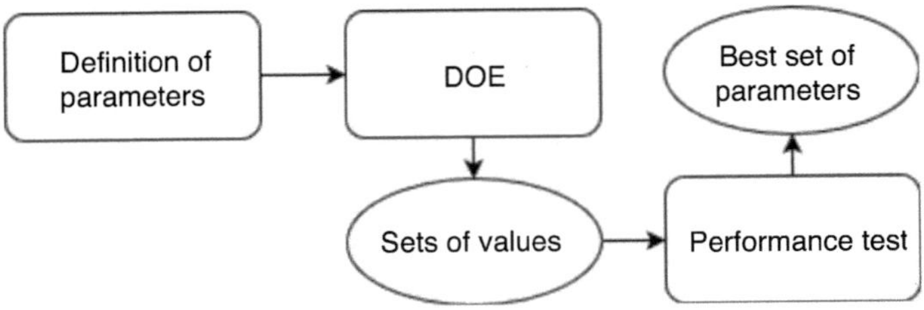

Fig. 11 Simplified tuning process for the model parameters

Table 6 Parameters for the SVM implementation

Parameter	Area of application
Standardized size of images	Dataset
No. classes	Dataset
C (Penalization for misclassification)	SVM
γ (Loss compensation)	SVM
No. epoch	Training
Learning rate	Training

According to the nature of the data required for training (the ImageNet dataset), the parameters are described in Table 6

The parameters utilized can be classified according to his area of governance in dataset parameters, SVM parameters, and training parameters.

The dataset parameters adapt the image dataset to the model altering its size, which is the resolution in pixels (width x height) and the number of classes required for the recognition model.

The SVM parameters tune the loss and cost functions to improve the classifier's performance by using the penalization and compensation parameters, which are represented by C and γ, respectively.

The training parameters can be used to manipulate the training process to reduce the like-hood of over-fitting or to achieve a good performance in a reasonable amount of time. A DOE was proposed to improve the tuning process and achieve an adequate set of parameters.

4 Results

4.1 Android App

After training and validation, the data fitting can be interpreted as the capacity of the model to make predictions about new inputs; otherwise it is the ability to classify images that were not part of the training and validation sets. In general, there are three cases of data fitting.

- **Under-fitting**: The model adjusts poorly to the data with a lot of misclassification cases. It needs more training or examples of data.
- **Acceptable fitting**: The model allows a reduced number of misclassifications and can make good predictions over new data.
- **Over-fitting**: The model adjusts very tightly to the training data, so the misclassification is greatly punished and can't make accurate predictions over new data.

A graphical user interface (GUI) in a desktop app was developed to provide a visual representation of the algorithm implementation and some performance metrics to evaluate the predictions made by the SVM model.

The GUI was developed using the Tkinter module in Python 3.9. This approach was selected because it allows to port the GUI app between the most common operative systems with access to a Python compiler. The desktop app can be executed in Windows, macOS, and most Linux distros.

The app was designed with a minimalist style to improve readability and provide a clean representation of the data obtained from the object recognition model. The basic app's GUI is shown in Fig. 12.

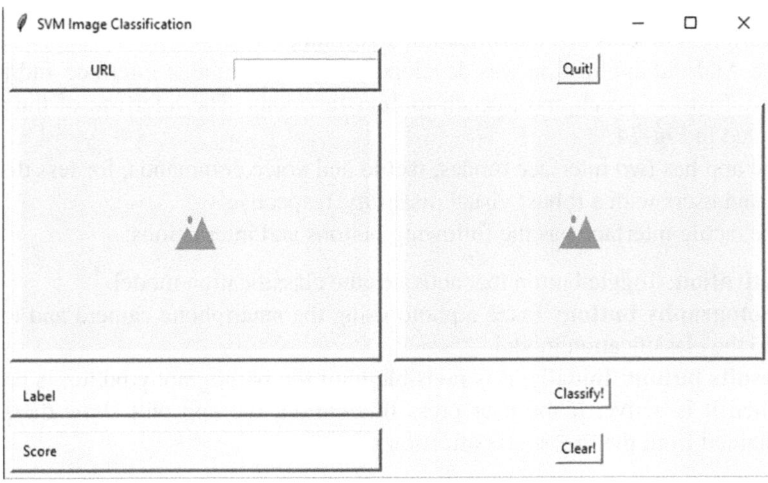

Fig. 12 GUI deployed in Windows environment

The user can interact with the app by selecting some images to test the object recognition algorithm. The results for each image are displayed in the interface using the label and confidence score generated by the model and a pie chart to visualize the score obtained by the other classes. The GUI's layout is composed of the following items:

- **Request box**: The user can indicate in a text box the parameters of the query and which image will be processed.
- **Left image box**: The image selected is displayed and scaled according to the app's window resolution.
- **Right image box**: A pie chart is displayed to provide information about the other classes detected in the image to compare the result obtained by the detected class with all the classes in the model.
- **Result box**: This box shows the class detected label corresponding to the type of object recognized in the image.
- **Score box**: Provides the accuracy score obtained in the classification process in percentage form.
- **Control buttons**: The user can use a set of buttons to control the app's behavior. The functions assigned to them are described as follows:

  - **Classify button:** If the user specifies the parameters and image to process, the model classifies the image selected and displays the results.
  - **Clear button:** Clear all the boxes in the GUI and allows to start a new classification process.
  - **Quit button:** Exits the app in secure mode without compromising the data obtained.

Several examples of the app classification process are shown in Fig. 13. The app was implemented with a general-purpose approach, so it can easily be configured with different models and classification algorithms.

The Android application was developed with a minimalist interface and implemented Text-to-Speech and Speech-to-Text functions. The interface of the app is presented in Fig. 14.

The app has two interface modes, tactile and voice commands, for less disabled users and users with a robust visual disability, respectively.

The tactile interface has the following buttons and interactions:

1. **Activation**: Toggle button that activates the classification model.
2. **Photography button**: Takes a photo using the smartphone camera and loads it into the classification model.
3. **Results button**: Initially, it is invisible until the photography button is pressed. When it is active, if the user press this button, the app will show the results obtained from the image classification process.

The voice command interface has the following interactions:

1. **Voice command "Start"**: The classification model is activated, and the app is ready for queries.

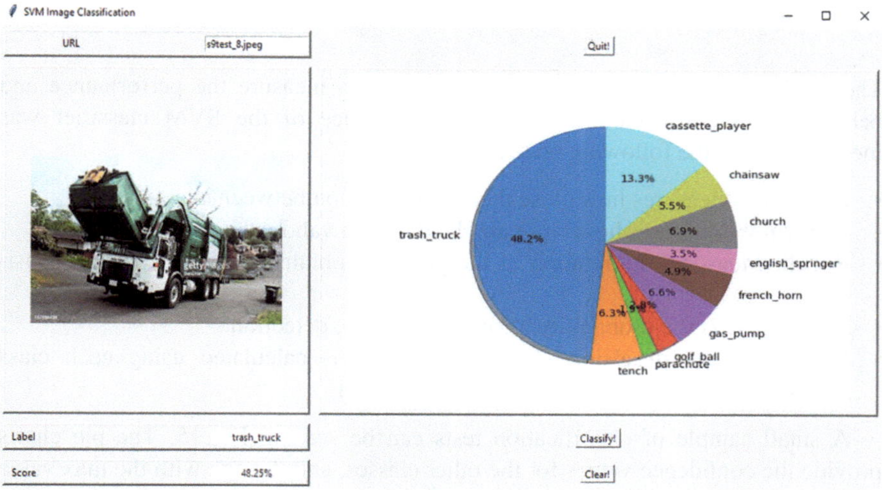

Fig. 13 Example of the classification process in the desktop app

Fig. 14 Interactions in the Android app

2. **Voice command "Search for –object_name–"**: The app will take a photo, and the classification model will process it to generate a label according to the class detected in the image. The results will be received by the user using audio feedback.
3. **Voice command "New query"**: The user can start a new query after completing the app's main process.

4.2 Classification Results

The model was tested using the validation set to measure the performance and behavior of the trained classes. The performance of the SVM classifier was measured using the following metrics:

- **Precision**: Measures how close the data dispersion between examples is.
- **Accuracy**: Measures how close are the data to a valid value.
- **Recall**: Measures the relation of the positives obtained and the expected positives.
- **Sensibility**: Is the probability for proper positive detection.
- **Mean average precision (MAP)**: This stat is calculated using each class precision value to obtain the mean for the model.

A small sample of classification tests can be seen in Fig. 15. The pie charts provide the confidence values for the other classes, and the one with the max value is the label generated by the model to classify the image.

The confusion matrix (Fig. 16) was constructed with the cases obtained for the object classes. The descriptions for each case are as follows:

- **True positive**: The object was correctly classified, and the label coincides with his type.
- **True negative**: The object was correctly classified and is not included in the classes. No label is generated.
- **False positive**: The object was incorrectly classified, and a label was assigned to a type of object not included in the original classes.
- **False negative**: The object was incorrectly classified, and his type coincides with a label but is detected as an object not included in the classes.

For each class, their respective performance parameters were obtained. According to Tables 7 and 8, the highest performing class was Class 8, and the lowest was Class 10. The value disparity could be explained by differences in images' quality for each class in the training set and the complexity of the object represented by the class. So, the worst performing classes correspond to difficult-to-classify objects with poor variation and quality in their corresponding training images. For the other part, the classes that surpassed the 0.6 value were composed of training images of better quality for classification tasks.

The main metrics of the model are presented in Table 9. The class metrics were used to obtain the mean average precision (mAP) of the model. According to the metric values obtained, the model's performance is better than a hypothetical luck-based classifier, so it can predict the corresponding class in an image more accurately than simply guessing it. The recall of the model represents how well the model classifies positive examples compared to the total of positive cases. In this case, a recall value of 0.5 represents that the model can classify correctly half of the positive cases. The mAP obtained was low compared to the accuracy and precision values due to the low-performing classes previously explained.

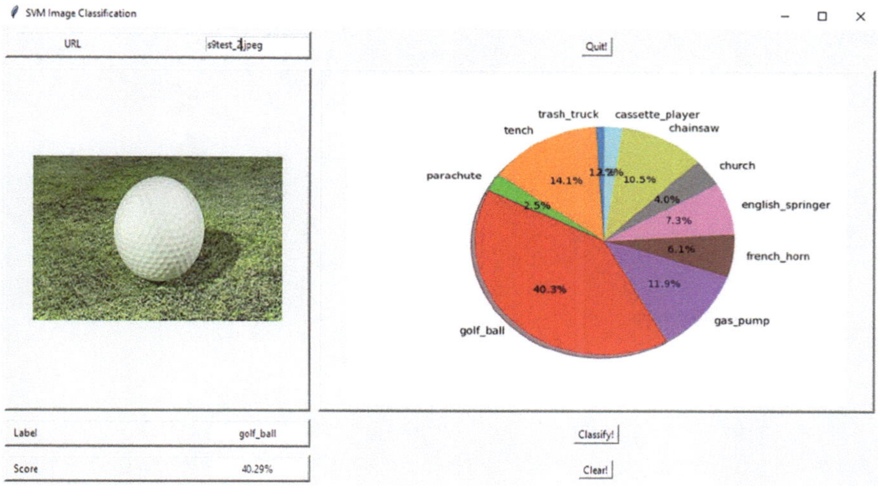

(a)

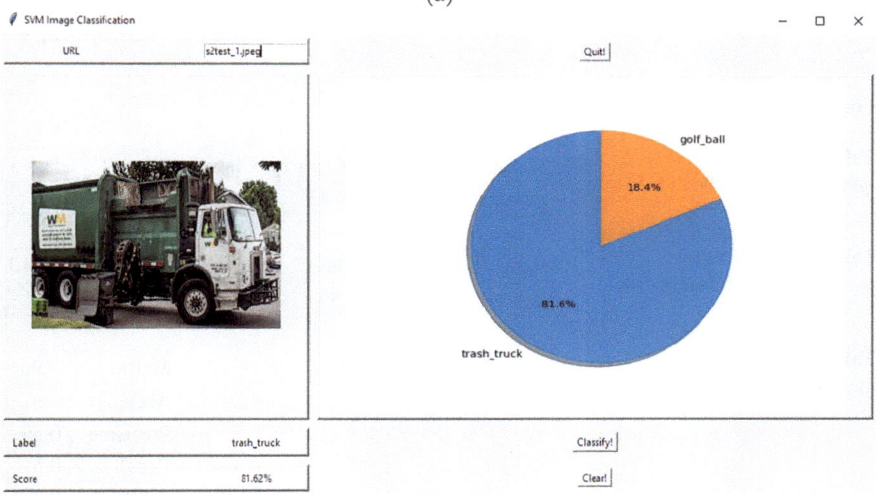

(b)

Fig. 15 Example of classification tests with the validation set. (**a**) Example with 10 classes. (**b**) Example with 2 classes

The precision and mAP over the complete training process are shown in Fig. 17.

The convergence of the model was achieved at the 6000 iterations mark. At this point, the model achieved the maximum precision value. The SVM implementation achieved a moderate accuracy value (60%) and can be deployed in a lightweight model, allowing its implementation in light apps and devices without dedicated computing resources.

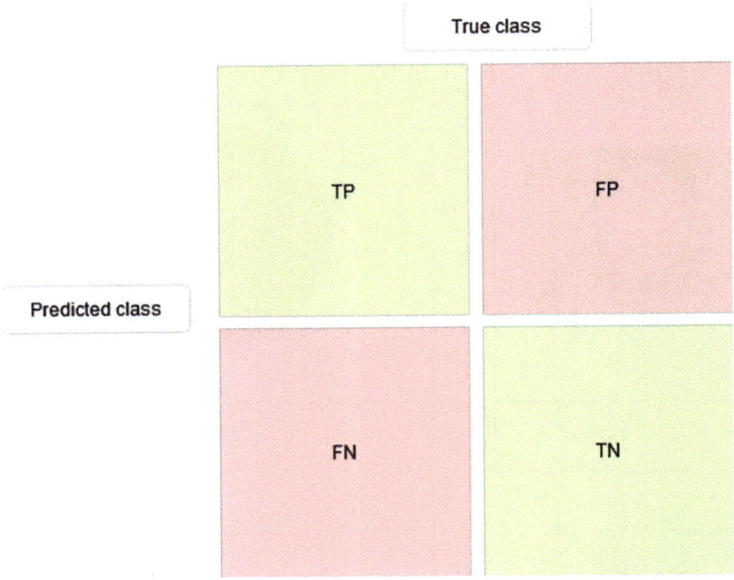

Fig. 16 Confusion matrix for the classes of the model

Table 7 Performance metrics for each class (1–5)

Metric	Class 1	Class 2	Class 3	Class 4	Class 5
Accuracy	0.56	0.43	0.7	0.68	0.52

Table 8 Performance metrics for each class (6–10)

Metric	Class 6	Class 7	Class 8	Class 9	Class 10
Accuracy	0.61	0.59	0.74	0.5	0.35

Table 9 Performance metrics for the model

Metric	Value
Accuracy	0.6
Precision	0.59
Recall	0.5
mAP	0.47

The approach's limitations include the small number of classes, the dataset, and the focus on classification-only tasks. The model can be easily scaled to process more complex data, like a significant number of classes, more populated images, and classification by features like color and shape, because the foundation of the SVM approach allows adjusting the model to a large number of features and classes.

According to the results obtained, the algorithm has a big room for improvement when combined with other machine learning techniques like deep learning. This combination can allow the implementation of classification models with the SVM classifier to achieve better accuracy and timely execution results.

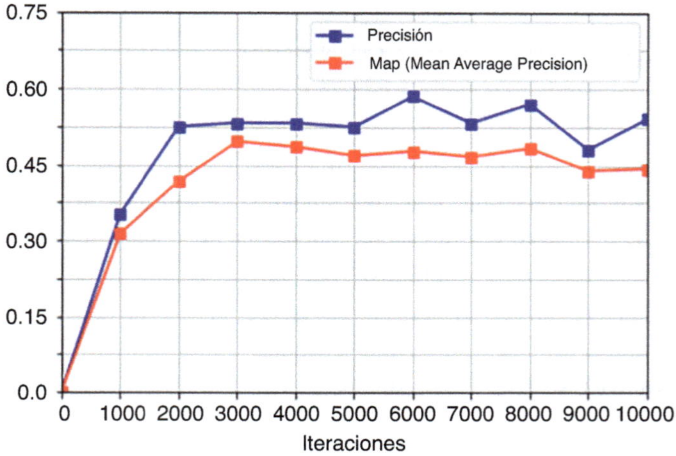

Fig. 17 mAP and precision in the training process

Table 10 Comparison between YOLO v4 classification model and the SVM implementation

Model	mAP	Accuracy	Dataset
SVM	0.47	0.6	ImageNet
YOLO v4	0.64	0.8	OpenImages

4.3 Metrics Comparison

The most representative metrics obtained for the SVM model were mean average precision (mAP) and accuracy because these measure the effectiveness of the model's predictions and the performance of the trained classes, respectively. The mAP metric is the mean of the individual precision means per class and it can tell how well the model adapts to the data.

Table 10 shows the mAP and accuracy stats for the classifier and the detection models. The stat values for the detection model (YOLO v4) are better than the classifier (SVM) due to the algorithm's robustness and its good adaptation to populated images. Nevertheless, the SVM is better suited for simple images for classification purposes.

The SVM model obtained a lower accuracy and mAP scores but a more reasonable execution time than the other model analyzed. The adobe means that the proposed model is faster than other classification algorithms but is compensated with lower accuracy.

A faster model is ideal for deployment in non-dedicated devices, like smartphones, or in more flexible environments, like web apps and desktop apps.

Another SVM model was trained and validated to compare the impact of the hyperparameter tuning optimization. This new model was manually tuned to adjust its hyperparameters. According to Table 11, the parameter tuning optimization produced a better model than one with arbitrary parameters. The performance

Table 11 Comparison between a manually and an optimized tuned SVM model

Tuning	Accuracy	Precision	mAP	Recall
Optimized	0.6	0.59	0.57	0.5
Manual	0.47	0.43	0.45	0.39

metrics were improved with the optimization with a significant gain compared to a model without optimization. Also, the model generated with manual tuning did not surpass the 50% accuracy threshold; that means the manual tuned model is not better than a classifier based on pure luck.

5 Conclusion and Future Scope

The algorithm was designed according to the limitations and characterization made in the project's starting stage. A graphical user interface was developed to visualize the algorithm performance better, and an SVM model was trained using a subset of the ImageNet dataset. The main research product is an Android application with voice command support, classification (SVM), and detection (YOLO v4) algorithms. According to the general results, the app could perform well in a controlled indoor environment with little perturbations and noise.

The results obtained in the classification and performance tests measured the overall precision of the algorithm at 60%, which is a good score for a bare-bones classifier without using techniques like data augmentation or specialized algorithm architectures. However, comparing the SVM classifier to the state-of-the-art algorithms, the current algorithm implementation provides less accurate results. It can fail in classification tasks when the input images contain complex scenes. The performance obtained could be a product of the dataset used and the parameter tuning or due to the nature of the classifier implemented.

Implementing the SVM classifier is not recommended for complex tasks. It can only achieve good results with simple data, like images with few objects and less-polluted background (e.g., simple colors as background). For more complex images, the YOLO v4 image detector must be used. According to his performance, this implementation has room for improvement in the following areas:

- **Algorithm**: The methodology process's main challenge was the algorithm's configuration to achieve the abovementioned scores. Thus, the architecture of the algorithm could be tuned to achieve better precision with more complex images. Also, a different algorithm could be combined with the existing one to reduce the weakness of the current implementation.
- **Dataset**: The dataset used can be augmented with more examples in the training and validation sets, more classes, better image resolution, and the implementation of techniques, like data augmentation, to improve the data quality.
- **Additional features**: Other features can be implemented in the system, like other vision computer-related tasks, such as pattern recognition and object

tracking, or more complex approaches, like face recognition and detection in three dimensions. Also, the SVM algorithm can be improved by combining it with more robust classifiers, like deep learning neural networks, to allow the implementation of object detection for more complex classes.
- **App deployment**: A critical point of the deployment of the application is the supported platforms. So, a web app can be developed to use the application in a large variety of systems, like smartphones, desktop PCs, laptops, and other intelligent devices.

Due to the nature of the algorithm, it can easily be adapted to complement another object recognition algorithm, so the next major step to improve it will be to implement the SVM in a deep learning algorithm to enhance a object detection model and optimize it to achieve better performance metrics compared with state-of-the-art implementations. The following projects and implementations can be derived from the SVM approach:

- **Feature extraction**: Includes the segmentation and extraction of features of interest in images, like colors, shapes, and patterns.
- **Object tracking**: It consists of tracking an object of interest across a sequence of frames to allow the system to "remember" the object for future tasks.
- **Object localization**: Requires the detection of an object in an image, indicating the object's position according to other items in the scene.

The SVM approach can be generalized for different image classification tasks as:

- **Identification of vegetables and fruits**: Implementation based on feature extraction and object classification. Identification of classes of different types of fruits and vegetables and their state of ripeness to separate the healthy specimens from those infected by plagues or rotten.
- **Identification of plants and animals species**: Classification of animal and plant specimens by their species. This approach requires a multiple relational class hierarchy. The classes need to be segmented into subclasses and sub-subclasses to generate a robust model to process this complex task.
- **Vision systems for arm robots**: Another related work is the implementation of an object classification algorithm to provide a robot with the capability to recognize objects by their features (like color or shape) and select the target classes.
- **Medical imaging**: Image processing, feature extraction, and classification models are often used to generate automatic diagnostic tools to detect illness and health problems in medical images.
- **Identification of traffic signs**: An approach related to the scope of the current SVM implementation is the detection and recognition of traffic signs to improve visually impaired people's awareness in urban environments. This implementation can reduce the risks of outdoor navigation.

References

1. J. Ribeiro, R. Lima, T. Eckhardt, S. Paiva, Robotic process automation and artificial intelligence in industry 4.0—a literature review. Proc. Comput. Sci. **181**, 51–58 (2021). CENTERIS 2020—International Conference on ENTERprise Information Systems / ProjMAN 2020—International Conference on Project MANagement/ HCist 2020—International Conference on Health and Social Care Information Systems and Technologies 2020, CENTERIS/ProjMAN/HCist 2020
2. X. Zhou, X. Xu, W. Liang, Z. Zeng, S. Shimizu, L.T. Yang, Q. Jin, Intelligent small object detection for digital twin in smart manufacturing with industrial cyber-physical systems. IEEE Trans. Ind. Inform. **18**(2), 1377–1386 (2022)
3. Z.M. Çınar, A. Abdussalam Nuhu, Q. Zeeshan, O. Korhan, M. Asmael, B. Safaei, Machine learning in predictive maintenance towards sustainable smart manufacturing in industry 4.0. Sustainability **12**(19), 8211 (2020)
4. W.H. Organization et al., World report on vision, 2019
5. B.K. Swenor, M.J. Lee, V. Varadaraj, H.E. Whitson, P.Y. Ramulu, Aging with vision loss: a framework for assessing the impact of visual impairment on older adults. Gerontologist **60**(6), 989–995 (2020)
6. B. Zhang, Computer vision vs. human vision, in *9th IEEE International Conference on Cognitive Informatics (ICCI'10)* (2010), pp. 3–3
7. P. Mungkaruna, P. Piyawongwisal, K. Ropkhop, U. Hatthasin, The talking color identifying device for the visually impaired, in *2016 13th International Conference on Electrical Engineering/Electronics, Computer, Telecommunications and Information Technology, ECTI-CON 2016* (2016)
8. P.S. Jamuni, S. Borkar, Colour and shape identification for assisting visually impaired person, in *Proceedings of the International Conference on I-SMAC (IoT in Social, Mobile, Analytics and Cloud), I-SMAC 2018* (2019), pp. 415–418
9. K. Srinivasan, V.R. Azhaguramyaa, Internet of Things (IoT) based object recognition technologies, in *Proceedings of the Third International Conference on I-SMAC (IoT in Social, Mobile, Analytics and Cloud) (I-SMAC 2019)* (2019), pp. 216–220
10. R. Phadnis, J. Mishra, S. Bendale, Objects talk—object detection and pattern tracking using tensorflow, in *Proceedings of the International Conference on Inventive Communication and Computational Technologies, ICICCT 2018* (2018), pp. 1216–1219
11. L. Valdes, *Accessibility on the Internet*, 2004
12. K. Manjari, M. Verma, G. Singal, A survey on assistive technology for visually impaired. Internet Things **11**, 100188 (2020)
13. M. Kandil, R. Albaghdadi, F. Alattar, I. Damaj, AmIE: an ambient intelligent environment for assisted living, in *2019 Advances in Science and Engineering Technology International Conferences, ASET 2019* (2019)
14. J. Connier, H. Zhou, C. De Vaulx, J.J. Li, H. Shi, P. Vaslin, K.M. Hou, Perception assistance for the visually impaired through smart objects: concept, implementation, and experiment scenario. IEEE Access **8**, 46931–46945 (2020)
15. R.A. Minhas, A. Javed, X-EYE: a bio-smart secure navigation framework for visually impaired people, in *2018 International Conference on Signal Processing and Information Security, ICSPIS 2018* (2019), pp. 2018–2021
16. D. Pawluk, N. Bourbakis, N. Giudice, V. Hayward, M. Heller, Haptic assistive technology for individuals who are visually impaired. IEEE Trans. Haptics **8**(3), 245–247 (2015)
17. L.G. Nick Antonopoulos, *Cloud Computing in a Nutshell*, vol. 15 (2011)
18. T. Chaurasia, P.K. Jain, Enhanced smart home automation system based on Internet of Things, in *Proceedings of the 3rd International Conference on I-SMAC IoT in Social, Mobile, Analytics and Cloud, I-SMAC 2019* (2019), pp. 709–713

19. A.K. Gupta, R. Johari, IOT based electrical device surveillance and control system, in *Proceedings—2019 4th International Conference on Internet of Things: Smart Innovation and Usages, IoT-SIU 2019* (2019)
20. R. Paradaa, J. Melia-seguib, A. Carrerasa, M. Morenza-cinosa, R. Pousa, Measuring user-object interactions in loT spaces, in *2015 IEEE International Conference on RFID Technology and Applications (RFID-TA) Measuring* (2015)
21. F.L. Barsha, Z. Tasneem, S. Mojib, M. Afrin, N. Jahan, M. Tasnim, U. Habiba, M.N. Islam, An IoT based automated door accessing system for visually impaired people, in *2019 5th IEEE International WIE Conference on Electrical and Computer Engineering, WIECON-ECE 2019—Proceedings* (2019), pp. 2–5
22. M.J. Junior, O.B. Maia, H. Oliveira, E. Souto, R. Barreto, Assistive technology through Internet of Things and edge computing, in *IEEE International Conference on Consumer Electronics - Berlin, ICCE-Berlin*, vol. 2019 (2019), pp. 330–332
23. R. Tapu, B. Mocanu, T. Zaharia, Wearable assistive devices for visually impaired: a state of the art survey. Pattern Recogn. Lett. **137**, 37–52 (2020). Learning and Recognition for Assistive Computer Vision
24. R. Tapu, B. Mocanu, T. Zaharia, DEEP-SEE: joint object detection, tracking and recognition with application to visually impaired navigational assistance. Sensors (Switzerland) **17**(11), 2473 (2017)
25. A. Pardasani, P.N. Indi, S. Banerjee, A. Kamal, V. Garg, Smart assistive navigation devices for visually impaired people, in *2019 IEEE 4th International Conference on Computer and Communication Systems, ICCCS 2019* (2019), pp. 725–729
26. C.T. Patel, V.J. Mistry, L.S. Desai, Y.K. Meghrajani, Environment for visually impaired people, in *2018 Second International Conference on Intelligent Computing and Control Systems (ICICCS)*, no. ICICCS (2018), pp. 1–4
27. I. Joe Louis Paul, S. Sasirekha, S. Mohanavalli, C. Jayashree, P. Moohana Priya, K. Monika, Smart eye for visually impaired-an aid to help the blind people, in *ICCIDS 2019—2nd International Conference on Computational Intelligence in Data Science, Proceedings* (2019)
28. A.S. Alon, R.M. Dellosa, N.U. Pilueta, H.D. Grimaldo, E.T. Manansala, EyeBill-PH: a machine vision of assistive Philippine bill recognition device for visually impaired, in *2020 11th IEEE Control and System Graduate Research Colloquium, ICSGRC 2020—Proceedings* (2020), pp. 312–317
29. A. Kuznetsova, H. Rom, N. Alldrin, J. Uijlings, I. Krasin, J. Pont-Tuset, S. Kamali, S. Popov, M. Malloci, A. Kolesnikov et al., The open images dataset v4. Int. J. Comput. Vis. **128**(7) (2020)
30. O. Russakovsky, J. Deng, H. Su, J. Krause, S. Satheesh, S. Ma, Z. Huang, A. Karpathy, A. Khosla, M. Bernstein, A. C. Berg, L. Fei-Fei, ImageNet large scale visual recognition challenge. Int. J. Comput. Vis. **115**(3), 211–252 (2015)
31. J. Howard, imagenette. https://github.com/fastai/imagenette/
32. A. Bochkovskiy, C. Wang, H.M. Liao, Yolov4: optimal speed and accuracy of object detection (2020). CoRR, vol. abs/2004.10934

Part II
Analytical Strategies for Productive Processes Based on Industry 4.0

Development and Evaluation of a Machine Learning Model for the Prediction of Failures in an Injection Moulding Process

A. Rojas-Rodríguez, F. S. Chiwo, H. Arcos-Gutiérrez, C. Ovando-Vázquez, and I. E. Garduño

1 Introduction

1.1 Injection Moulding

Injection moulding (IM) is the top manufacturing process in the industry due to its technical and economic advantages. Its main advantage over other manufacturing processes is the improved quality of the moulded parts at lower costs and production times [1]. The operating principle begins by injecting molten plastic material into the mould cavity using an injection moulding machine. When the plastic material enters the injection moulding machine through a hopper, the material is molten before being pressed by the screw, thereby generating a large amount of heat due to friction. The molten plastic is collected at the front of the cylinder and is continuously heated to maintain the injection temperature. The screw then pushes the material into the sealed cavity of the mould in a process known as an injection. Once the initial injection has been completed and the mold cavity has been filled with molten polymer, additional plastic is injected under high pressure

A. Rojas-Rodríguez
Universidad Anáhuac, Naucalpan de Juárez, Edo. de México, México

F. S. Chiwo
CIATEQ A.C, San Luis Potosí, México

H. Arcos-Gutiérrez · I. E. Garduño (✉)
CONACYT – CIATEQ A.C, San Luis Potosí, México
e-mail: isaias.garduno@ciateq.mx

C. Ovando-Vázquez
CONACYT – Instituto Potosino de Investigación Científica y Tecnológica, Centro Nacional de Supercómputo, San Luis Potosí, México

L. C. Méndez-González et al. (eds.), *Innovation and Competitiveness in Industry 4.0 Based on Intelligent Systems*, EAI/Springer Innovations in Communication and Computing, https://doi.org/10.1007/978-3-031-29775-5_5

to compensate for the volume reduction caused by cooling. The last task is to compensate for the reduction in the volume of the plastic due to cooling. At this point, checking that the mould cavity is perfectly filled guarantees that the piece solidifies in a process known as packing. Finally, the moving part moves back to push the pin that reaches the back plate to push the moulded product, the slide system and the waste out. The cycle is called the injection moulding cycle [2].

In the case of the plastic injection process, the quality of a moulded part depends on the plastic material properties and the process parameters. When these parameters are optimal, it immediately reduces the injection cycle time and improves product quality. However, in practice, the configuration of the process parameters is mainly based on the operator's experience. Therefore, this method does not always guarantee appropriate process parameters or reproducibility values. Additionally, plastic has thermoviscoelastic properties, so establishing suitable process parameters to obtain the desired quality of the product is a challenge. Although this entire injection moulding cycle is fully automated and runs on a single machine, its defect rate is relatively high due to the somewhat complicated nature of the process [3]. If any control parameters deviate slightly from their standard value, defective parts can result. The control parameters include the charge duration, material/mould temperature, filling duration, injection speed, injection pressure, filling-compaction transition speed, holding pressure, holding time and cooling temperature, among others. As a result, these parameters are often selected manually and then adjusted through trial and error, which is costly and time-consuming.

1.2 Artificial Intelligence

Along with the Internet of Things (IoT), Big Data Analytics (BDA) and cyber-physical systems (CPS), artificial intelligence (AI) is one of the core techniques of the current industrial revolution, known as Industry 4.0. AI allows learning from previous experiences to develop a connected system with ubiquitous features that aims to create smart manufacturing systems and assemblies with fault robustness and self-control. These ideas are called smart factories or intelligent manufacturing processes, IoT-enabled manufacturing and cloud manufacturing [4]. Initially proposed in 1956, AI corresponds to the science in which computers mimic human behaviour or improve it.

The term generally comprises computer science concepts that deal with data processing systems to perform tasks like decision-making, learning, prediction, self-improvement and self-control. They are based on multiple mathematical and statistical concepts to solve real-life problems and the total chain value [5]. AI is in every sector of society, for example, in the health sector for early diagnosis and disease management [6], in the construction industry for monitoring and risk management [7] and in the education for online learning with video conferences, web-based learning applications and cloud services [8]. The primary focus of this research is on the injection molding industry [2, 9–11].

1.3 Machine Learning

Machine learning (ML) is a computing paradigm based on the capacity to solve a problem employing previous examples [12, 13]. In other words, it is the study, design and application of computer algorithms with the ability to auto-improve through the experience of previous results and by using and processing the analysed data [14]. ML is a branch of artificial intelligence that enables machines to perform their tasks independently and more efficiently [15]. The methodology uses data analysis and computing technologies to establish classification, regression or clusterisation to extract information from raw data by proposing a model for prediction and optimisation. Technically speaking, ML refers to all those techniques that enable learning on a computer without being programmed [16–18].

The use of machine learning in production processes is a tool based on estimating the future values of some variables that characterise a system, generally a machine, a plant or a production process, through mathematical models for identifying anomalies and potential failures early. The concepts of ML algorithms usually rely on supervised and unsupervised learning and combinations of both. The tasks associated with supervised learning are classification and regression. Clustering [19] is unsupervised learning since the output is unknown, like a label or a value of interest related to the process [20, 21]. Generally, applied ML aims to find the optimal mapping of the inputs to outputs given a set of data in a specific mathematical model [11, 22, 23].

1.4 Unsupervised Machine Learning Algorithms

Unsupervised learning deals with the optimisation process for clustering data with unlabelled data. In this methodology, the inputs and outputs of the problem are unknown, and a priori information about the data is unknown [11, 24]. These features are the unsupervised machine learning algorithms suitable for double-blind tests to make inferences and determine the optimal hidden features to apply for predicting or splitting variables. The most unsupervised machine learning algorithms are for data reduction and clusterisation [25–28]. Cluster analysis is a technique used for grouping data in disjoint groups, known as clusters, where similarities or dis-similarities rule the differentiation of each cluster in the observations contained in each cluster. Generally, cluster analysis is an approach for data separation and uncovering the natural structure of the data analysed [29, 30]. One of the most critical features of the clustering techniques is the possibility of performing the analysis with unlabelled data to find relevant information about the nature of the data. Among the clustering algorithms for dealing with unlabelled data is Density-Based Spatial Clustering of Applications with Noise (DBSCAN) [31, 32]. Such an algorithm was reported initially by Ester et al. [33], whose purpose is to find density regions in the datasets that can be considered independent clusters. This algorithm only requires information about the distance between the observations in the dataset, that is, the Euclidean distance between the observations and the minimal points to consider in the cluster. Furthermore, the DBSCAN algorithm does not require information about the number of clusters; it just calculates the parameter from the metrics presented in the data [34].

Recent applications of the unsupervised machine learning algorithms include fault detection in predictive maintenance [35], the prognosis of industrial equipment [36, 37], noninvasive assessment of laser-based surgeries for skin diseases and their classification [38, 39], material characterisation [40], hyper-spectral classification of recycled plastic parts [41], Internet services and mobile network improvement [42], water environment studies [43] and recently mental health diseases identification [44] and COVID-19 detection [44, 45].

1.5 Machine Learning Algorithm Applied to Injection Moulding Process for Fault Diagnosis

Energy efficiency, strict environmental regulations and global competition are the most challenging aspects of the manufacturing process, leading to the development of emergent technologies and complex control systems for its management that are, at first sight, the most appropriate approaches for dealing with these drawbacks. However, when the complexity of a system increases, the cost will increase, also leading to an inefficient system, so a more suitable approach is required for dealing with the data of the production results. In other words, it is a better approximation to perform an algorithm to optimise the production line than to acquire new equipment or develop a novel technique [46]. In this new scenario, when data analysis is crucial for the manufacturing process, fault classification and detection are vital for process tracking and control, aiming to determine which fault has occurred. Also known as anomaly detection, the concept of fault classification and detection is a set of mathematical and computational techniques applied for detecting observations that deviate from the usual pattern of the process analysed [47]. Thus, it is possible to propose a complete and reasonable answer for fault clearance and recovery. Fault classification is considered a multi-class classification. In this context, machine learning and deep learning methods are helpful for fault classification.

The control process of the IM input parameters is a complex task since it depends on the optimal response of the sensors and their location within the IM machine. Its electronic configuration is essential since it can be a simple analogy system or even a complex embedded digital system with several signal-processing techniques. This configuration will depend on the IM machine and the produced part. With this new feature, the cost of an intelligent IM system will increase with the number of sensors required and the computational resources needed. This clear drawback can be improved by selecting only the necessary resources and carefully selecting only those sensors of the critical parameters. The main challenge in applying

ML algorithms is the availability of viable datasets with meaningful information regarding all the IM processes, control parameters and the quality of the part produced. In this regard, workers are in charge of gathering the documentation of the datasets; consequently, the possibility of having an unmeaningful dataset increases due to the lack of precision in the data acquisition process. The primary purpose of ML is not to replace handmade human work but to improve the manufacturing process by optimising the relationship between the inputs, like control parameters, and the quality of produced parts, considered as the outputs of the manufacturing process [3]. Once the process is optimised, it is necessary to train the operators in the production line to efficiently record the information by establishing a protocol of manual data recording and developing an online platform capable of pre-evaluating the quality of the data recorded. Once the correct data are gathered, the system will upload the dataset to a cloud-based platform to provide access to any data scientist to develop the ML algorithms or a data engineer for real-time data testing of the failure detection performance. Developing the optimal platform with ML algorithms is beyond this work's scope.

Defined as the set of techniques used in complex manufacturing environments where the detection of failures is difficult [48, 49], the optimisation via machine learning techniques is based on estimating future values of variables that characterise a system. The most common examples are machines, factories and production processes, where mathematical models and computational algorithms can identify anomalies and potential failures in advance [4]. In the case of the plastic injection moulding process, the quality of a moulded part depends on the plastic material properties and the process parameters. Optimal process parameters reduce injection cycle time and increase product quality. In practice, the setting of the process parameters is mainly based on the mastery of experienced plastic engineers and is expensive and time-consuming. The appropriate process parameters for injection moulding are set up through analytical methods using the formulated mathematical equations. However, reliable solutions obtained do not always describe in their entirety the process due to the complexity of the injection moulding process; this is because the generated models go through simplifications. Although publications focused primarily on optimising the injection moulding process by known parameters, some are still academic and difficult to apply in practice. The main drawback is the lack of viable data. Another drawback is the labelling of variables that is almost constant in the literature by determining the pressure and temperature during the injection cycle as predictors and data related to the contraction, the scrap and the quality of the final product as possible responses. Although useful for machine learning applications in the manufacturing process, this aspect is not always possible because of having data without labelling or even physical meaning to the process. The latest leads to the necessity of viability of reliable datasets crucial to implementing the machine learning algorithms. In this work, the application of unsupervised ML algorithms into IM processes deals with the pressure observed in two cavities within an IM machine. First, failures are characterised, detected and predicted in the injection moulding by selecting the optimal pressure values required for each machine. The atypical observations, known as outliers, were detected and subsequently used as a reference for failure detection to isolate those values to associate and compare the predictions with the direct effect of the failures in the manufacturing line, like the contraction, scrap or component total loss.

2 Methodology

The analysed dataset comes from a JSW model J650AD injection equipment. The dataset contains pressure values in pounds per square gauge (psig) units for two injection cavities measured at the end of the cavity (EOC), labelled as EOC_1_2 and EOC_3_4, with the time stamp of the acquisition, configured to take measurements in a lapse of 2–4 s per observation. 162,132 records compounded the complete dataset corresponding to 1 year of operation per cavity recorded in a comma-separated value (CSV) file. To keep the integrity of the dataset with all its features, an exploratory data analysis, based on a numerical and graphical study of location and dispersion measurements, was performed only to assess the validity of the data recorded by monitoring the presence of the outliers and its contribution to the entire dataset.

Since the main objective of the research is to cluster and characterise failures in a noisy dataset and then predict them, a Density-Based Spatial Clustering of Applications with Noise (DBSCAN) was applied. Due to its simplicity and minimal input parameters requirement, the DBSCAN was selected in this research to cluster the input parameters according to the observed responses. In the case of this research, since there was no previous information about the dataset, the time stamp used, and the IM process DBSCAN is the right choice. This algorithm was applied to cluster the unlabelled data and extract the contribution of the outliers.

The data correspond to a wrong measurement due to the configuration of the sensor during the IM process. Other effects regard an external influence that modified the IM machine performance, all considered failures in a manufacturing line, overshadowing accurate and optimal measurements. Therefore, it is proposed that there are correlations between outliers with failures in the injection moulding process or even wrong functionality of the machine. The dataset was processed using the general-purpose programming language Python and its essential functions for NaN and outliers' suppression in a Google Colab notebook to perform the preliminary data cleaning and then the exploratory data analysis to assess the quality of the data acquired. The licensed matrix-based software MATLAB™ 2022a [50], particularly the Machine Learning and Statistics Toolbox, was used for the unsupervised machine learning study. This study also used the online MATLAB Online [51], particularly the Machine Learning and Statistics Toolbox [51].

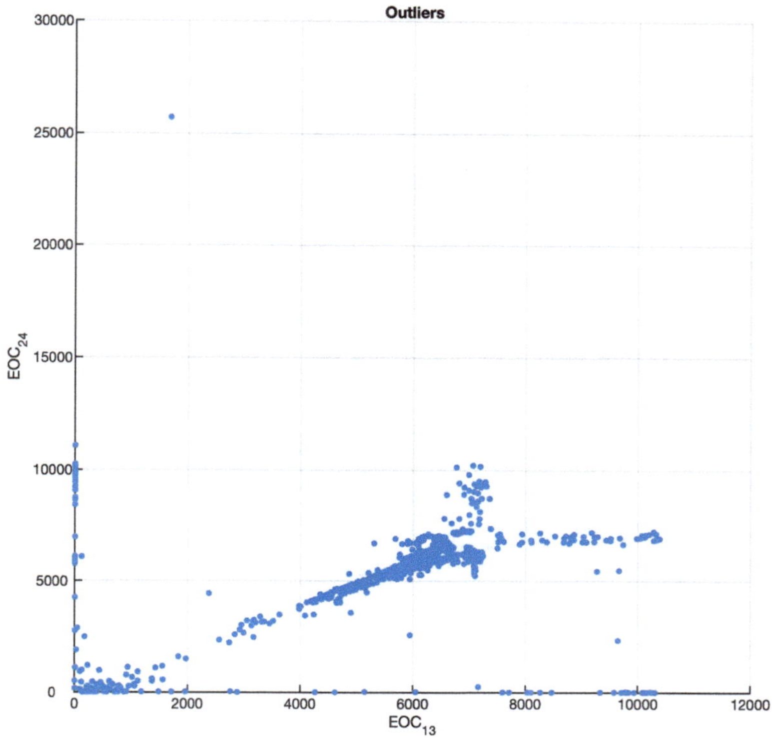

Fig. 1 The full dataset of the pressure values observed in each cavity. No relevant information was kept in this form, leading to more robust data analysis and processing algorithms

3 Results and Discussion

The data recorded by hand presented several deficiencies due to the errors induced by the personnel. In this way, after performing a missing analysis, it was determined that there is a total of 7.67% of data with no record. For example, regarding the EOC_1_3, the missing data are around 2251, representing approximately 1.38% of the total dataset. On the other hand, in the EOC_2_4, there are 4578 missing records, which means 2.82% of the dataset. Finally, both measurements showed 5632 missing records or 3.47% of the dataset. Since there was a high missing data volume in the dataset, the performed EDA was used to check the primary difference between the original and cleaned datasets. Figure 1 shows the pressure data, where the strength of atypical observations, or outliers, can be observed.

Clustering can be inferred in the range from 0 to 1000 in both measurements. This effect is because the manually captured dataset included several registers with "NaN" or "Not a Number" values, indicating a wrong acquisition. Here, it was applied a substitution for zeros to those values, which can be considered the human error factor included in the dataset.

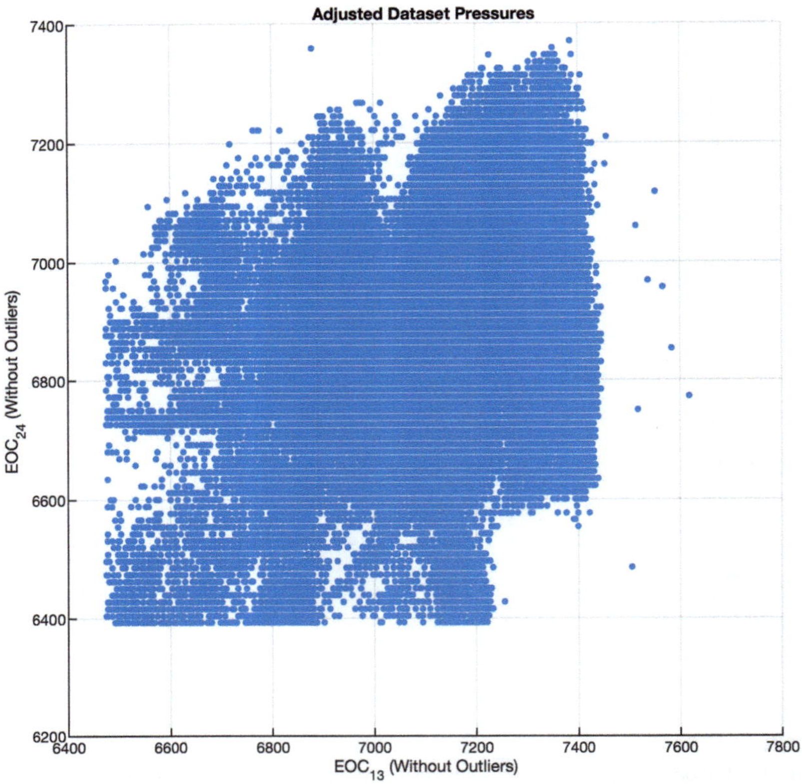

Fig. 2 Datasets without the presence of outliers and NaN values. A general cluster shows atypical values that overshadow the dataset, so this machine learning study cannot rely on exploratory data analysis (EDA)

By applying cleaning work to the analysis, those values located outside of the range of 6400–7600 psig for EOC_1_3 and 6400–7400 psig for EOC_2_4, where the data clustering is more concentrated than in other regions, were discarded in a process known as outlier detection and suppression [52–56]. In this stage, a considerable reduction in the observations was necessary due to most of the values being outliers, as observed in Fig. 1. Since the main objective of the research is to assess the outliers and correlate them with failures, exploratory data analysis was used only to draw the differences between the original data and the cleaned data. Figure 2 shows the resulting pressure values of 6400–7400 psig in the dataset without NaN values and outliers.

Figure 3 shows the outlier's contribution to the dataset before and after performing a data cleaning stage. This observation can be considered an indicator of the uselessness of the outlier detection stage, and some discarded observations are not outliers. Accordingly, another more robust algorithm is required to study the nature

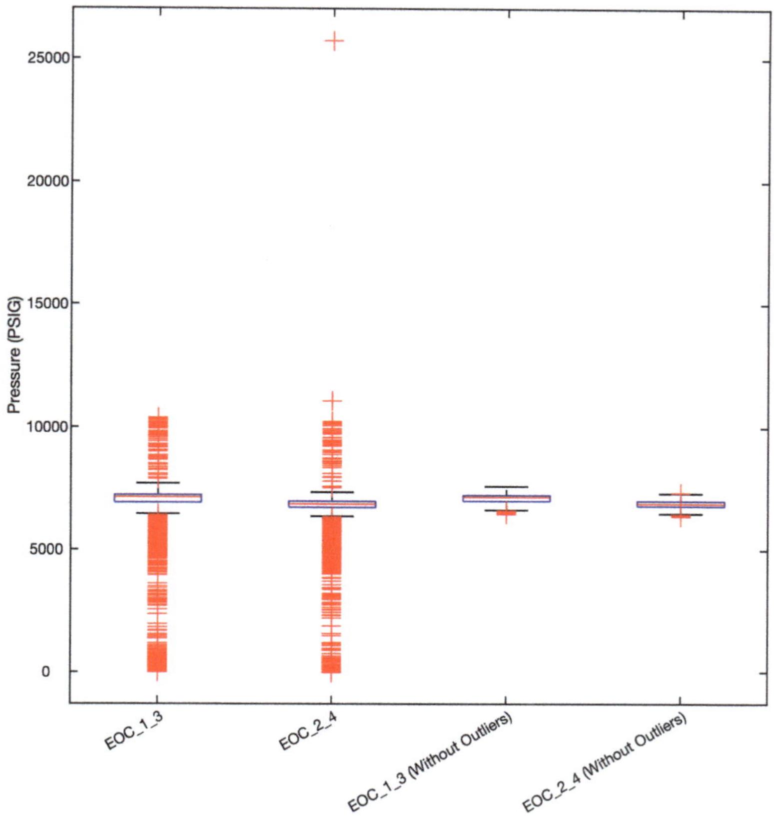

Fig. 3 Pressure data boxplots with and without outliers show these values' contribution beyond the normal range. In the cleaned data, several outliers are still present and can be misclassified and used as valid data for the analysis

of those outliers and their relationship with the manufacturing failures, the human error factor and the optimal functionality of the machine.

As a final part of the EDA, the values observed in the dataset in the histograms are shown (see Fig. 4). Figure 4a, c show the substitution of NaN values for zeros in the dataset histograms. Figure 4b, d show higher values with a negatively skewed distribution in the cleaned dataset, with the difference that the values observed showed lower values. Pressure data boxplots with and without outliers show these values' contribution beyond the normal range. In the cleaned data, several outliers are still present and can be misclassified and used as valid data for the analysis.

A clusterisation algorithm was applied to the original dataset to improve outliers' detection and association with physical failures in the manufacturing process. This clustering algorithm defines the dataset's outliers and the human error factor and explains those observations. Consequently, it indicates the optimal values for the manufacturing process and those with lower values related to the experiment's

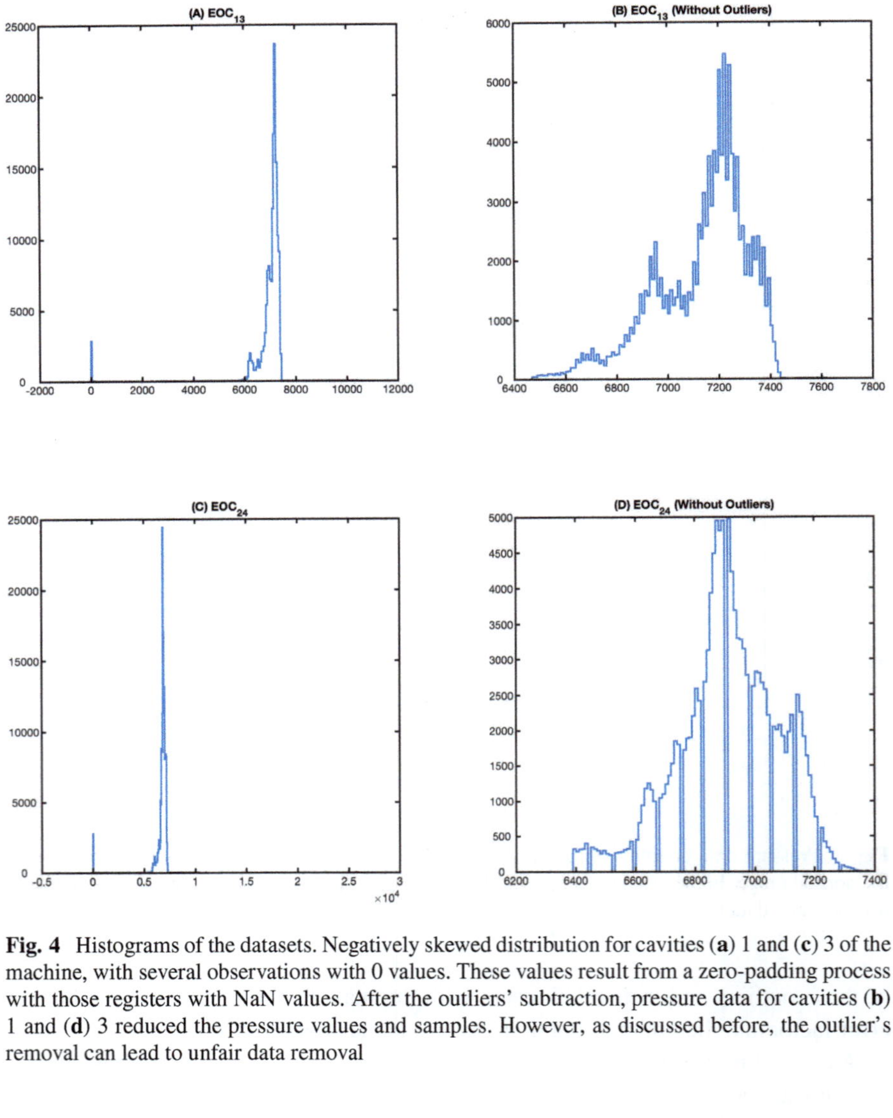

Fig. 4 Histograms of the datasets. Negatively skewed distribution for cavities (**a**) 1 and (**c**) 3 of the machine, with several observations with 0 values. These values result from a zero-padding process with those registers with NaN values. After the outliers' subtraction, pressure data for cavities (**b**) 1 and (**d**) 3 reduced the pressure values and samples. However, as discussed before, the outlier's removal can lead to unfair data removal

stabilisation. Finally, the results represent the lower pressure values or the required time for the optimal performance of the injection moulding machine analysed.

Density-Based Spatial Clustering of Applications with Noise (DBSCAN) [57] is an algorithm that clusters the unlabelled datasets into several groups with arbitrary shapes, depending on the radius of the distances between each value, and with the minimal a priori information about the dataset. These features, along with the minimal input requirements, offer a significant advantage over other clusterisation techniques used in machine learning applications, like the K-Means, where the number of clusters is known, or the hierarchical clustering, where the performance

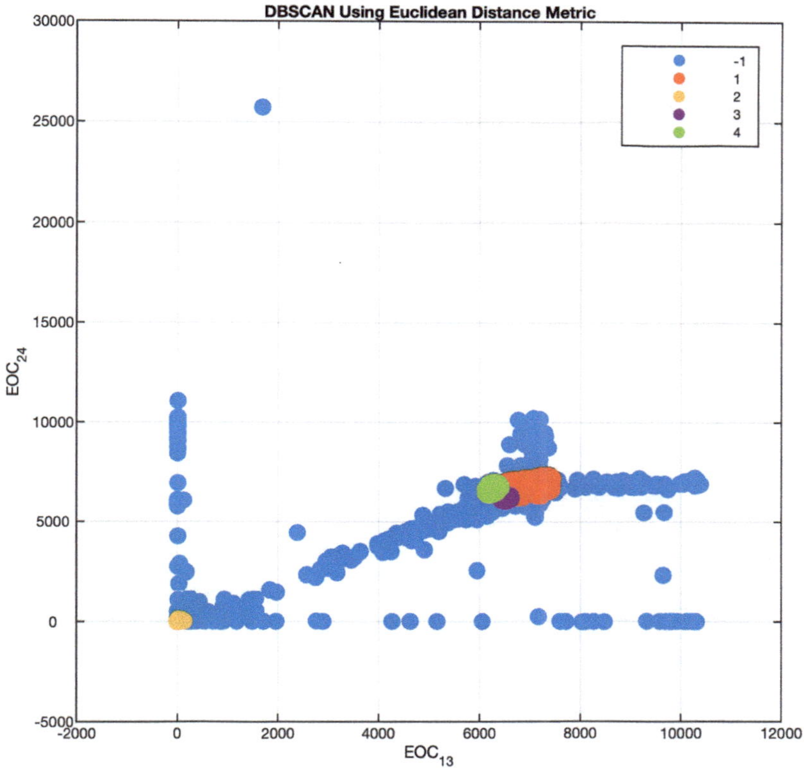

Fig. 5 Visual representation of the clusters obtained by the application of the DBSCAN algorithm to the original pressure data by calculating the radius and minimal points for each cluster from the dataset dimensions; the four obtained clusters are labelled as 1 (orange), 2 (yellow), 3 (purple) and 4 (green). The remaining values labelled with −1 (blue) do not belong to any of the four clusters obtained using the algorithm

of the algorithm fails for big datasets. This research selected this algorithm by following the stated objective of failure detection and prediction. Since DBSCAN is used for noise and outliers' detection, that is, detecting those observations that do not belong to any calculated cluster, it is possible to separate the failure detection and human error contribution from the meaningful observations. The result is that there is not any data reduction, filtering or even regression between the variables [9, 10, 20, 22, 33]. Figure 5 shows the N clusters obtained after applying the DBSCAN algorithm.

The original number of observations in the dataset was 162,132 observations per 2 cavities, so it was transformed into an $M \times N$ matrix with $M = 162{,}132$ and $N = 2$; the samples in cluster 1 were 132,976, 2569 for cluster 2, 1872 for cluster 3 and 5666 for cluster 4. Figure 6 presents the scatter plots of each cluster.

Cluster 1, Fig. 6a, presents a similarity in shape with the EDA result after learning and subtracting outliers in the dataset (see Fig. 2). In this new cluster, there are

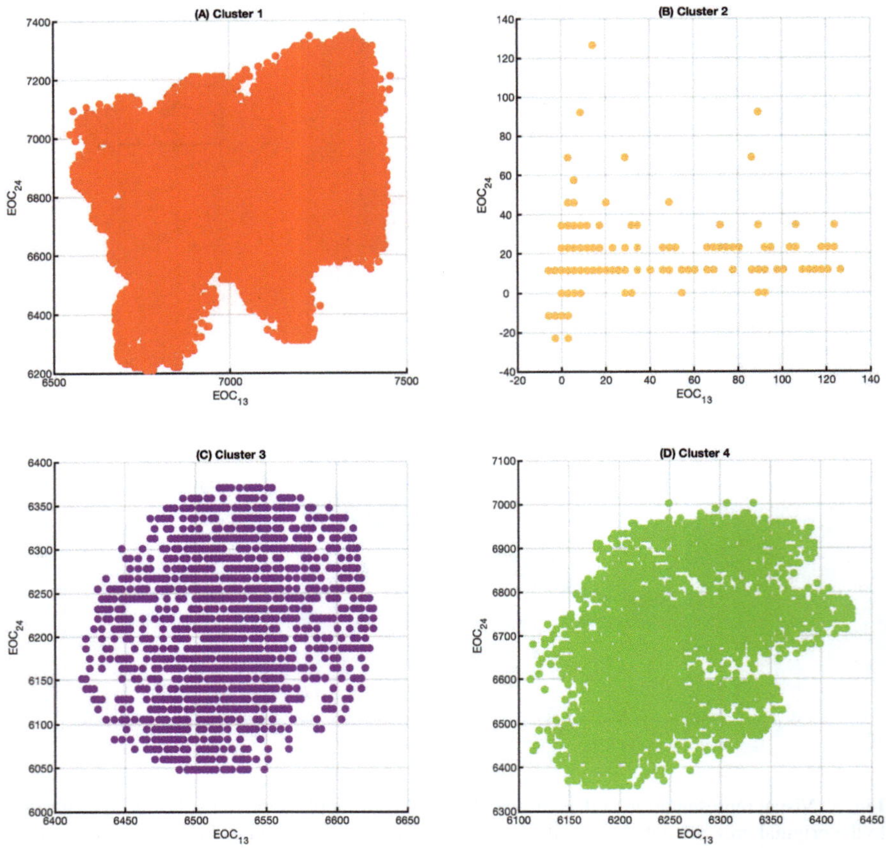

Fig. 6 Scatter plots for the pressure data observed in each cluster. (**a**) Cluster 1 shows the optimal values of the pressure data with a shape similar to the one obtained with the EDA. (**b**) Although cluster 2 visually shows fewer observations than the obtained value of 2569, it represents the contribution of noise and missing values in the dataset. (**c**) Cluster 3 shows values around 6000–66,000 psig. The observed lower values compared with those of cluster 1 can be labelled as the transitory response of the machine; in other words, cluster 3 shows values related to the stabilisation time required for an optimal operation. (**d**) Finally, cluster 4 shows values around 6100–7000 psig, which are closer to the values of cluster 1, a response attributed to the location of the sensors in the mould

no observations outside the limits of the cluster itself, contrary to the EDA, where several observations were still present after the cleaning. Those observations outside this cluster are not outliers but observations of another well-defined cluster by the DBSCAN algorithm. The typical pressure values range from 6500 to 7500 psig for cavities 1–3 and 6200 to 7400 psig for cavities 2–4.

This range of values can be considered optimal for the injection moulding process performed in the machine analysed. In this part of the analysis, it is necessary to recall that the optimum parameter values depend exclusively on the product manufactured. The IM parameters will vary depending on the machine's typical values, the manufacturing process' efficiency and the quality of the produced parts.

Cluster 2, Fig. 6b, presents values around 0, with mean values of 4.75 and 10.76 for cavities 1–3 and 2–4, respectively. The values' interpretation corresponds to the contribution of the error, mainly to the human element or a bad lecture in the sensors. In conclusion, it is the noise contribution to the dataset.

Clusters 3 and 4, shown in Fig. 6c, d, respectively, showed pressure values with lower magnitudes than in cluster 1. Since the measured pressure values are at lower times, these data are considered the response observed in a stabilisation time required for the machine to take the optimal pressure values. In comparison with cluster 3, the density of cluster 4 is more substantial, so that it can be an indicator of high-pressure values observed in these cavities. It is important to remark that the pressure sensors were in different locations of the injection moulding machine, so their values will differ in the several stages of the injection moulding process. With this observation, it is clear that clusters 3 and 4 are associated with the injection stages, although not with optimal pressure values, as in the case of cluster 1.

In Fig. 7, the scatter plots of clusters 1, 3 and 4 present pressure values to inspect the clusters obtained with the DBSCAN algorithm. After removing the outliers, a shape similar to the dataset was observed (see Fig. 2). Regardless, the values from 6000 to 6400 psig for the EOC_1_3 were not labelled as outliers and were considered independent clusters. Likewise, values assessed in the EOC_2_4 from 6000 to 6400 psig in a different cluster were not considered outliers in the EDA as performed at the beginning. The main contribution of the performed approach is the detection and isolation of those data related to the optimal operating values in the analysed machine.

Figure 8 shows the pressure distributions in cluster 1, where the negatively skewed distribution is still present and might be related to the required time to reach the optimal pressure values. The opposite case is in Fig. 9, where the cluster 2 data values are zero, indicating a noise contribution to the dataset. Figure 10 shows the skewed distributed pressure data of cluster 3, which can be related to the stabilisation time for the pressure in the IM machine. Although there is no linear relationship between the pressure and the time, this observation needs to be assessed with different techniques in the same IM process to justify the observed variations. Finally, Fig. 11 shows the pressure data for cluster 4, where a positively skewed distribution in cavities 1–3 is observed. The latest indicates a strong presence of data at the beginning of the injection moulding process. Meanwhile, slightly positively skewed data in cavities 2–4, located at the end of the mould, indicate the final measurements for the injection moulding process.

Outliers' treatment is crucial for detecting failures and anomalies in the manufacturing process due to different mechanical, electrical or human factors [58, 59]. Other terms are rare events, noise, one-class classification, novelty detection, abnormal data and singular points [60]. These observations have been a topic in failure detection and monitoring research. Several statistical and machine learning approaches could be helpful for its treatment and evaluation as potential failures in the manufacturing processes. Also, the outliers are related to mechanical and

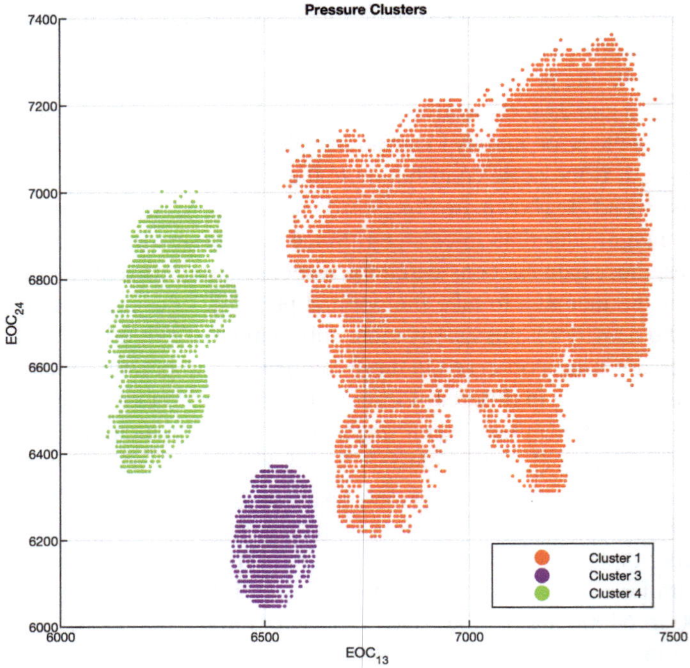

Fig. 7 Pressure data observed in clusters 1 (orange), 3 (purple) and 4 (green), all of them with values in the range from 6000 to 7500 psig, were considered in this work as the pressure values without the contribution of outliers

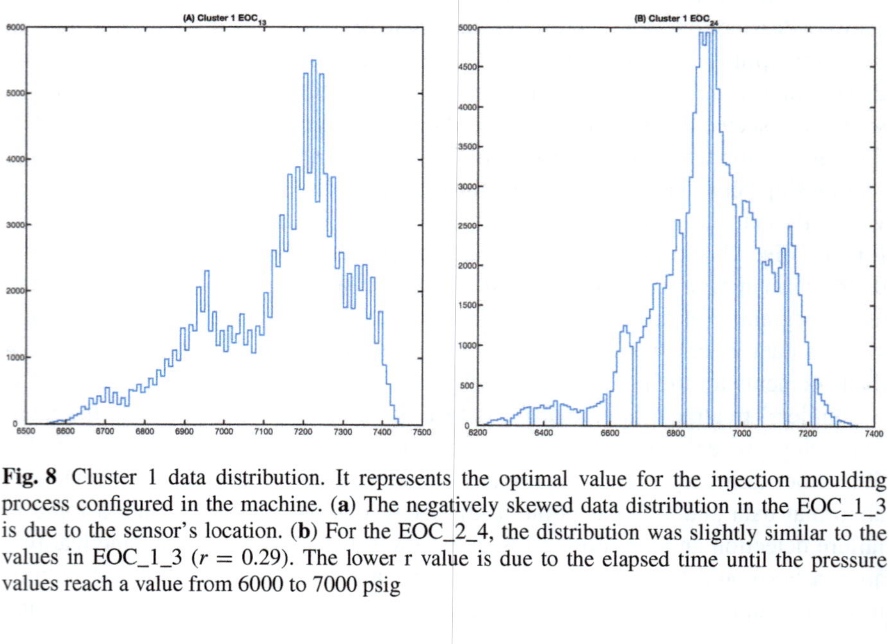

Fig. 8 Cluster 1 data distribution. It represents the optimal value for the injection moulding process configured in the machine. (**a**) The negatively skewed data distribution in the EOC_1_3 is due to the sensor's location. (**b**) For the EOC_2_4, the distribution was slightly similar to the values in EOC_1_3 ($r = 0.29$). The lower r value is due to the elapsed time until the pressure values reach a value from 6000 to 7000 psig

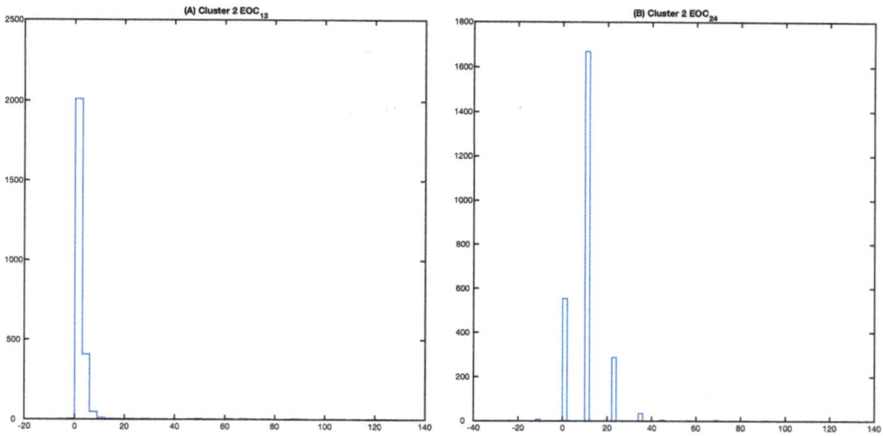

Fig. 9 Cluster 2 data distribution with the noise contribution in the dataset. (**a**) There are three values in the EOC_1_3; 0 is the highest frequency. The latest is due to the 0 substitution of the NaN values. (**b**) For the EOC_2_4, the 0 values are still present but with a lower frequency. The values with 10 have a higher frequency and contribute to the noise during the measurements

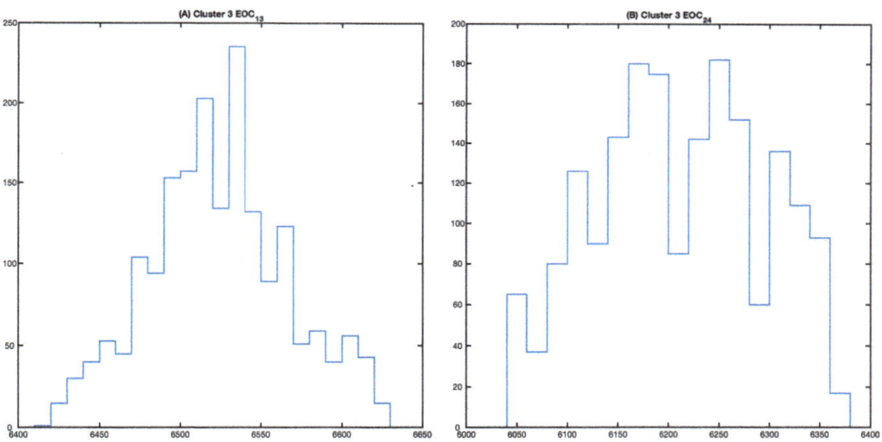

Fig. 10 Cluster 3 data distributions for the (**a**) EOC_1_3 and (**b**) EOC_2_4. Although low, both measurements show positive skewness values in their distributions (0.0862 and 0.0022), which can be related to a standardisation time during the IM process

electrical failures in the production line, where the only improvement factor is the human contribution to monitor the production parameters. This task seems to be quite simple, but it requires a lot of human and technical resources to avoid the presence of outliers. By adding intelligent control systems to the production line, such as batteries, the result will be a completely intelligent system, which is the primary motivation of this research and all the related works in Industry 4.0.

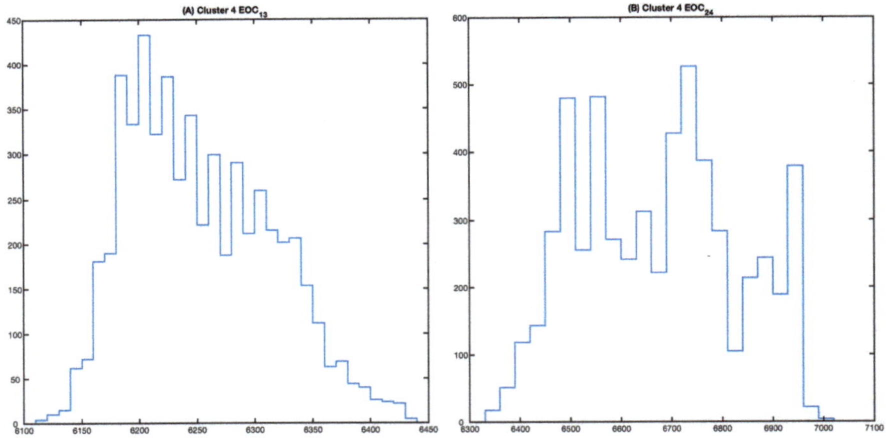

Fig. 11 Cluster 4 data distributions for (**a**) EOC_1_3 and (**b**) EOC_2_4. The skewed distributions are related to the beginning and finish of the injection moulding process. In the EOC_1_3, the positively skewed distribution (skewness = 0.41) is possibly due to the sensor's location and measurement at the beginning of the IM process. The EOC_2_4 data distribution shows another but slowly positive skewness (skewness = 0.09), possibly related to the measurement performed at the end of the IM process

In the proposed research, a new approach using the DBSCAN algorithm for clusterisation and subsequent outliers labelling is being developed. Here, DBSCAN uses no-label data corresponding to pressure values in an injection moulding machine. The time stamp used was recorded in a time-lapse of 2 years of production, but with the limitation of several registers with null values, indicating an electrical failure, an inappropriate sensor measurement or even a manual shutdown. These situations lead to a delay in the production schedule and financial losses, which can be avoided by assessing the features of those measurements considered as outliers to find a cluster of interest. As previously reported [60, 61], the limitation of machine learning in the manufacturing process is that the algorithms implemented are limited to specific parts of the process analysed instead of the complete analysis of the whole manufacturing process. In this research, there is an identification of each part of the injection moulding process reflected as the clusters obtained using the DBSCAN algorithm. Therefore, it is not only those values labelled as the optimal values for obtaining an efficient production line and failure detection.

Along with this observation, this research's main contribution is using unsupervised machine learning algorithms for failure detection and prediction. As the DBSCAN is an algorithm for clusterisation, the obtained five clusters are related to different stages in the injection moulding process. Therefore, the optimal values required for an efficient production line were detected and classified as cluster 1. Cluster 2 shows those values related to the noise contribution in the production line due to wrong sensor measurements and can be related to a failure. Clusters 3 and 4 contain the values obtained during the injection moulding machine's required stabilisation time and cooling, respectively. Also, neither of these two clusters can be considered a failure. Finally, the remaining data were identified as failures since the values mainly were outliers during the injection moulding process because of mechanical and electrical factors in the machine or, most importantly, due to a wrong combination of the control parameters in the process.

With the meaningful data characterised and labelled, the outlier's treatment is now the remaining issue in this research. Figure 12 shows the data distribution of the outliers in the measurements with negatively skewed distributions, −3.18 for EOC_1_3 and −3.46 for EOC_2_4, related to the failures observed in the dataset. The main reason is a wrong acquisition, a bad sensor function or even an anomaly detected in the material mixing during the injection process. Compared with the original dataset, this number of failure observations is relatively low, and using a simple clustering algorithm and thresholding will remove or mislabel failures. Therefore, the use of unsupervised machine learning algorithms is justified. This clustering isolates the noise contributions, detects the failures and helps predict the optimal parameters and pressure ranges for controlling the injection moulding process.

As reported by Selvaraj et al. (2019) [2], machine learning algorithms applied to the injection moulding processes deal with the repetitive experimentation in the injection process and the computational time consumed by simulations by iterating its control parameters. Therefore, special attention is required to the dataset size, as it requires as much information as possible. Also, validation is required by including different datasets to guarantee the algorithm's robustness. Most of the reported research in the machine learning applications in the injection moulding process deals with supervised algorithms and deep neural networks for more complex cases [9]. Recently, an approach to assess the quality of the plastic parts produced by an

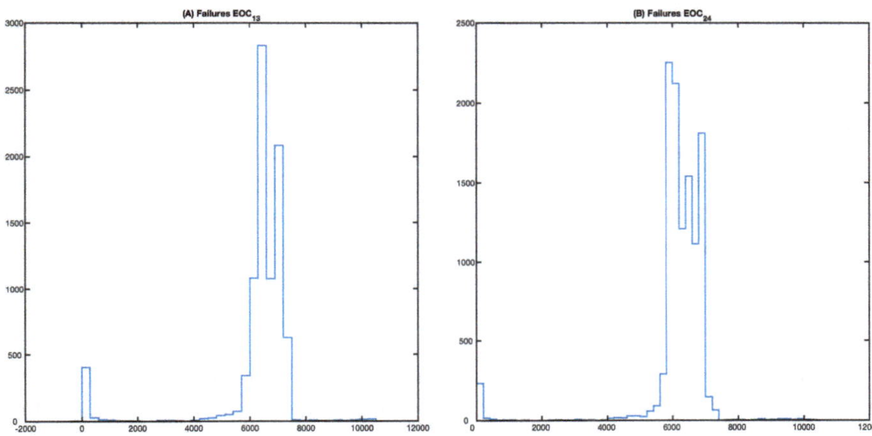

Fig. 12 Outliers' distribution in the dataset for the (**a**) EOC_1_3 and (**b**) EOC_2_4. Both measurements show a negatively skewed distribution with values closer to those observed in cluster 1, where the pressure values range from 6000 to 7000 psig

IM machine with optical techniques was reported [40]. This novel approach deals with the unsupervised classification of spectroscopic measurements but does not deal with failure detection.

In another research, Kolokas et al. [36] successfully applied an unsupervised algorithm for failure detection. However, the analysed datasets do not consider outliers' clusterisation and their corresponding relationship with different stages of the manufacturing process. Also, in another recent research [62], the unsupervised algorithm approach was applied to an IM to find the optimal features in the dataset for prediction. This study could be applied for failure detection since it only remarks the optimal values of the IM process. Finally, Liu et al. [63] reported a novel approach for defect parts in an IM process with small datasets. This fact in AI research and its applications, including the IM process, is highlighted as one limitation. The reported research deals with applying deep convolutional networks and transfer learning to detect the defect products by image recognition. This research yields a new scenario that can be improved by adding other classes to the study, such as the clusters related to the noise, the failures and the standardisation times.

In most cases, previous research on machine learning algorithms applied to the injection moulding process deals with supervised learning. If not the case, deep learning is the most accurate technique for finding hidden parameters in the dataset. However, using unsupervised algorithms can obtain fast and straightforward solutions to optimise values or detect failures in terms of outliers.

4 Conclusions

This analysis showed that the fault diagnosis of the injection process is a complex clusterisation problem. Conventional machine learning techniques help to cluster and define the fault class. In this research, an unsupervised machine learning algorithm was used for failure detection and optimal parameter identification in an injection moulding process to evolve Industry 4.0 paradigm.

This analysis shows that the fault diagnosis of the injection process is a complex classification problem. Conventional machine learning techniques help to cluster and define the fault class. However, this implies the need for no labelled data, like predictors or responses, to establish a classification algorithm and evaluate the system's efficiency. In future research, it would be interesting to define several threshold values for the pressure data and the resulting clusters and then train a classification model.

The use of unsupervised algorithms in failure detection is not a novel approach. However, with the clusterisation tasks, the optimal values of an injection moulding machine can be identified and the outliers' detection as well. The presence of outliers can result from the different stages of the injection moulding process and not only failure. Therefore, adding different unsupervised algorithms to correlate the findings and assess the manufactured objects' quality will be of interest to standardise the operation's optimal control parameters. The lack of labelling data and some response variables to evaluate the clusterisation algorithm represents the main limitations of the research. However, this issue has been identified as a critical parameter in Industry 4.0, requiring optimal data and robust machine learning algorithms for classification, regression and clusterisation. Another limitation of this research is the reduced dataset since a higher dimension will considerably improve the unsupervised algorithm's robustness.

In this research, an unsupervised machine learning algorithm was used for failure detection and optimal parameter identification in an injection moulding process to evolve Industry 4.0 paradigm. The following steps concern associating the cluster with stages and the manufactured product's quality. This approach reduces the dataset's sparsity, and the proposed algorithm's advantages lie on the use of ML techniques, ease of implementation and increased detection accuracy through concurrent analysis.

Another proposal concerns implementing more sensors to capture other process variables in the injection moulding process. Consequently, test the machine learning algorithms reported in this research and others considered for clusterisation, classification and regression, and use the results of the optimised parameters to predict and improve the efficiency and quality of the manufactured components.

Acknowledgements I. E. Garduño and H. Arcos–Gutiérrez gratefully acknowledge the support from the Investigadoras e Investigadores por México CONACYT programme through project No. 674.

I. E. Garduño and F. S. Chiwo acknowledge the support from the COPOCYT Fideicomiso 23871 Multas Electorales Convocatoria 2021–01 through the project Optimización de Parámetros en Procesos de Moldeo por Inyección de Plásticos con Enfoque Hacia Manufactura 4.0.

References

1. M.R. Khosravani, S. Nasiri, Injection molding manufacturing process: Review of case-based reasoning applications. J. Intell. Manuf. **31**, 847–864 (2020). https://doi.org/10.1007/s10845-019-01481-0
2. S.K. Selvaraj, A. Raj, R. Rishikesh Mahadevan, U. Chadha, V. Paramasivam, A review on machine learning models in injection molding machines. Adv. Mater. Sci. Eng. **2022**, 1–28 (2022). https://doi.org/10.1155/2022/1949061
3. M.R. Khosravani, S. Nasiri, T. Reinicke, Intelligent knowledge-based system to improve injection molding process. J. Ind. Inf. Integr. **25**, 100275 (2022). https://doi.org/10.1016/j.jii.2021.100275
4. R.Y. Zhong, X. Xu, E. Klotz, S.T. Newman, Intelligent manufacturing in the context of industry 4.0: A review. Engineering **3**(5), 616–630 (2017). https://doi.org/10.1016/J.ENG.2017.05.015
5. C. Zhang, Y. Lu, Study on artificial intelligence: The state of the art and future prospects. J. Ind. Inf. Integr. **23**, 100224 (2021). https://doi.org/10.1016/j.jii.2021.100224
6. N. Schwalbe, B. Wahl, Artificial intelligence and the future of global health. Lancet **395**(10236), 1579–1586 (2020). https://doi.org/10.1016/S0140-6736(20)30226-9
7. S.O. Abioye, L.O. Oyedele, L. Akanbi, A. Ajayi, J.M. Delgado, M. Bilal, O.O. Akinade, A. Ahmed, Artificial intelligence in the construction industry: A review of present status, opportunities and future challenges. J. Build. Eng. **44**, 103299 (2021). https://doi.org/10.1016/j.jobe.2021.103299

8. L. Chen, P. Chen, Z. Lin, Artificial intelligence in education: A review. IEEE Access **8**, 75264–75278 (2020). https://doi.org/10.1109/ACCESS.2020.2988510
9. H. Jung, J. Jeon, D. Choi, A.J.Y. Park, Application of machine learning techniques in injection molding quality prediction: Implications on sustainable manufacturing industry. Sustainability **13**(8), 4120 (2021). https://doi.org/10.3390/su13084120
10. R.D. Párizs, D. Török, T. Ageyeva, J.G. Kovács, Machine learning in injection molding: An industry 4.0 method of quality prediction. Sensors **22**(7), 2704 (2022). https://doi.org/10.3390/s22072704
11. A. Polenta, S. Tomassini, N. Falcionelli, P. Contardo, A.F. Dragoni, P. Sernani, A comparison of machine learning techniques for the quality classification of molded products. Information **13**(6), 272 (2022). https://doi.org/10.3390/info13060272
12. T. Jo, *Machine Learning Foundations* (Springer, 2021). https://doi.org/10.1007/978-3-030-65900-4
13. G. Taranto-Vera, P. Galindo-Villardón, J. Merchán-Sánchez-Jara, J. Salazar-Pozo, A. Moreno-Salazar, V. Salazar-Villalva, Algorithms and software for data mining and machine learning: A critical comparative view from a systematic review of the literature. J. Supercomput. **77**(10), 11481–11513 (2021). https://doi.org/10.1007/s11227-021-03708-5
14. S.L. Mirtaheri, R. Shahbazian, *Machine Learning: Theory to Applications* (CRC Press, 2022). https://doi.org/10.1201/9781003119258
15. M. Mohammed, M.B. Khan, E.B. Bashier, *Machine Learning: Algorithms and Applications* (CRC Press, 2016). https://doi.org/10.1201/9781315371658
16. A. Géron, *Hands-on Machine Learning with Scikit-Learn, Keras and TensorFlow* (O'Reilly Media, Inc, 2022)
17. H. Jiang, *Machine Learning Fundamentals: A Concise Introduction* (Cambridge University Press, 2021)
18. T.T. Teoh, Z. Rong, *Artificial Intelligence with Python* (Springer, 2022). https://doi.org/10.1007/978-981-16-8615-3
19. M. Mittal, L.M. Goyal, D.J. Hemanth, J.K. Sethi, Clustering approaches for high-dimensional databases: A review. Wiley Interdiscip. Rev.: Data Min. Knowl. Discov. **9**(3), e1300 (2019). https://doi.org/10.1002/widm.1300
20. B. Mahesh, Machine learning algorithms – A review. Int. J. Sci. Res. (IJSR). [Internet] **9**, 381–386 (2020). https://doi.org/10.21275/ART20203995
21. S. Ray, A quick review of machine learning algorithms, in *2019 International Conference on Machine Learning, Big Data, Cloud and Parallel Computing (COMITCon)*, (IEEE, 2019), pp. 35–39. https://doi.org/10.1109/COMITCon.2019.8862451
22. P. Agrawal, C. Gupta, A. Sharma, V. Madaan, N. Joshi, *Machine Learning and Data Science* (2022). https://doi.org/10.1002/9781119776499
23. S. Rafatirad, H. Homayoun, Z. Chen, S.M. Dinakarrao, *Machine Learning for Computer Scientists and Data Analysts: From an Applied Perspective* (Springer, 2022). https://doi.org/10.1007/978-3-030-96756-7
24. M.W. Berry, A. Mohamed, B.W. Yap (eds.), *Supervised and Unsupervised Learning for Data Science* (Springer, 2019). https://doi.org/10.1007/978-3-030-22475-2
25. J. Alzubi, A. Nayyar, A. Kumar, Machine learning from theory to algorithms: An overview, in *Journal of Physics: Conference Series*, Vol. 1142, No. 1, (IOP Publishing, 2018), p. 012012. https://doi.org/10.1088/1742-6596/1142/1/012012
26. M. Khanum, T. Mahboob, W. Imtiaz, H.A. Ghafoor, R. Sehar, A survey on unsupervised machine learning algorithms for automation, classification and maintenance. Int. J. Comput. Appl. **119**(13) (2015). https://doi.org/10.5120/21131-4058
27. B.K. Tripathy, S. Anveshrithaa, S. Ghela, *Unsupervised Learning Approaches for Dimensionality Reduction and Data Visualization* (CRC Press, 2021). https://doi.org/10.1201/9781003190554
28. M. Verkerken, L. D'hooge, T. Wauters, et al., Towards model generalization for intrusion detection: Unsupervised machine learning techniques. J. Netw. Syst. Manag. **30**, 12 (2022). https://doi.org/10.1007/s10922-021-09615-7

29. R. Scitovski, K. Sabo, F. Martínez-Álvarez, Š. Ungar, *Cluster Analysis and Applications* (Springer, 2021). https://doi.org/10.1007/978-3-030-74552-3
30. S.T. Wierzchoń, M.A. Kłopotek, *Modern Algorithms of Cluster Analysis* (Springer, 2018). https://doi.org/10.1007/978-3-319-69308-8
31. M. Amiruzzaman, R. Rahman, M.R. Islam, R.M. Nor, Evaluation of DBSCAN algorithm on different programming languages: An exploratory study, in *2021 5th International Conference on Electrical Engineering and Information & Communication Technology (ICEEICT)*, (IEEE, 2021), pp. 1–6. https://doi.org/10.1109/ICEEICT53905.2021.9667925
32. K. Khan, S.U. Rehman, K. Aziz, S. Fong, S. Sarasvady, DBSCAN: Past, present and future, in *The Fifth International Conference on the Applications of Digital Information and Web Technologies (ICADIWT 2014)*, (IEEE, 2014), pp. 232–238. https://doi.org/10.1109/ICADIWT.2014.6814687
33. M. Ester, H.P. Kriegel, J. Sander, X. Xu, A density-based algorithm for discovering clusters in large spatial databases with noise, in *KDD*, Vol. 96, No. 34, (1996), pp. 226–231
34. F. Ros, S. Guillaume (eds.), *Sampling Techniques for Supervised or Unsupervised Tasks* (Springer, 2020). https://doi.org/10.1007/978-3-030-29349-9
35. N. Amruthnath, T. Gupta, A research study on unsupervised machine learning algorithms for early fault detection in predictive maintenance, in *2018 5th International Conference on Industrial Engineering and Applications (ICIEA)*, (IEEE, 2018), pp. 355–361. https://doi.org/10.1109/IEA.2018.8387124
36. W. Nelson, C. Culp, Machine learning methods for automated fault detection and diagnostics in building systems – A review. Energies **15**(15), 5534 (2022). https://doi.org/10.3390/en15155534
37. N. Kolokas, T. Vafeiadis, D. Ioannidis, D. Tzovaras, A generic fault prognostics algorithm for manufacturing industries using unsupervised machine learning classifiers. Simul. Model. Pract. Theory **103**, 102109 (2020). https://doi.org/10.1016/j.simpat.2020.102109
38. F.S. Chiwo, E. Guevara, M.G. Ramírez-Elías, C.C. Castillo-Martínez, C.E. Osornio-Martínez, R. Cabrera-Alonso, F. Pérez-Atamoros, F.J. González, Use of Raman spectroscopy in the assessment of skin after CO_2 ablative fractional laser surgery on acne scars. Skin Res. Technol. **25**(6), 805–809 (2019). https://doi.org/10.1111/srt.12722
39. M.A. Kassem, K.M. Hosny, R. Damaševičius, M.M. Eltoukhy, Machine learning and deep learning methods for skin lesion classification and diagnosis: A systematic review. Diagnostics **11**(8), 1390 (2021). https://doi.org/10.3390/diagnostics11081390
40. S. Masubuchi, T. Machida, Classifying optical microscope images of exfoliated graphene flakes by data-driven machine learning. npj 2D Mater. Appl. **3**(1), 1–7 (2019). https://doi.org/10.1038/s41699-018-0084-0
41. M.L. Henriksen, C.B. Karlsen, P. Klarskov, M. Hinge, Plastic classification via in-line hyperspectral camera analysis and unsupervised machine learning. Vib. Spectrosc. **118**, 103329 (2022). https://doi.org/10.1016/j.vibspec.2021.103329
42. M. Usama, J. Qadir, A. Raza, H. Arif, K.L. Yau, Y. Elkhatib, A. Hussain, A. Al-Fuqaha, Unsupervised machine learning for networking: Techniques, applications and research challenges. IEEE Access **7**, 65579–65615 (2019). https://doi.org/10.1109/ACCESS.2019.2916648
43. M. Zhu, J. Wang, X. Yang, Y. Zhang, L. Zhang, H. Ren, B. Wu, L. Ye, A review of the application of machine learning in water quality evaluation. Eco-Environ. Health (2022). https://doi.org/10.1016/j.eehl.2022.06.001
44. B. Kung, M. Chiang, G. Perera, M. Pritchard, R. Stewart, Unsupervised machine learning to identify depressive subtypes. Healthc. Inform. Res. **28**(3), 256–266 (2022). https://doi.org/10.4258/hir.2022.28.3.256
45. A. de Hoffer, S. Vatani, C. Cot, G. Cacciapaglia, M.L. Chiusano, A. Cimarelli, F. Conventi, A. Giannini, S. Hohenegger, F. Sannino, Variant-driven early warning via unsupervised machine learning analysis of spike protein mutations for COVID-19. Sci. Rep. **12**(1), 1–4 (2022). https://doi.org/10.1038/s41598-022-12442-8
46. C. Aldrich, L. Auret, *Unsupervised Process Monitoring and Fault Diagnosis with Machine Learning Methods* (Springer, London, 2013). https://doi.org/10.1007/978-1-4471-5185-2

47. K.P. Tran, *Control Charts and Machine Learning for Anomaly Detection in Manufacturing*, Springer Series in Reliability Engineering (Springer, 2022). https://doi.org/10.1007/978-3-030-83819-5
48. J.A. Harding, M. Shahbaz, A. Kusiak, Data mining in manufacturing: A review. J. Manuf. Sci. Eng. **128**, 969–976 (2006). https://doi.org/10.1115/1.2194554
49. T. Wuest, *Identifying Product and Process State Drivers in Manufacturing Systems Using Supervised Machine Learning* (Springer, 2015). https://doi.org/10.1007/978-3-319-17611-6
50. MathWorks Inc., MATLAB. Natick, Massachusetts, US (2022)
51. MathWorks Inc., MATLAB Online (2022)
52. A. Blázquez-García, A. Conde, U. Mori, J.A. Lozano, A review on outlier/anomaly detection in time series data. ACM Comput. Surv. (CSUR) **54**(3), 1–33 (2021). https://doi.org/10.1145/3444690
53. A. Gaddam, T. Wilkin, M. Angelova, J. Gaddam, Detecting sensor faults, anomalies and outliers in the Internet of things: A survey on the challenges and solutions. Electronics **9**(3), 511 (2020). https://doi.org/10.3390/electronics9030511
54. W.L. Martinez, A.R. Martinez, J. Solka, *Exploratory Data Analysis with MATLAB* (Chapman and Hall/CRC, 2017)
55. F. Meng, G. Yuan, S. Lv, Z. Wang, S. Xia, An overview on trajectory outlier detection. Artif. Intell. Rev. **52**(4), 2437–2456 (2019). https://doi.org/10.1007/s10462-018-9619-1
56. A. Smiti, A critical overview of outlier detection methods. Comput. Sci. Rev. **38**, 100306 (2020). https://doi.org/10.1016/j.cosrev.2020.100306
57. E. Schubert, J. Sander, M. Ester, H.P. Kriegel, X. Xu, DBSCAN revisited, revisited: Why and how you should (still) use DBSCAN. ACM Trans. Database Syst. (TODS) **42**(3), 1–21 (2017). https://doi.org/10.1145/3068335
58. M.E. Celebi, K. Aydin (eds.), *Unsupervised Learning Algorithms* (Springer, Berlin, 2016). https://doi.org/10.1007/978-3-319-24211-8
59. B. Wang, Z. Mao, Outlier detection based on a dynamic ensemble model: Applied to process monitoring. Inf. Fusion. **51**, 244–258 (2019). https://doi.org/10.1016/j.inffus.2019.02.006
60. J. Carrasco, D. López, I. Aguilera-Martos, D. García-Gil, I. Markova, M. García-Barzana, M. Arias-Rodil, J. Luengo, F. Herrera, Anomaly detection in predictive maintenance: A new evaluation framework for temporal unsupervised anomaly detection algorithms. Neurocomputing **462**, 440–452 (2021). https://doi.org/10.1016/j.neucom.2021.07.095
61. S. Doltsinis, P. Ferreira, N. Lohse, Reinforcement learning for production ramp-up: A Q-batch learning approach, in *2012 11th International Conference on Machine Learning and Applications*, Vol. 1, (IEEE, 2012), pp. 610–615. https://doi.org/10.1109/ICMLA.2012.113
62. A.S. Struchtrup, D. Kvaktun, R. Schiffers, Comparison of feature selection methods for machine learning based injection molding quality prediction, in *AIP Conference Proceedings*, Vol. 2289, No. 1, (AIP Publishing LLC, 2020), p. 020052. https://doi.org/10.1063/5.0028546
63. J. Liu, F. Guo, H. Gao, M. Li, Y. Zhang, H. Zhou, Defect detection of injection molding products on small datasets using transfer learning. J. Manuf. Process. **70**, 400–413 (2021). https://doi.org/10.1016/j.jmapro.2021.08.034

An Approach to Select an Open Source ERP for SMEs Based on Industry 4.0 and Digitization Considering the SHERPA and WASPAS Methods

Juan Vicente Barraza de la Paz, Luis Alberto Rodríguez-Picón, Iván Juan Carlos Pérez-Olguín, and Luis Carlos Méndez-González

1 Introduction

The fourth industrial revolution or Industry 4.0 (I4.0) represents a new paradigm of work, in which small- and medium-sized enterprises (SMEs) have trouble integrating, mainly due to the inherent limitations that SMEs have, but I4.0, supported by a growing number of technologies, is closing this gap every day.

Overall, I4.0 is being driven by the following technological advances: autonomous robots, simulation, horizontal and vertical integration systems, the Internet of Things, cybersecurity, cloud, additive manufacturing, augmented reality, and at the same time big data analytics [1], which, coupled with the technological advances of the last decade, contribute to producing low-cost, high-speed electronic circuits that have changed the way signals are processed and have achieved greater efficiency between suppliers, producers, and customers, as well as between humans and machines [1, 2].

SMEs with aspirations to be part of the I4.0 must be more competitive; this can be achieved through the improvement of their processes and continuous evaluation and data exchange in real time, thereby achieving a flexible and quality business management, so a first approach to the I4.0 is through the integration of information systems or digitization of their processes, thereby allowing the acceleration of their resources.

One way to accelerate their resources is through Enterprise Resource Planning (ERP) systems, which are a set of unified software from which all the information of the operation of an organization is obtained [3]. Through these systems, continuous

J. V. B. de la Paz (✉) · L. A. Rodríguez-Picón · I. J. C. Pérez-Olguín · L. C. Méndez-González
Department of Industrial Engineering and Manufacturing, Autonomous University of Ciudad Juárez, Chihuahua, Mexico
e-mail: al216655@alumnos.uacj.mx

L. C. Méndez-González et al. (eds.), *Innovation and Competitiveness in Industry 4.0 Based on Intelligent Systems*, EAI/Springer Innovations in Communication and Computing, https://doi.org/10.1007/978-3-031-29775-5_6

improvement of processes such as purchasing, finance, manufacturing, sales, and inventories, among others, is achieved. This type of solutions achieves a complete integration of business processes with an integrated view of operations from a single information technology (IT) architecture [4].

For these reasons, most global companies have already adopted an ERP system, but for small- and medium-sized enterprises (SMEs), the adoption of an existing ERP system is more complicated, mainly due to the high cost of licensing, infrastructure, the complexity of implementation, and after-sales services [5], even with these limitations, and thanks to all the benefits already mentioned, SMEs increasingly find it more profitable and a competitive need to follow suit through the acquisition of this type of management systems [6].

Considering the above reasons, and mainly to eliminate high licensing costs, SMEs can analyze options such as open source ERP systems, which are softwares whose source code and some other rights are available for use or modification by others and that have been developed mainly by communities, promising significant benefits when adopted.

Choosing an ERP system, including open source systems, is not an easy task, making this stage a very critical process and even more so for SMEs, due to the large number of multicriteria decisions that exist when selecting an ERP system that suits the needs of the SME, so this research proposes the use of the methodology of Systematic Help for ERP Acquisition (SHERPA) [7] complemented with the multi-criteria decision making (MCDM) model of Weighted Aggregated Sum Product (WASPAS) [8].

SMEs that wish to be part of this industrial revolution must find a solution that is in line with their business model and even beyond their facilities. In the following section, a literature review is conducted in which topics related to I4.0, SMEs, ERP systems, the SHERPA method, and the WASPAS method of MCDM are discussed to support and contextualize the following sections.

2 Literature Review

The term of the fourth industrial revolution or Industry 4.0 derives from a proposal of the German government in 2011, which is realized by means of communication between industrial components along the entire value chain [9], which are constantly achieving continuous improvement in companies [1].

Companies are using more and more information on a large scale, increasing their dependence on information technologies [10]. Consequently, I4.0 is being driven by the consolidation of information systems along the value chain including with scopes beyond their own installation, thereby understanding the organization as a single system composed of multiple complex systems, whereby a first approach for SMEs to I4.0 can be brought through this integration.

The automation model of technological systems, which constitute the concept of I4.0, integrated with information technologies through big data analysis or artificial

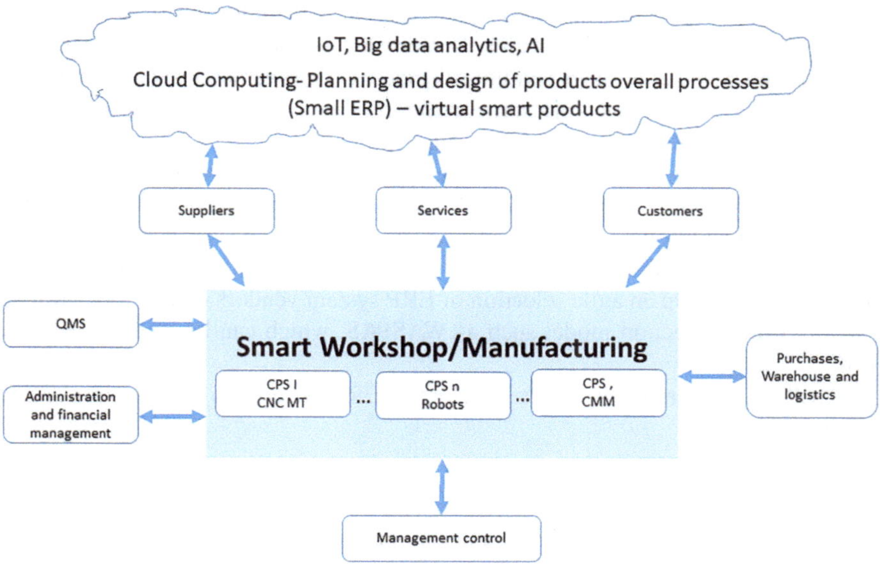

Fig. 1 Framework of smart manufacturing system of I4.0 model, adopted according to [11]

intelligence, has brought excellent opportunities to manufacturing systems, such as the management of relationships between supply and customer demand, allowing to cover aspects of quantity, variety, time-to-market appearance, price, quality and design, and the control and planning of supply chains [11].

For SMEs, a comprehensive smart manufacturing and ERP model based on the fundamentals of Industry 4.0 will be considered. Cyber Physical Systems (CPS), M2M-Human Machine Communication, I4.0 Operations Management, Digital Twins, Horizontal and Vertical Integration, IoT (Internet of Things), IIoT (Industrial Internet of Things), real time communications, big data, cloud, automation pyramid, supplier relationships, smart logistics, and customer relationships are shown in Fig. 1 [11].

Digitized and networked technological systems via IoT have the ability to transfer production control tasks to intelligent objects. This makes the manufacturing system itself more flexible and adaptable and defines a new paradigm for production planning and control [11].

ERP systems offer the strategy team of organizations' support in dealing with uncertainty by helping to identify specific potential failures, clarify their possible outcomes, assess possible consequences, and finally help to evaluate strategic alternatives and make decisions that reflect a wide range of aspects and circumstances [12].

SMEs need an ERP system to improve their operations; however, the investment limit of SMEs makes them decide for open source ERP rather of some proprietary version [13]. The main advantage of open source ERP systems include allowing organizations to take the lead in system selection and implementation regardless of the manufacturer, ability to tailor the software to the needs of the SME, and lower or no licensing cost, in addition to process improvements such as reducing raw material costs and inventories, increasing productibility, improving operational efficiency, and achieving competitive advantages [14].

Faced with the need to implement an ERP, the right selection of an open source ERP is an important problem faced by SMEs, due to the multiple options available in the market, in addition to the different criteria involved in this selection. Therefore, tools such as SHERPA proposed by Burgés et al. (2000) [15], and reviewed by Oltra et al. [16], help an agile selection of ERP system vendors complemented with a multi-criteria selection model such as WASPAS, which can be of great help for SMEs.

SHERPA's different phases for selecting an ERP system are briefly summarized below [15]:

- Phase 0: Learning business tactics and deciding to get or not an ERP.
  In phase 0, two activities are determined; the first one is to know in depth the business (mission, strategy, etc.) and its operations, in order to select the most appropriate option. In the second activity, the evaluators must make the decision if the company requires an ERP, reviewing the existing alternatives and selecting those that best suit the business [15].
- Phase 1: Alternatives and first filter.
  Once the business is thoroughly known and the candidate ERPs are known, the evaluators must discard those options that do not meet the minimum requirements. By analyzing each of the alternatives, the team should gather the necessary information to narrow down the options to between 5 and 8 [15].
- Phase 2: In-depth analysis of the alternatives, second filter.
  In this phase, as much information as possible is obtained from the alternatives, applying more exhaustive criteria to select two or three alternatives that fit the business. After this phase, a formal notation is more adequate [15].
- Phase 3: Validation of alternatives.
  In this phase, candidates must prove to the evaluation team that their product is the most suitable. This can be done through validation of real data by reviewing the functionality and adaptability of the business. The evaluators will analyze each alternative and present their proposal to the IT team and senior management [15].
- Phase 4: Last word and preparation.
  The evaluators will draw up the contract with the selected supplier, determining the general terms and conditions. Both IT and senior management will approve the resolution in order to sign the contract. Throughout the SHERPA evaluation, the selected tables and criteria will improve the evaluations in each phase [15].

The SHERPA model by Ilia et al. [15] uses the NoFun language designed to meet ISODEC (Integration of Social Development Center) software measurement guidelines. It decomposes the main software quality characteristics into measurable software sub-characteristics and divides the language into three distinct types, modules can introduce domains and attributes, and attributes can be defined in relation to other attributes [15].

On the other hand, MCDM methods are based on mathematical models with different levels of complexity and with the ability to take into consideration multiple criteria and their weights. Most multi-criteria methods are subject to constant improvements and changes by researchers.

Some previous researches where MCDM methods have been used in ERP system selection are "A multi-criteria decision making model for ERP selection: An empirical case in industrial organization of Iran" [17], "An Integrated Approach for Identification of critical factor for ERP implementation using Entropy and WASPAS method," [18], "An Integrated Fuzzy Multi Criteria Group Decision Making Approach for ERP System Selection" [19], "Applications of MCDM Methods in the ERP System Selection Process in Enterprises" [12], "ERP software selection with MCDM: application of TODIM method" [20], "Open Source ERP Selection for Small and Medium Enterprises by Using Analytic Hierarchy Process" [13], "Selection among ERP outsourcing alternatives using a fuzzy multi-criteria decision making methodology" [6], and "Development of a hybrid methodology (MCDM) for ERP system selection" [21], and that is why Brzozowski and Birfer [12] consider that MCDM have been widely used as a solution tool for the ERP system selection problem [12].

An MCDM problem consists of the attributes or criteria, the relative importance (weight) of each attribute, and the performance measures of the alternatives with respect to the various attributes. The main objective of the MCDM is to select the greatest option from a set of viable alternatives when faced with multiple conflicting criteria [22].

The WASPAS method has been included in the MCDM and was developed by Zavadskas et al. [8], the proposed method achieved a higher accuracy than the weighed product model (WPM) and 1.6 times more than the weighted sum model (WSM), suggesting that it is more accurate and advantageous than just using one of the previous methods, and to date, it has been applied in various industrial applications [23, 24].

The steps of the WASPAS method are as follows [22]:

- Step 1. Determine the type of criterion whether it is direct or inverse.
- Step 2. Obtain the normalized data matrix using the following two equations:

For direct criteria, use

$$\overline{x}_{ij} = \frac{x_{ij}}{x_j^{\text{Max}}} \tag{1}$$

For inverse criteria, use

$$\overline{x}_{ij} = \frac{x_j^{\mathrm{Min}}}{x_{ij}} \tag{2}$$

- Step 3. Calculate the normalized decision matrix for WPM and WSM, by means of the formulas:

$$A_i^{\mathrm{WPM}} = \prod\nolimits_{j=1}^{n} x_{ij}^{w_j} \tag{3}$$

$$A_i^{\mathrm{WSM}} = \sum\nolimits_{j=1}^{n} w_j x_{ij} \tag{4}$$

- Step 4. Obtain the preference scores and hierarchy of alternatives for WPM and WSM.
- Step 5. Calculate the preference score using.

$$Q_i = \lambda Q_i^1 + (1 - \lambda)\, Q_i^2 \tag{5}$$

$$Q_i^1 \text{ : weighted sum model}$$

$$Q_i^2 \text{ : weighted product model}$$

$$\lambda \text{ : value defined by the desicsion maker}$$

$$Q_i = Q_i^1 \quad \text{When } \lambda = 1$$

$$Q_i = Q_i^2 \quad \text{When } \lambda = 0$$

There is no ERP on the market that can fully meet the needs of enterprises. Since each company has different strategies and objectives, the selection of evaluation criteria becomes an important process, and the multidisciplinary content reveals the multi-criteria decision making [25].

Some criteria used by different authors are shown in Table 1, which provide support for the selection and definition of the criteria established later by the SME specialists based on the needs and characteristics of the SME, which gives rise to not considering the NoFun language used in the SHERPA method, for the determination of the criteria.

Table 1 Possible evaluation criteria according to different authors

Kazancogle and Burmaoglu [20]	Efe [26]	Tasnawijitwong and Samamchuen [13]
Strategy adjustment	Cost	Cost
Technology	Supplier specifications	Implementation time
Change management	Technical specifications	Vendor support
Risk	Usability	Functionality
Implementability		Flexibility
Business functionality		Reliability
Vendor credentials		User-friendly
Flexibility		
Cost		
Benefits		

The weights define the relative importance and influence of the input parameters on the final justification. Therefore, a combination of entropy and AHP methods will be used to determine the objective and subjective importance of the criteria.

The entropy method is an objective method for determining the weight of individual responses without the choice of the decision maker [27]; this method works on the principle that information from indicators of greater weight is more constructive than information from indicators of lesser weight [28].

The steps to be followed to determine the weights through the entropy method are as follows:

- Step 1. Normalization of the decision matrix array (performance indices) to obtain the project results through the formula:

$$P_{ij} = \frac{Xij}{\sum_{i=1}^{m} X_{ij}} \tag{6}$$

- Step 2. Calculation of entropy, a measure of the project results using the following formula:

$$E_j = -k\sum_{i=1}^{m} P_{ij} \ln P_{ij} \tag{7}$$

where $k = 1/\ln(\mathrm{m})$

- Step 3. Define the objective weights based on the concept of entropy, with the following formula:

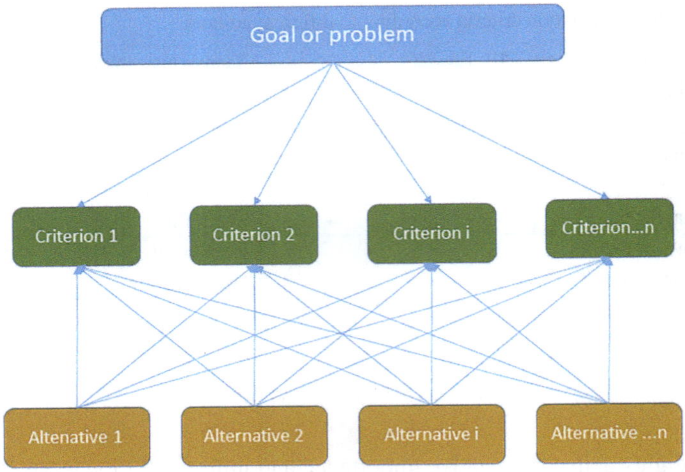

Fig. 2 Decision model

Table 2 Paired comparison matrix, which will be called A1

	C_1	CC_2	CC_3	……	CC_n
C_1	1	$C_{1/}\ C_2$	$C_{1/}\ C_3$	...	$C_{1/}\ C_n$
C_2	$C_{2/}\ C_1$	1	$C_{2/}\ C_3$	…	$C_{2/}\ C_n$
C_3	$C_{3/}\ C_1$	$C_{3/}\ C_2$	1	…	$C_{3/}\ C_n$
…	…	…	…	1	…
C_n	$C_{n/}\ C_1$	$C_{n/}\ C_2$	$C_{n/}\ C_3$	…	1

$$W_j = \frac{1 - E_j}{\sum_{j=1}^{n} \left(1 - E_j\right)} \tag{8}$$

The AHP method is a subjective method proposed to compare each criterion with others and derive the weights of the criteria based on the subjective opinions of the experts [29]. The AHP can be evaluated between qualitative and quantitative criteria. In addition, AHP methods tolerate small discrepancies because human decisions are not always consistent [13].

The steps to be followed to determine the weights through the AHP method are as follows:

- Step 1. Build the decision model, which is a hierarchical model, which at the top is the goal or problem, followed by the criteria or sub-criteria and the alternatives at the bottom, as shown below in Fig. 2.
- Step 2. Construct the paired comparison matrix, shown in Table 2, using Saaty's relative importance scale [30], which is shown in Table 3.
- Step 3. Obtaining the weights

Table 3 Possible evaluation criteria according to different authors

Scale	Definition	Explanation
1	Equally important	Two elements contribute equally
3	Moderately important	Slight preference of one element over another
5	Strongly important	Strong preference of one element over another
7	Very strong or demonstrated importance	Much more preference of one element over another, demonstrated preference
9	Extremely strong importance	Clear and absolute preference of one element over another
2, 4, 6, and 8	Intermediate value	Intermediate of above values

To obtain the weights of each of the criteria, it is necessary to multiply the value of each of the criteria of the paired comparison matrix and raise it to 1/n as indicated in Eq. 9; the result of this will be called A2.

$$A2 = \begin{bmatrix} (C_{11} * C_{12} * C_{13} * C_{....} * C_{1n})^{\frac{1}{n}} \\ (C_{21} * C_{12} * C_{13} * C_{....} * C_{1n})^{\frac{1}{n}} \\ (C_{31} * C_{12} * C_{13} * C_{....} * C_{1n})^{\frac{1}{n}} \\ (C_{...1} * C_{12} * C_{13} * C_{....} * C_{1n})^{\frac{1}{n}} \\ (C_{n1} * C_{n2} * C_{n3} * C_{....} * C_{nn})^{\frac{1}{n}} \end{bmatrix} = \begin{bmatrix} R_1 \\ R_2 \\ R_3 \\ R_{...} \\ R_n \end{bmatrix} \tag{9}$$

Next, the column is summed with the data obtained from A2, as shown in Eq. 10.

$$\text{Sum} = (R_1 + R_2 + R_3 + R_{...} + R_n) \tag{10}$$

To obtain the weights, matrix A2 is divided by the value of the sum resulting from Eq. 10, obtaining the matrix of the weights of each of the criteria A3, where the sum of these must be equal to 1.

$$A3 = \begin{bmatrix} R_1 / \text{Sum} \\ R_2 / \text{Sum} \\ R_3 / \text{Sum} \\ R_{...} / \text{Sum} \\ R_n / \text{Sum} \end{bmatrix} = \begin{bmatrix} W_1 \\ W_2 \\ W_3 \\ W_{...} \\ W_n \end{bmatrix} \tag{11}$$

- Step 4. Finally, the consistency test is performed, the acceptance of the pairwise comparison matrix is accepted when the consistency radius is less than 0.1, and to do this, it is necessary to perform the sum of the multiplication of each of the criteria by the result of A2, as expressed in Eq. 12.

$$A4 = \begin{bmatrix} ((C_{11} * R_1) + (C_{12} * R_1) + (C1_3 * R_1) + \cdots + (C_{1n} * R_1)) \\ ((C_{21} * R_2) + (C_{22} * R_2) + (C_{23} * R_2) + \cdots + (C_{2n} * R_2)) \\ ((C_{31} * R_3) + (C_{32} * R_3) + (C_{33} * R_3) + \cdots + (C_{3n} * R_3)) \\ ((C_{\ldots 1} * R_{\ldots}) + (C_{\ldots 2} * R_{\ldots}) + (C_{\ldots 3} * R_{\ldots}) + \cdots + (C_{\ldots n} * R_{\ldots})) \\ ((C_{nn} * R_n) + (C_{nn} * R_n) + (C_{nn} * R_n) + \cdots + (C_{nn} * R_n)) \end{bmatrix} = \begin{bmatrix} RC_1 \\ RC_2 \\ RC_3 \\ RC_{\ldots} \\ RC_n \end{bmatrix} \tag{12}$$

The result obtained in A4 is divided by the matrix of the obtained weights A3, as expressed in Eq. 13.

$$A5 = \begin{bmatrix} RC_1 / W_1 \\ RC_2 / W_2 \\ RC_3 / W_3 \\ RC_{\ldots} / W_{\ldots} \\ RC_n / W_n \end{bmatrix} = \begin{bmatrix} CW_1 \\ CW_2 \\ CW_3 \\ CW_{\ldots} \\ CW_n \end{bmatrix} \tag{13}$$

Then, the average of A5 is obtained, which will be represented by the symbol lambda λ_{max}:

$$\lambda_{max} = \left[(CW_1 + CW_2 + CW_3 + CW_{\ldots} + CW_n) / n\right] \tag{14}$$

Then, the consistency index is obtained by means of Eq. 15.

$$CI = \frac{\lambda_{max} - n}{n - 1} \tag{15}$$

Finally, the consistency ratio is obtained by dividing the consistency index by the random consistency index (RI) shown in Table 4, as shown in Eq. 16.

Table 4 Random consistency index (RI)

Elements	1	2	3	4	5	6	7	8	9	10	11	12
RI	0	0	0.58	0.09	1.12	1.24	1.32	1.41	1.45	1.49	1.51	1.58

$$\mathrm{CR} = {}^{\mathrm{CI}}\!/_{\mathrm{RI}} \tag{16}$$

If CR <0.1 the paired comparison matrix is considered acceptable.

Equation 17 is applied to combine the objective importance W_k and the subjective importance λ_k to evaluate the integrated importance of the indicator parameter W_k^* used by Taheriyoun et al. [27].

$$\mathrm{W}_{\mathrm{k}}^{*} = \frac{\lambda_{\mathrm{k}} * \mathrm{W}_{\mathrm{k}}}{\sum_{\mathrm{j}=1}^{\mathrm{n}} \lambda_{\mathrm{j}} \mathrm{W}_{\mathrm{j}}} \tag{17}$$

3 Methodology

The methodology of this research begins with the search and selection of an open source ERP system for an SME, in which it is proposed to use the method proposed by Ilia, Franch, and Pastor in 2000, called SHERPA [15], which include five phases detailed in Fig. 3, which is presented below.

In addition to complementing the SHERPA method from phase 3, with the WASPAS method, which is detailed in Fig. 4 below, in order to perform the analysis of the candidates to obtain the most feasible solution for the SME from a multicriteria decision making, decide the weights of each of the criteria through

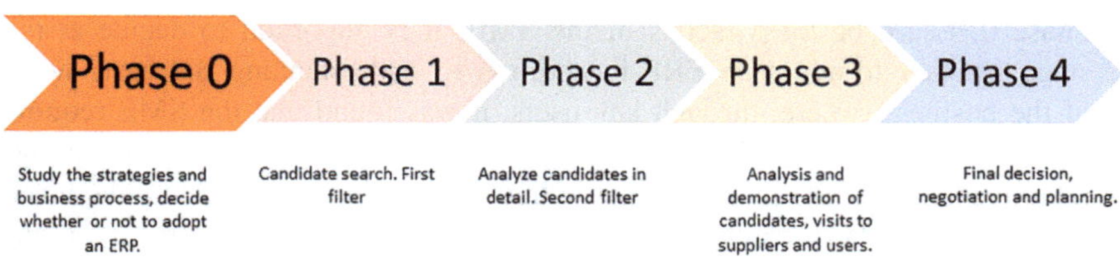

Fig. 3 Phases of SHERPA method [15]

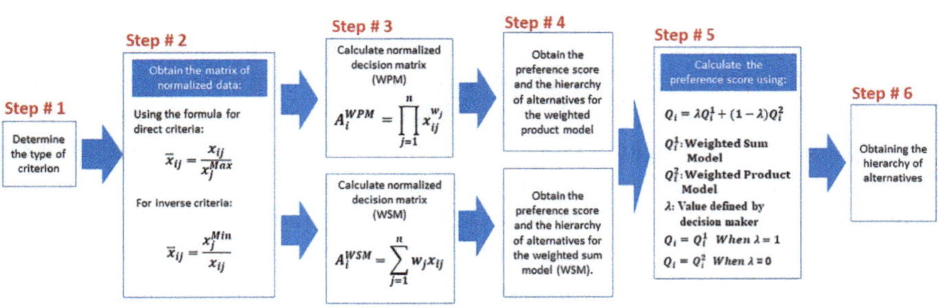

Fig. 4 WASPAS

a combination of the entropy and AHP methods in order to consider the points of view of decision makers both subjectively and objectively.

In order to visualize the methods in Fig. 5, a graphical representation of the steps to follow for the selection of an open source ERP system through the SHERPA and WASPAS methods is shown.

In order to ground the methodology used in a graphical representation using the hierarchical model, Fig. 6 is established as a summary of Fig. 5.

4 Case Study

The SME under study is a company located in Ciudad Juarez, Chihuahua, which seeks to standardize processes and manage its administrative and production processes more objectively, in addition to having general knowledge of the operation of ERP systems, thanks to previous experience due to the handling of the Oracle system by its personnel.

For system selection, the SME has an information technology (IT) department, which can evaluate the technical requirements of the different options presented, in addition to selecting a working group of key users with a high degree of knowledge of operations in areas such as finance, purchasing, warehouse, manufacturing, and sales, which will determine the needs of the aforementioned areas.

The following results were obtained through the SHERPA and WASPAS methodologies:

- Phase 0: Learn business tactics of the SME; it is important to decide at this stage whether to adopt an ERP in addition to understanding the basic needs of the business, where, through key users, it was found that the SME requires the following modules: customer relationship management or CRM (customer relation management), purchasing, warehouse, manufacturing, sales, and finance.
- Phase 1: Candidate search (first filter), work organization, review the organization, evaluate options, and select options from the market. In this phase, the following candidates were detected: ERPNext, ERP5, Odoo, Tryton, Openbravo, Axelor, and Flectra; this first filter criteria such as costs were evaluated, since the system had to have at least one open source version available for evaluation.
- Phase 2: Analyze in detail the candidates (second filter); in this stage, the number of candidates is reduced, analyzing which of them best suits the company. In this phase, it is important to know the technical and support requirements of the platforms; it is vital to know the programming language and the type of database used. In this phase, the IT department chose to continue with ERPNext, ERP5, and Odoo, since they are based on the Python programming language and fit the needs of the company, with modules such as CRM, purchasing, warehouse, manufacturing, sales, and finance.
- Phase 3: Validation of alternatives, in this stage, the preliminary evaluation of the characteristics of the systems is carried out, preparing and analyzing

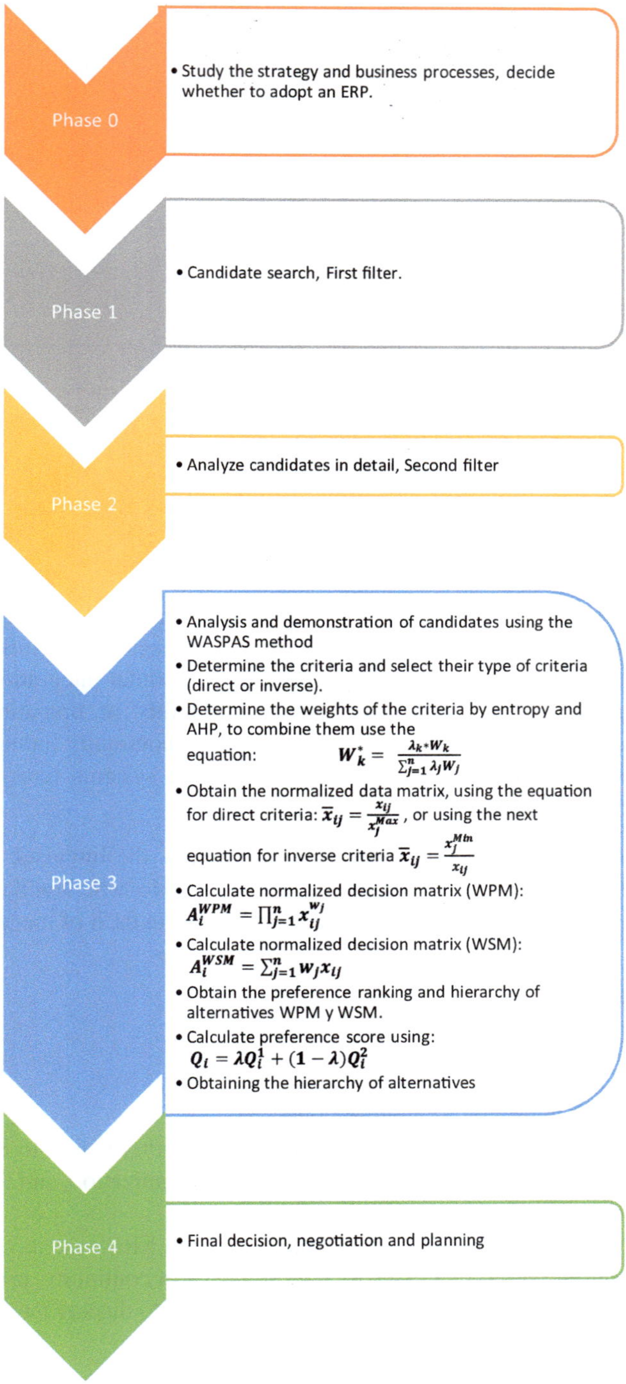

Fig. 5 SHERPA + WASPAS

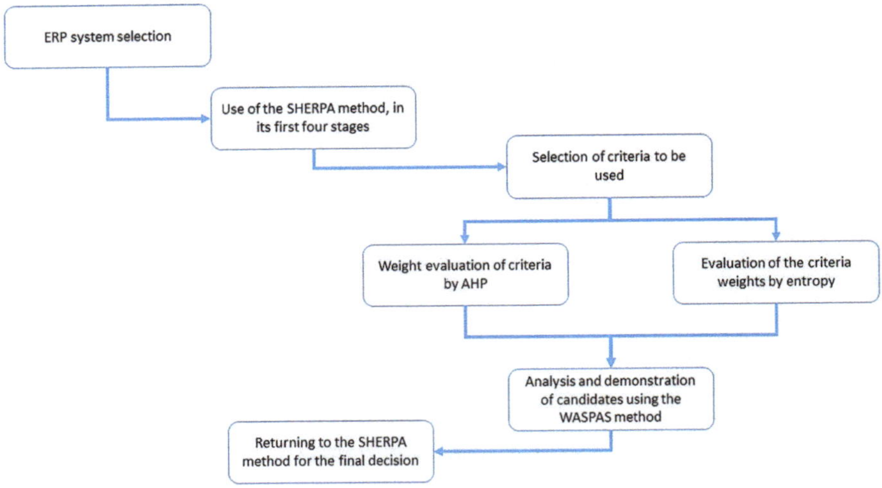

Fig. 6 SHERPA and WASPAS summary

demonstrations. In this stage, the WASPAS model was used in order to obtain the best alternative considering the criteria general characteristics (reliability, user-friendliness, customization, training), flexibility (modular integration, licensing, programming language, installation, and availability of operating systems), support (collaborators, community, and enterprise), continuity (number of installations and community size), and maturity of the systems (current versions, supported versions, and future versions).

As all the terms use several types of measurement, the linguistic values were converted to numerical values through the following scale, which will be applied to each of the criteria and will be expressed as results (R) in each of the tables.

Low = 1
Below average = 2
Average = 3
Good = 4
Excellent = 5

Table 5 compares the general characteristics of the alternatives to be evaluated, such as reliability of use, user-friendliness, level of customization, and whether there is training in the use of the system.

Table 6 reviews the flexibility or adaptation of the ERP system options to be evaluated, focused on the processes to achieve business continuity, prioritizing the programming language and operating systems, in order to display the data required by the SME.

Table 5 General characteristics (prepared by the company's experts, 2022)

ERP	Reliability	User-friendly	Customization	Training	R
ERPNext	Can work on an Ubuntu server and can be easily updated through its command line; it uses MariaDB as a database.	✓	2	Across communities	2
ERP5	The system operates under Linux and Unix, performance monitoring from the application.	✓	3	Easy structural design that guides the user	3
Odoo	It has a large number of modules to use and simple backups.	✓	4	Through communities and website, in addition to offering more training in payment options	4

Table 6 Flexibility (prepared by the company's experts, 2022)

ERP	Modular integration	License	Programming language/DB	Installation/SO	R
ERPNext	X	Open source License: GNU General Public License version 3.0 (GPLv3)	Python/MySQL	It is managed by the Frappe organization and can be used by purchasing a hosting plan, or by downloading and installing a virtual machine, Unix, Linux, or Mac.	4
ERP5	X	License Opensource: GNU General Public License version 3.0 (GPLv3)	ZOPE y Python/ZODB	It is managed by the Frappe organization and can be used by purchasing a hosting plan, or by downloading and installing a virtual machine, Unix, Linux, or Mac.	4
Odoo	✓	License: Odoo Community Version is Opensource distributed under GNU LGPL v3; and Odoo Enterprise Edition License v1.0 (paid at competitive price).	Python/PostgreSQL	An initial configuration of the system is performed in which a database is configured, available in Ubuntu, Debian, RPM, Sources, and Windows environment.	4

Table 7 Support (prepared by the company's experts, 2022)

EPR	Collaborators	Community	Enterprise	R
ERPNext	India, United Arab Emirates, Philippines, Tanzania, Egypt, Germany, and Oman	✓	✓	5
ERP5	N/A	Through distribution mails	✓	3
Odoo	Worldwide	✓	✓	5

Table 8 Continuity (prepared by the company's experts, 2022)

ERP	Number of installations	Community Size	R
ERPNext	+500 companies	+1000 people	5
ERP5	N/A	N./A	3
Odoo	+5,000,000 users	+1500 people	4

Table 9 Maturity (prepared by the company's experts, 2022)

ERP	Actual Version	Supported version	Future release	R
ERPNext	14	14, 13, 12	N/A	3
ERP5	5.5 (November, 2014)	5.5(November 2014), 5.4.7(January 2013)	N/A	4
Odoo	15 (October 2021)	15 (October 2021), 14 (October 2020), 13 (October 2019), 12 (October 2018), 11 (October 2017)	N/A	5

Table 10 Weight of alternatives

	General characteristics	Flexibility	Support	Continuity	Maturity	
Entropy weights	0.030	0.005	0.020	0.020	0.925	
AHP weights	0.055	0.256	0.465	0.142	0.082	CR = 0.07
Weights (W_j)	0.018	0.013	0.102	0.031	0.835	

Table 7 shows a review of the support provided by the options being evaluated, contemplating the maintenance of the ERP systems and possible improvements, from the community version to an enterprise version.

Table 8 reviews the continuity of the systems through the communities, in order to assess the support that could be obtained from them.

Table 9 represents the maturity of the options evaluated, through the versions that the systems have in operation, as well as the support of previous versions or if they are working on any future version.

Once the scale to be used has been established, the weighting weights will be obtained through the entropy method and AHP, obtaining the results shown in Table 10.

Once the weighting weights are obtained and the type of criteria is determined, Step 1, in this case, all were direct as shown in Table 11, which is presented below.

Step 2 is to obtain the normalized data matrix expressed in Table 12.

Table 11 Criterial types

		Direct	Direct	Direct	Direct	Direct
Weight (W_j)		0.018	0.013	0.102	0.031	0.835
Criterial names						
		General characteristics	**Flexibility**	**Support**	**Continuity**	**Maturity**
Alternatives	ERPNext	2	4	5	5	3
	ERP5	3	4	3	3	4
	Odoo	4	4	5	4	5

Table 12 Normalized data matrix

		General characteristics	Flexibility	Support	Continuity	Maturity
Normalized matrix	ERPNext	0.500	1.000	1.000	1.000	0.750
	ERP5	0.750	1.000	0.600	0.600	1.000
	Odoo	1.000	1.000	1.000	0.800	1.000

Table 13 Decision matrix de WSM and WPM

		General characteristics	Flexibility	Support	Continuity	Maturity
WSM	ERPNext	0.0091	0.0129	0.1023	0.0313	0.5012
	ERP5	0.0136	0.0129	0.0614	0.0188	0.6683
	Odoo	0.0182	0.0129	0.1023	0.0250	0.8354
WPM	ERPNext	0.9875	1.0000	1.0000	1.0000	0.6526
	ERP5	0.9948	1.0000	0.9491	0.9841	0.8299
	Odoo	1.0000	1.0000	1.0000	0.9930	1.0000

Table 14 Preference score WSM and WPM

		Preference score	Rank
WSM	ERPNext	0.6568	3
	ERP5	0.7750	2
	Odoo	**0.9937**	**1**
WPM	ERPNext	0.6445	3
	ERP5	0.7712	2
	Odoo	**0.9930**	**1**

Step 3 is to obtain the decision matrix of the sum model (WSM) using Eq. (3) and weighted product (WPM) using Eq. (4); the result of the operations is expressed in Table 13.

In Step 4, the preference score of the sum model (WSM) and weighted product (WPM) is obtained using Eq. (5); the results of the operations are shown in Table 14.

In Steps 5 and 6, we obtain the preference score and the hierarchies of the alternatives of the WASPAS model, as shown in Table 15.

Table 15 WASPAS preference score

		Score WASPAS	Rank WASPAS
Alternatives	ERPNext	0.6506	3
	EPR5	0.7731	2
	Odoo	**0.9934**	**1**

Phase 4, which involved final decision, negotiation and planning, review, and approval, resulted in the selection of Odoo ERP in its community version. This system fully met all the criteria preselected by the SME specialists.

5 Conclusion

When the processes in technological changes, especially in the changing environment, involve uncertainty, current technologies allow SMEs to benefit from open source platforms, which through communities generate efficient solutions of great benefit, achieving with these applications to be able to synchronize administrative and productive processes in order to compete in a globalized business world.

The problem of selecting among the different options that exist in the ERP systems market can be solved through a multi-criteria analysis, where one of the main activities to be performed is to determine which the best criteria to evaluate are. In the case study, it was determined not to consider such an important criterion as cost, since the SME was looking to evaluate open solutions, which would not generate any initial cost.

Therefore, SMEs can begin the path to I4.0 through the integration of information systems, by means of a correct selection and implementation of an ERP system, which will help them to a great extent to achieve success in the transformation and digitization of their processes.

Through the conducted analysis, it was decided to select Odoo as ERP system. Some of the most influential factors were that it contemplates the business needs of the SME, through the wide variety of modules in its community version, allowing the gradual implementation of these modules, thus dissolving the great complexity that the implementation of these types of systems represents. In addition, it proposes a lasting value in terms of scalability, expansion, robustness, security, and affordable prices in case the system needs to be upgraded to an enterprise version.

The Odoo system implemented in the SME is in constant review and updating of information through the different departments of the company, in order to achieve the objectives of the company, in addition to ensuring that the information it contains is accurate, and continues to assist in making decisions such as what can be faster and more effective budgets, ensure that there are no duplicate payments to suppliers, ensure the existence of materials for the formation of finished products, and ensure timely billing of products sold and the collection of the same.

This is achieved through the following:

Table 16 Improvement indicators

Indicator	Improvement (%)
Response time in quotations	80%
Improved customer management	90%
Platform stability	90%
Warehouse management	85%
Management of orders received	92%
Sales	15%

- Continuous review of raw material prices.
- Revaluation of finished products.
- Review of delivery times to customers and suppliers.
- The analysis of future modules that could be useful for the company.

Table 16 shows the improvement indicators obtained through the implementation of Odoo; indicators such as response time in budgets, improved customer management, and warehouse management will continue to increase over time, thanks to the capture of more part numbers and bills of materials, as for sales information accumulated by the company was taken to establish this parameter.

Although in the case study, Odoo was selected as the best option, this is not an indication that it is the best option for another SME; that is why the replicability of the methods described in this research could be given from a previous analysis of the company in question in order to know their needs and specific characteristics and thus select the criteria that best fit the SME by determining their weights through Entropy and AHP.

Given the results obtained in the selection, development, and implementation of Odoo, it can be said that in the case study, the SME now has more administrative tools, which help to resolve customer and supplier requests more efficiently. Therefore, the SME feels more prepared to take on new challenges with future customers and with these tools to confront the challenges that may arise.

References

1. M. Rubmann, M. Lorenz, P. Gerbert, et al., Industry 4.0: The future of productivity and growth in manufacturing industries. Bost Consult Gr.
2. Alasdair Gilchrist. Industry 4.0 : The Industrial Internet of Things. 2016.
3. T.B. Sambe, S. Maag, A. Cavalli, A methodology for enterprise resource planning automation testing application to the open source ERP-ODOO. ICSOFT 2019 – Proc 14th Int Conf Softw Technol, 407–415 (2019)
4. Y.M. Cheng, What drives cloud ERP continuance? An integrated view. J. Enterp. Inf. Manag. **31**, 724–750 (2018)
5. S.S. Shahawai, R. Idrus, Research methodology for assessing Malaysian SMEs perspective on ERP, in *Proceedings – 2009 3rd Asia International Conference on Modelling and Simulation, AMS 2009*, (2009. Epub ahead of print 2009). https://doi.org/10.1109/AMS.2009.122
6. C. Kahraman, A. Beskese, I. Kaya, Selection among ERP outsourcing alternatives using a fuzzy multi-criteria decision making methodology. Int. J. Prod. Res. **48**. Epub ahead of print 2010. https://doi.org/10.1080/00207540903175095

7. X.B. Ilia, X. Franch, J.A. Pastor, Formalising ERP selection criteria, in *10th International Workshop on Software Specification and Design, IWSSD 2000*, (2000. Epub ahead of print 2000). https://doi.org/10.1109/IWSSD.2000.891132
8. E.K. Zavadskas, Z. Turskis, J. Antucheviciene, et al., Optimization of weighted aggregated sum product assessment. Elektron ir Elektrotechnika **122**, 3–6 (2012)
9. S.H. Bonilla, H.R.O. Silva, M.T. da Silva, et al., Industry 4.0 and sustainability implications: A scenario-based analysis of the impacts and challenges. Sustainability **10**. Epub ahead of print 2018. https://doi.org/10.3390/su10103740
10. T.P. Raptis, A. Passarella, M. Conti, Data management in industry 4.0: State of the art and open challenges. IEEE Access **7**. Epub ahead of print 2019. https://doi.org/10.1109/ACCESS.2019.2929296
11. V. Majstorovic, S. Stojadinovic, ERP model for industry 4.0 concept. Proc. Manuf. Syst. **15**
12. M. Brzozowski, I. Birfer, Applications of MCDM methods in the ERP system selection process in enterprises. Handel Wewn.trzny **3**, 40–52 (2017)
13. S. Tasnawijitwong, T. Samanchuen, Open source ERP selection for small and medium enterprises by using analytic hierarchy process, in *Proceedings of 2018 5th International Conference on Business and Industrial Research: Smart Technology for Next Generation of Information, Engineering, Business and Social Science, ICBIR 2018*, (2018. Epub ahead of print 2018). https://doi.org/10.1109/ICBIR.2018.8391226
14. Aversano L, Di Brino M, Guardabascio D, et al. Understanding enterprise open source software evolution. In: Procedia Comput. Sci.. 2015. Epub ahead of print 2015. DOI: https://doi.org/10.1016/j.procs.2015.08.609.
15. X.B. Ilia, X. Franch, J.A. Pastor, Formalising ERP selection criteria. 10th Int Work Softw Specif Des IWSSD 2000, 115–122 (2000)
16. R. Oltra-Badenes, H. Gil-Gomez, V. Guerola-Navarro, Metodología Para La Selección De Sistemas Erp Para Pymes. 3C Empres Investig y Pensam crítico **7**. Epub ahead of print 2018. DOI: https://doi.org/10.17993/3cemp.2018.070436.10-33/
17. A. Alizadeh, B. Jahan, M. Chehrehpak, A multi-criteria decision making model for ERP selection: An empirical case in industrial organizations of Iran, in *In Proceedings of 2016 12th International Conference on Industrial Engineering, ICIE 2016*, (2016)
18. S. Shukla, An integrated approach for identification of critical factors for ERP implementation using entropy and WASPAS method. Int. J. Mech. Eng. **6**, 80–89 (2016)
19. B. Efe, An integrated fuzzy multi criteria group decision making approach for ERP system selection. Appl. Soft. Comput. J. **38**. Epub ahead of print 2016. https://doi.org/10.1016/j.asoc.2015.09.037
20. Y. Kazancoglu, S. Burmaoglu, ERP software selection with MCDM: Application of TODIM method. Int. J. Bus. Inf. Syst. **13**. Epub ahead of print 2013. https://doi.org/10.1504/IJBIS.2013.055300
21. H.S. Kilic, S. Zaim, D. Delen, Development of a hybrid methodology for ERP system selection: The case of Turkish Airlines. Decis. Support Syst. **66**. Epub ahead of print 2014. https://doi.org/10.1016/j.dss.2014.06.011
22. S. Chakraborty, E.K. Zavadskas, Applications of WASPAS method in manufacturing decision making. Informatica **25**. Epub ahead of print 2014. DOI: https://doi.org/10.15388/Informatica.2014.01
23. R. Kumar, A. Bhattacherjee, A.D. Singh, et al., Selection of portable hard disk drive based upon weighted aggregated sum product assessment method: A case of Indian market. Meas Control (United Kingdom) **53**. Epub ahead of print 2020. https://doi.org/10.1177/0020294020925841
24. K. Rudnik, G. Bocewicz, A. Kucińska-Landwójtowicz, et al., Ordered fuzzy WASPAS method for selection of improvement projects. Expert Syst. Appl. **169**. Epub ahead of print 2021. https://doi.org/10.1016/j.eswa.2020.114471
25. Y. Kazancoglu, S. Burmaoglu, ERP software selection with MCDM: Application of TODIM method. Int. J. Bus. Inf. Syst. **13**, 435–452 (2013)
26. B. Efe, An integrated fuzzy multi criteria group decision making approach for ERP system selection. Appl. Soft. Comput. J. **38**, 106–117 (2016)

27. M. Taheriyoun, M. Karamouz, A. Baghvand, Development of an entropy-based Fuzzy eutrophication index for reservoir water quality evaluation. Iran J. Environ. Health Sci. Eng. **7**
28. R. Kumar, S. Singh, P.S. Bilga, et al., Revealing the benefits of entropy weights method for multi-objective optimization in machining operations: A critical review. J. Mater. Res. Technol. **10**. Epub ahead of print 2021. https://doi.org/10.1016/j.jmrt.2020.12.114
29. M. Lin, C. Huang, Z. Xu, et al., Evaluating IoT platforms using integrated probabilistic linguistic MCDM method. IEEE Internet Things J. **7**. Epub ahead of print 2020. https://doi.org/10.1109/JIOT.2020.2997133
30. T.L. Saaty, A scaling method for priorities in hierarchical structures. J. Math. Psychol. **15**. Epub ahead of print 1977. https://doi.org/10.1016/0022-2496(77)90033-5

The Technological Role of Steepest Ascent Optimization in Industry 4.0 Modeling

Paulo Eduardo García-Nava, Luis Alberto Rodríguez-Picón, Luis Carlos Méndez-González, Iván Juan Carlos Pérez-Olguín, and Roberto Romero-López

1 Introduction

Industrial competitiveness is growing all around the globe everyday; therefore, productivity is a very important factor in organizations.

According to Trojanowska et al. [1], one of the most important aspects of manufacturing assessment is the determination of productivity. A variety of professionals from all around the world consider that the main objective of productivity consists in doing more things with less resources.

This is crucial for most companies and a deep analysis to improve productivity is vital. Irfan et al. [2] mention some of the most significant labor productivity factors such as the shortage of skilled labor, payment delay by owner, unrealistic scheduling, or even a lack of clarity of technical specification.

Industrial revolutions have brought enormous changes in production strategies. As said by Nuvolari [3], these revolutions have offered a more accurate historical account for providing insights to innovative practices and sustainability policies. As mentioned by Melnyk [4], humans as labor force are not notably modified in the first three industrial revolutions; nonetheless, the fourth revolution intends to replace most of physical jobs and open the gate for creative economy and decisions-making. According to Popkova et al. [5], the fourth industrial revolution (or Industry 4.0) foresees full elimination of human from the production system, guaranteeing automatization of the production process, developing global industrial networks, changing the essence of industrial patents, and creating the possibility

P. E. García-Nava (✉) · L. A. Rodríguez-Picón · L. C. Méndez-González · I. J. C. Pérez-Olguín · R. Romero-López
Department of Industrial Engineering and Manufacturing, Autonomous University of Ciudad Juárez, Ciudad Juárez, Chihuahua, Mexico
e-mail: al206600@alumnos.uacj.mx

L. C. Méndez-González et al. (eds.), *Innovation and Competitiveness in Industry 4.0 Based on Intelligent Systems*, EAI/Springer Innovations in Communication and Computing, https://doi.org/10.1007/978-3-031-29775-5_7

of quick direction changing of industrial production specialization. In the same way, Leong et al. [6] ensure that Industry 4.0 has offered many advantages (such as interconnection) and challenges (such as the design of the interconnection) for industry members. Organizations take advantage of this industrial paradigm shift; therefore, improving sustainability of the enterprise becomes a positive consequence.

There are many factors such as talent development, machinery advancement, and infrastructure development which involve huge investment that need to be considered. In this scenario, one of the most important factors is the technological strategy applied. Technology has become a notable important issue for organizations due to the benefits it brings once it is totally well adopted by companies.

Leech [7] provides information related by presenting estimations of the productivity impact of intelligent machine technology and the rate of return to research and development and productivity growth in the automotive, aerospace, and capital construction industries as generators of substantial productivity growth. One of the most notable technological tools are those related to the computational software packages that assist human intervention and decision-making.

Khan et al. [8] mention that most of the software development organizations are adopting a global software and programming language development, because of the significant return on investment and process improvement it produces through statistics. In this scenario, a very important and popular programming language for statistical computing, analysis, and human intervention is the one known as *R*. The researchers Shim and Kim [9] explore general approaches for intervention meta-analysis for quantitative data using the *R* software. In specific, the authors suggested methods for evaluating the effect sizes of the target studies in statistical intervention for meta-analysis.

For example, Grömping [10] advices that *R* package for Design of Experiments (DOE) is used to create full factorial designs and general factorial experiments based on orthogonal arrays. In addition to this functionality, the package offers features for analyzing experimental designs and the infrastructure for a suite of further packages on designing and analyzing experiments. Miranda-Ackerman and García-Lechuga [11] remark that DOE as a subfield of statistics has its origins trying to understand systems and making decisions. Something similar is reported by Jacyna et al. [12] who mention that DOE is a very popular methodology for planning and conducting experiments, where the effect of each tested factor on the studied responses is systematically examined and documented. Similarly, Yolmeh and Jafari [13] report the use of response surface methodology (RSM) as a technique widely used to optimize several processes. In their case, for the optimization of food processes such as extraction, drying, blanching, enzymatic hydrolysis and clarification, production of microbial metabolites, and formulation. Further, Challadurai [14] applies RSM to optimize parameters in casting, welding, and machinability studies of composite materials to design the experiments and minimize the numbers of experiments for specific number of factors and its levels with many advantages over, for example, Taguchi method.

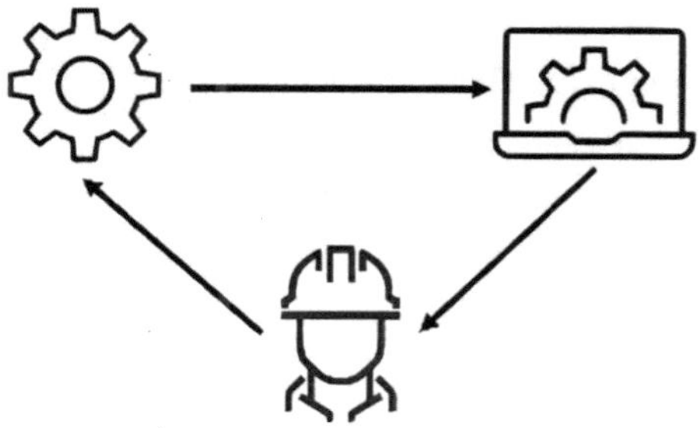

Fig. 1 Interaction of analyzed elements in the procedure

Law [15] reports that these models often have many input factors, determining which ones have a significant impact on performance responses of interest. The author discusses classical experimental designs such as full factorial, fractional factorial, and central composite.

Everything previously mentioned has the possibility to use the *R* software capability by providing functions to generate designs, fit first- and second-order models, and even obtain the path of steepest ascent to determine the optimum value of processes. The number of applications using the *R* software is huge because it not only has a numerical approach but also a plot design potential. It allows to easily visualize the information provided.

Comparable with this capability, Allen [16] presents many useful statistical techniques powered by *Minitab*® software for human intervention and decision-making because most scientific research generates data and its analysis helps the understanding of natural behavior just as [17] mentioned.

In his scenario, statistical software such as *Minitab*® is also a useful tool. Lesik [18] provides evidence about the huge benefit of the application that this software gives focusing on the applications of statistics. Same as Akers [19] assures, *Minitab*® has become the premier statistics and analytics software package widely used in automotive, aerospace, engineering, pharmaceutical, oil and gas production, telecoms, banking, insurance, or finance organizations.

Furthermore, Blackburn [20] mentions that the primary objective of analysis is to determine root causes with evidence. They mostly are supported by statistical software packages such as *Minitab*®.

In this chapter, a strategy based on statistical methods is illustrated through DOE. The proposed method considers the data of a simulated process with different considered factors to design an experiment. The analysis is done automatically applying a previously built code using a macro of Excel. The aim of this approach is to present, in a comprehensive way, an innovative scenario of appropriate human intervention for decision-making. The development is done in a production process simulation using both of the referenced statistical packages, *Minitab*® and *R softwares*. The interaction of production system, statistical software packages, and human interaction is next shown in Fig. 1.

The succeeding parts of the manuscript are explained by the automation sequence proposed. This sequence is formed by the following:

A. The designed specialized code for the applied macro for the response drop automation.
B. The subsequent activity assisted by the *R* software to obtain statistical information.
C. The optimization procedure through specific methodologies.

In this analysis, optimization of the output was possible with a $Y_{max} = 11.14$. The next sections will explain the complete procedure to get to the improved value.

The specific novelty that makes this study interesting is the way that elements are together well arranged to design a valuable, practical, and usable strategy for continuous improvement in terms of optimization when large amount of data take place in any industrial duty.

After this first section of introduction, the rest of the document is divided in three more sections. Section 2 is dedicated to the proposed method including five different stages (production simulated data considerations, design experimentation development, automation process modeling, statistical analysis, and human intervention). In Sect. 3, results of the experiment using both of the software packages are shown. Finally, Section 4 presents the discussion and conclusion of the entire document.

2 Method

The method is illustrated as a flowchart, and it basically consists of five main stages shown in the flow of Fig. 2.

2.1 Production Simulated Data Consideration

The analyzed information relates a simulated operation that monitors seven characteristics of a process as shown in Fig. 4. Each of these manufacturing operating conditions (MOC) manages a lower specification limit (LSL) and an upper specification limit (USL). Of course, the center points (CP) are those arithmetically in the center of that range. Each condition goes as follows:

(a) The first condition is illustrated with letter P. Its LSL is 11.0 units, a CP of 13.5 units, and a USL of 16.0 units.
(b) The second condition is represented by letter Q. Its LSL is 68.0 units, a CP of 84.0 units, and a USL of 100.0 units.
(c) The third condition is shown as letter R. Its LSL is 67.0 units, a CP of 92.0 units, and a USL of 117.0 units.

Fig. 2 The method applied

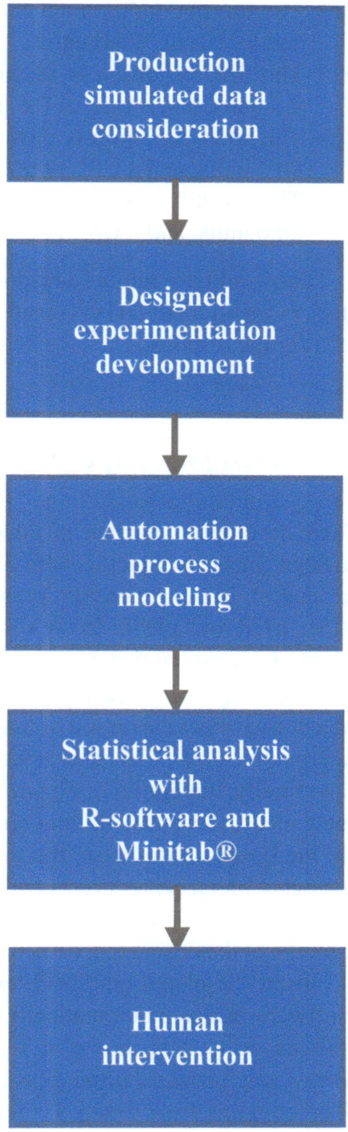

(d) The fourth condition has letter S. Its LSL is 275.0 units, a CP of 300.0 units, and a USL of 325.0 units.
(e) The fifth condition is illustrated with letter T. Its LSL is 0.4 units, a CP of 0.5 units, and a USL of 0.6 units.
(f) The sixth condition is illustrated with letter U. Its LSL is 70.0 units, a CP of 80.0 units, and a USL of 90.0 units.
(g) The seventh condition is illustrated with letter V. Its LSL is 16.0 units, a CP of 20.0 units, and a USL of 24.0 units.

This simulator returns a response value *Y* that obeys the combination of the seven factors mentioned; it also considers a LSL and a USL. Additionally, it shows a target (T) for the response.

2.2 Designed Experimentation Development

The developed analysis is done using designed experimentation. The factorial design considers 7 factors with 2 levels each, 2 replicates, and 2 center points, giving a total of 262 experimental runs. As it is previously explained, the factors show a LSL, a CP, and a USL. In the simulator used, this factorial design considers the LSL as the lower value (level − 1) and the CP as the higher value (level 1).

The design of the factorial arrangement was built using *Minitab*® capability. It is needed to select the option of S*tatistics* in the top main menu, into this menu, there is an option called *DOE*, after selecting it, other menu displays the option of *Factorial*, and finally, the option of *Create Factorial Design* shall be followed. Next, into this new menu, there are different options. It is necessary to select the *number of factors* in the study (in this case, there are seven factors). In the option of *Designs***,** a full design with its center points and replicates previously mentioned are considered. In the section of *Options*, it is important to confirm the randomization of runs and the storage of the design in a worksheet. In the button of *Factors*, a little box is displayed to insert the name, type, and levels of factors. After all these options are set, the software package automatically displays the factorial arrangement. The design of this experiment is shown in Table 1. The mismatch between the column of *Run order* and *Standard order* illustrates the applied randomization to the experiment. The columns from letters P to V represent each of the factors. Finally, column *Y* denotes the response variable. The way this column was obtained is the core of the automation stage of the method; it is called *automation process modeling* and it is explained next.

Table 1 Factorial design in Minitab® with responses

Run order	Std order	P	Q	R	S	T	U	V	Y
1	249	11	68	67	300	0.5	80	20	**4.57**
2	132	13.5	84	67	275	0.4	70	16	**2.25**
3	256	13.5	84	92	300	0.5	80	20	**2.30**
4	210	13.5	68	67	275	0.5	70	20	**2.33**
5	248	13.5	84	92	275	0.5	80	20	**2.17**
6	193	11	68	67	275	0.4	70	20	**2.02**
7	195	11	84	67	275	0.4	70	20	**3.48**
8	252	13.5	84	67	300	0.5	80	20	**3.47**
9	212	13.5	84	67	275	0.5	70	20	**2.88**
10	215	11	84	92	275	0.5	70	20	**3.59**
11	34	13.5	68	67	275	0.4	80	16	**3.30**
12	110	13.5	68	92	300	0.4	80	20	**3.27**
13	207	11	84	92	300	0.4	70	20	**3.90**
14	137	11	68	67	300	0.4	70	16	**3.97**
15	11	11	84	67	300	0.4	70	16	**3.95**
16	228	13.5	84	67	275	0.4	80	20	**4.23**
17	243	11	84	67	275	0.5	80	20	**3.38**
18	7	11	84	92	275	0.4	70	16	**3.55**
19	2	13.5	68	67	275	0.4	70	16	**4.59**
20	131	11	84	67	275	0.4	70	16	**3.70**
21	25	11	68	67	300	0.5	70	16	**3.95**
22	177	11	68	67	275	0.5	80	16	**3.87**
23	185	11	68	67	300	0.5	80	16	**4.13**
24	8	13.5	84	92	275	0.4	70	16	**3.61**
25	1	11	68	67	275	0.4	70	16	**3.69**
26	107	11	84	67	300	0.4	80	20	**3.95**
27	179	11	84	67	275	0.5	80	16	**3.39**
28	23	11	84	92	275	0.5	70	16	**3.33**
29	163	11	84	67	275	0.4	80	16	**3.15**
30	130	13.5	68	67	275	0.4	70	16	**4.59**
31	116	13.5	84	67	275	0.5	80	20	**5.35**
32	133	11	68	92	275	0.4	70	16	**4.51**
33	150	13.5	68	92	275	0.5	70	16	**5.72**
34	260	12.25	76	79.5	287.5	0.45	75	18	**4.62**
35	151	11	84	92	275	0.5	70	16	**4.62**
36	159	11	84	92	300	0.5	70	16	**3.82**
37	200	13.5	84	92	275	0.4	70	20	**3.48**
38	42	13.5	68	67	300	0.4	80	16	**3.06**
39	192	13.5	84	92	300	0.5	80	16	**3.84**
40	183	11	84	92	275	0.5	80	16	**4.29**
41	259	12.25	76	79.5	287.5	0.45	75	18	**5.20**
42	191	11	84	92	300	0.5	80	16	**4.70**
43	217	11	68	67	300	0.5	70	20	**4.64**

(continued)

Table 1 (continued)

Run order	Std order	P	Q	R	S	T	U	V	Y
44	189	11	68	92	300	0.5	80	16	**3.94**
45	168	13.5	84	92	275	0.4	80	16	**5.65**
46	47	11	84	92	300	0.4	80	16	**4.60**
47	79	11	84	92	300	0.4	70	20	**4.15**
48	9	11	68	67	300	0.4	70	16	**4.64**
49	101	11	68	92	275	0.4	80	20	**4.62**
50	27	11	84	67	300	0.5	70	16	**4.17**
51	112	13.5	84	92	300	0.4	80	20	**4.71**
52	245	11	68	92	275	0.5	80	20	**4.69**
53	58	13.5	68	67	300	0.5	80	16	**3.80**
54	230	13.5	68	92	275	0.4	80	20	**4.33**
55	186	13.5	68	67	300	0.5	80	16	**4.05**
56	97	11	68	67	275	0.4	80	20	**4.67**
57	173	11	68	92	300	0.4	80	16	**4.64**
58	174	13.5	68	92	300	0.4	80	16	**4.14**
59	68	13.5	84	67	275	0.4	70	20	**4.57**
60	202	13.5	68	67	300	0.4	70	20	**3.07**
61	218	13.5	68	67	300	0.5	70	20	**4.27**
62	172	13.5	84	67	300	0.4	80	16	**3.68**
63	141	11	68	92	300	0.4	70	16	**3.67**
64	54	13.5	68	92	275	0.5	80	16	**4.32**
65	139	11	84	67	300	0.4	70	16	**4.41**
66	65	11	68	67	275	0.4	70	20	**4.08**
67	37	11	68	92	275	0.4	80	16	**3.74**
68	127	11	84	92	300	0.5	80	20	**4.49**
69	32	13.5	84	92	300	0.5	70	16	**3.62**
70	134	13.5	68	92	275	0.4	70	16	**3.78**
71	140	13.5	84	67	300	0.4	70	16	**4.82**
72	124	13.5	84	67	300	0.5	80	20	**3.63**
73	158	13.5	68	92	300	0.5	70	16	**5.39**
74	85	11	68	92	275	0.5	70	20	**4.64**
75	3	11	84	67	275	0.4	70	16	**4.49**
76	257	12.25	76	79.5	287.5	0.45	75	18	**4.06**
77	227	11	84	67	275	0.4	80	20	**4.43**
78	169	11	68	67	300	0.4	80	16	**4.54**
79	161	11	68	67	275	0.4	80	16	**4.50**
80	224	13.5	84	92	300	0.5	70	20	**3.52**
81	171	11	84	67	300	0.4	80	16	**4.48**
82	50	13.5	68	67	275	0.5	80	16	**4.66**
83	109	11	68	92	300	0.4	80	20	**4.56**
84	46	13.5	68	92	300	0.4	80	16	**4.18**
85	234	13.5	68	67	300	0.4	80	20	**3.61**
86	17	11	68	67	275	0.5	70	16	**4.67**

(continued)

Table 1 (continued)

Run order	Std order	P	Q	R	S	T	U	V	Y
87	184	13.5	84	92	275	0.5	80	16	**3.79**
88	82	13.5	68	67	275	0.5	70	20	**4.59**
89	48	13.5	84	92	300	0.4	80	16	**3.83**
90	258	12.25	76	79.5	287.5	0.45	75	18	**4.00**
91	166	13.5	68	92	275	0.4	80	16	**4.30**
92	59	11	84	67	300	0.5	80	16	**3.79**
93	66	13.5	68	67	275	0.4	70	20	**5.62**
94	29	11	68	92	300	0.5	70	16	**3.87**
95	214	13.5	68	92	275	0.5	70	20	**5.03**
96	135	11	84	92	275	0.4	70	16	**4.48**
97	81	11	68	67	275	0.5	70	20	**4.17**
98	100	13.5	84	67	275	0.4	80	20	**3.98**
99	240	13.5	84	92	300	0.4	80	20	**3.81**
100	41	11	68	67	300	0.4	80	16	**3.69**
101	244	13.5	84	67	275	0.5	80	20	**3.27**
102	239	11	84	92	300	0.4	80	20	**4.63**
103	13	11	68	92	300	0.4	70	16	**4.21**
104	253	11	68	92	300	0.5	80	20	**4.52**
105	213	11	68	92	275	0.5	70	20	**4.48**
106	117	11	68	92	275	0.5	80	20	**3.51**
107	220	13.5	84	67	300	0.5	70	20	**4.46**
108	125	11	68	92	300	0.5	80	20	**4.69**
109	33	11	68	67	275	0.4	80	16	**4.77**
110	70	13.5	68	92	275	0.4	70	20	**3.63**
111	6	13.5	68	92	275	0.4	70	16	**4.58**
112	43	11	84	67	300	0.4	80	16	**3.90**
113	4	13.5	84	67	275	0.4	70	16	**3.52**
114	229	11	68	92	275	0.4	80	20	**4.65**
115	31	11	84	92	300	0.5	70	16	**4.03**
116	63	11	84	92	300	0.5	80	16	**4.19**
117	52	13.5	84	67	275	0.5	80	16	**4.54**
118	262	12.25	76	79.5	287.5	0.45	75	18	**5.04**
119	105	11	68	67	300	0.4	80	20	**4.72**
120	83	11	84	67	275	0.5	70	20	**3.86**
121	20	13.5	.84	67	275	0.5	70	16	**4.04**
122	129	11	68	67	275	0.4	70	16	**4.64**
123	211	11	84	67	275	0.5	70	20	**4.50**
124	178	13.5	68	67	275	0.5	80	16	**4.04**
125	118	13.5	68	92	275	0.5	80	20	**4.00**
126	93	11	68	92	300	0.5	70	20	**4.59**
127	146	13.5	68	67	275	0.5	70	16	**4.05**
128	203	11	84	67	300	0.4	70	20	**4.17**
129	16	13.5	84	92	300	0.4	70	16	**3.99**
130	15	11	84	92	300	0.4	70	16	**4.13**

(continued)

Table 1 (continued)

Run order	Std order	P	Q	R	S	T	U	V	Y
131	99	11	84	67	275	0.4	80	20	**4.27**
132	87	11	84	92	275	0.5	70	20	**4.09**
133	175	11	84	92	300	0.4	80	16	**4.33**
134	223	11	84	92	300	0.5	70	20	**4.55**
135	201	11	68	67	300	0.4	70	20	**4.50**
136	111	11	84	92	300	0.4	80	20	**4.58**
137	120	13.5	84	92	275	0.5	80	20	**4.36**
138	76	13.5	84	67	300	0.4	70	20	**4.42**
139	226	13.5	68	67	275	0.4	80	20	**5.44**
140	21	11	68	92	275	0.5	70	16	**4.57**
141	180	13.5	84	67	275	0.5	80	16	**4.73**
142	235	11	84	67	300	0.4	80	20	**4.65**
143	254	13.5	68	92	300	0.5	80	20	**4.48**
144	149	11	68	92	275	0.5	70	16	**4.65**
145	206	13.5	68	92	300	0.4	70	20	**3.07**
146	80	13.5	84	92	300	0.4	70	20	**3.41**
147	208	13.5	84	92	300	0.4	70	20	**5.04**
148	30	13.5	68	92	300	0.5	70	16	**4.37**
149	157	11	68	92	300	0.5	70	16	**4.26**
150	241	11	68	67	275	0.5	80	20	**4.12**
151	114	13.5	68	67	275	0.5	80	20	**4.44**
152	155	11	84	67	300	0.5	70	16	**4.38**
153	196	13.5	84	67	275	0.4	70	20	**2.78**
154	236	13.5	84	67	300	0.4	80	20	**3.37**
155	113	11	68	67	275	0.5	80	20	**3.81**
156	106	13.5	68	67	300	0.4	80	20	**3.54**
157	115	11	84	67	275	0.5	80	20	**4.05**
158	238	13.5	68	92	300	0.4	80	20	**5.24**
159	122	13.5	68	67	300	0.5	80	20	**4.33**
160	36	13.5	84	67	275	0.4	80	16	**3.91**
161	199	11	84	92	275	0.4	70	20	**3.87**
162	40	13.5	84	92	275	0.4	80	16	**4.01**
163	251	11	84	67	300	0.5	80	20	**4.59**
164	144	13.5	84	92	300	0.4	70	16	**4.27**
165	92	13.5	84	67	300	0.5	70	20	**4.22**
166	14	13.5	68	92	300	0.4	70	16	**5.52**
167	95	11	84	92	300	0.5	70	20	**4.59**
168	102	13.5	68	92	275	0.4	80	20	**5.22**
169	209	11	68	67	275	0.5	70	20	**4.70**
170	121	11	68	67	300	0.5	80	20	**4.55**
171	69	11	68	92	275	0.4	70	20	**4.56**
172	222	13.5	68	92	300	0.5	70	20	**4.14**
173	181	11	68	92	275	0.5	80	16	**4.54**
174	75	11	84	67	300	0.4	70	20	**4.35**

(continued)

Table 1 (continued)

Run order	Std order	P	Q	R	S	T	U	V	Y
175	73	11	68	67	300	0.4	70	20	**3.92**
176	89	11	68	67	300	0.5	70	20	**4.00**
177	176	13.5	84	92	300	0.4	80	16	**4.58**
178	10	13.5	68	67	300	0.4	70	16	**3.32**
179	167	11	84	92	275	0.4	80	16	**3.96**
180	18	13.5	68	67	275	0.5	70	16	**4.21**
181	103	11	84	92	275	0.4	80	20	**4.09**
182	84	13.5	84	67	275	0.5	70	20	**3.22**
183	128	13.5	84	92	300	0.5	80	20	**5.53**
184	154	13.5	68	67	300	0.5	70	16	**3.66**
185	153	11	68	67	300	0.5	70	16	**4.82**
186	246	13.5	68	92	275	0.5	80	20	**4.47**
187	123	11	84	67	300	0.5	80	20	**4.86**
188	182	13.5	68	92	275	0.5	80	16	**6.17**
189	188	13.5	84	67	300	0.5	80	16	**4.66**
190	147	11	84	67	275	0.5	70	16	**4.08**
191	62	13.5	68	92	300	0.5	80	16	**4.37**
192	162	13.5	68	67	275	0.4	80	16	**3.93**
193	194	13.5	68	67	275	0.4	70	20	**5.81**
194	232	13.5	84	92	275	0.4	80	20	**4.53**
195	247	11	84	92	275	0.5	80	20	**4.67**
196	90	13.5	68	67	300	0.5	70	20	**4.19**
197	119	11	84	92	275	0.5	80	20	**4.16**
198	45	11	68	92	300	0.4	80	16	**3.98**
199	143	11	84	92	300	0.4	70	16	**3.71**
200	187	11	84	67	300	0.5	80	16	**4.70**
201	77	11	68	92	300	0.4	70	20	**4.63**
202	78	13.5	68	92	300	0.4	70	20	**4.28**
203	35	11	84	67	275	0.4	80	16	**4.69**
204	221	11	68	92	300	0.5	70	20	**4.08**
205	53	11	68	92	275	0.5	80	16	**4.63**
206	205	11	68	92	300	0.4	70	20	**4.48**
207	39	11	84	92	275	0.4	80	16	**4.32**
208	71	11	84	92	275	0.4	70	20	**3.79**
209	56	13.5	84	92	275	0.5	80	16	**5.75**
210	19	11	84	67	275	0.5	70	16	**3.48**
211	198	13.5	68	92	275	0.4	70	20	**4.37**
212	57	11	68	67	300	0.5	80	16	**4.73**
213	38	13.5	68	92	275	0.4	80	16	**5.06**
214	88	13.5	84	92	275	0.5	70	20	**4.86**
215	138	13.5	68	67	300	0.4	70	16	**5.00**
216	255	11	84	92	300	0.5	80	20	**4.69**
217	28	13.5	84	67	300	0.5	70	16	**3.88**
218	165	11	68	92	275	0.4	80	16	**4.13**

(continued)

Table 1 (continued)

Run order	Std order	P	Q	R	S	T	U	V	Y
219	55	11	84	92	275	0.5	80	16	**4.11**
220	160	13.5	84	92	300	0.5	70	16	**3.68**
221	108	13.5	84	67	300	0.4	80	20	**3.93**
222	24	13.5	84	92	275	0.5	70	16	**4.90**
223	22	13.5	68	92	275	0.5	70	16	**4.32**
224	74	13.5	68	67	300	0.4	70	20	**4.36**
225	170	13.5	68	67	300	0.4	80	16	**4.09**
226	197	11	68	92	275	0.4	70	20	**4.79**
227	164	13.5	84	67	275	0.4	80	16	**4.56**
228	67	11	84	67	275	0.4	70	20	**4.66**
229	72	13.5	84	92	275	0.4	70	20	**4.85**
230	261	12.25	76	79.5	287.5	0.45	75	18	**4.19**
231	86	13.5	68	92	275	0.5	70	20	**3.65**
232	60	13.5	84	67	300	0.5	80	16	**3.82**
233	49	11	68	67	275	0.5	80	16	**4.67**
234	104	13.5	84	92	275	0.4	80	20	**4.84**
235	61	11	68	92	300	0.5	80	16	**4.58**
236	204	13.5	84	67	300	0.4	70	20	**2.84**
237	91	11	84	67	300	0.5	70	20	**3.93**
238	12	13.5	84	67	300	0.4	70	16	**4.85**
239	26	13.5	68	67	300	0.5	70	16	**4.12**
240	142	13.5	68	92	300	0.4	70	16	**4.85**
241	94	13.5	68	92	300	0.5	70	20	**3.53**
242	250	13.5	68	67	300	0.5	80	20	**5.23**
243	231	11	84	92	275	0.4	80	20	**4.52**
244	242	13.5	68	67	275	0.5	80	20	**4.29**
245	126	13.5	68	92	300	0.5	80	20	**3.86**
246	190	13.5	68	92	300	0.5	80	16	**4.28**
247	219	11	84	67	300	0.5	70	20	**3.92**
248	216	13.5	84	92	275	0.5	70	20	**5.04**
249	145	11	68	67	275	0.5	70	16	**4.58**
250	64	13.5	84	92	300	0.5	80	16	**3.90**
251	98	13.5	68	67	275	0.4	80	20	**4.52**
252	44	13.5	84	67	300	0.4	80	16	**3.32**
253	152	13.5	84	92	275	0.5	70	16	**3.92**
254	148	13.5	84	67	275	0.5	70	16	**4.69**
255	156	13.5	84	67	300	0.5	70	16	**3.51**
256	51	11	84	67	275	0.5	80	16	**4.47**
257	237	11	68	92	300	0.4	80	20	**4.03**
258	136	13.5	84	92	275	0.4	70	16	**4.63**
259	233	11	68	67	300	0.4	80	20	**4.51**
260	96	13.5	84	92	300	0.5	70	20	**3.93**
261	5	11	68	92	275	0.4	70	16	**4.58**
262	225	11	68	67	275	0.4	80	20	**4.68**

2.3 Automation Process Modeling

Several software packages have the capability to develop statistical analysis automatically and immediately; nevertheless, the insertion of the data could be a time-consuming issue. The way the analyst inserts the levels for each run to obtain the response variable is often tedious. Overall, for cases like this where many runs must be managed. The time that it takes to run 262 iterations could be considerably huge. Different solutions have been proposed and applied. For this chapter, the proposed automation sequence has a strong relation between the process simulator applied and the *R software* analysis capability with an additional extension of results using *Minitab*®. The innovative designed sequence is given as shown in the following:

(a) The process simulator is linked to the *Excel macro* to take all of the 262 runs automatically. It brings the advantage to avoid the user to insert all of them individually.
(b) *R software* automatically takes the Excel worksheet with the information from simulator to statistically analyze the information to obtain the model (the regression equation) that allows optimization.
(c) Using that model, a *SADM* path can be built through Excel or *R software*. After developing the path, levels of factors are ready to be inserted again into the simulator to obtain new responses.

The Excel macro takes part again to automatically insert the new levels in the simulator to obtain the responses from the path to maximize Y.

Next, Fig. 3 shows the flow of the information including the automation sequence of the operation.

The only previous activity that has to be performed only one time is the creation of the macro. In this case, a specialized code for a macro that performs the task is applied. In general terms, this task drops 262 response values in each of the experimental lines according to the levels presented and of course the value offered by the simulator used. The macro used is next presented in Table 2.

This macro basically takes the seven values (or levels) of each of the factors of the individual experimentation line (treatment) from the factorial design and it inserts it in each of the seven inputs of the simulator as shown in Fig. 4. This simulator drops a unique response due to the combination of the levels of the factors. The value is copied in other cells as illustrated in Table 3. The macro performs this action 262 times for each of the 262 response values. The column Y of Table 1 shows the total amount of response dropped automatically by the mentioned macro.

Once all the values are obtained; they are ready to be analyzed using any statistical software.

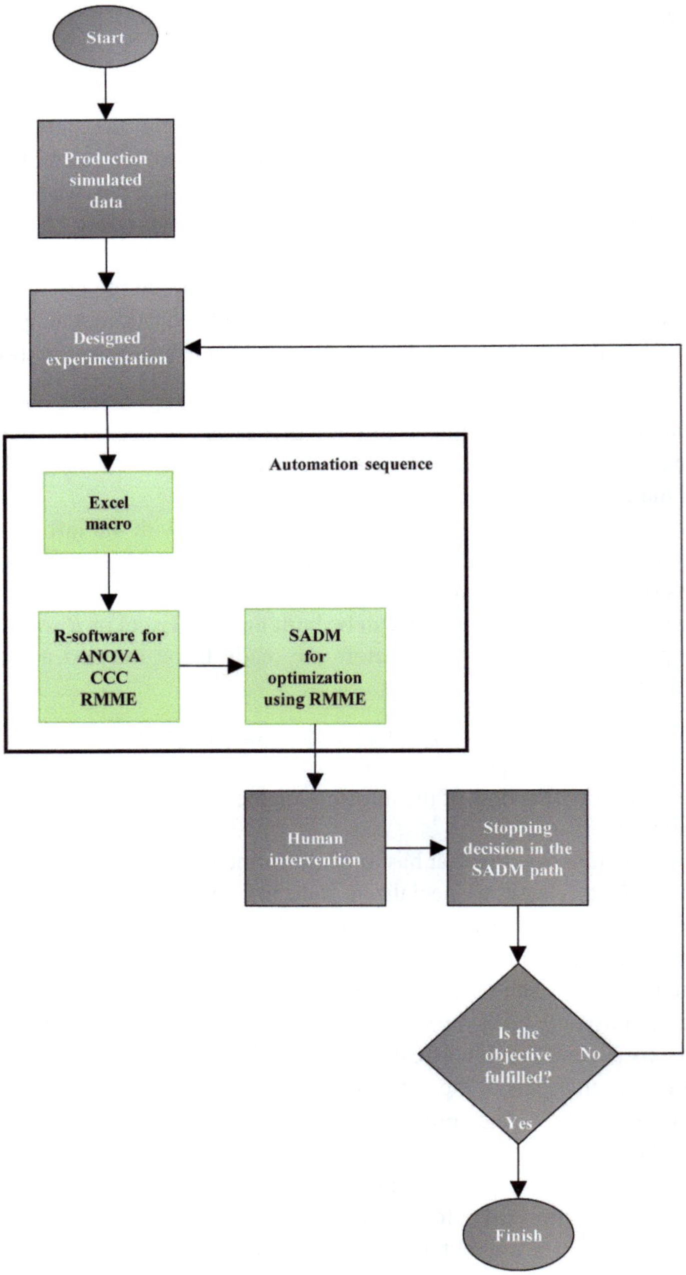

Fig. 3 The automation sequence of the operation

Table 2 Specialized code for the applied macro to obtain the response drop automation

```
Sub Test ()
Dim Range As Range
Dim Cell As Object
Dim B(26) As String
B(0) = ""
B(1) = "A"
B(2) = "B"
B(3) = "C"
B(4) = "D"
B(5) = "E"
B(6) = "F"
B(7) = "G"
B(8) = "H"
B(9) = "I"
B(10) = "J"
B(11) = "K"
B(12) = "L"
B(13) = "M"
B(14) = "N"
B(15) = "O"
B(16) = "P"
B(17) = "Q"
B(18) = "R"
B(19) = "S"
B(20) = "T"
B(21) = "U"
B(22) = "V"
B(23) = "W"
B(24) = "X"
B(25) = "Y"
B(26) = "Z"
j = 13
VC = 0
Dim limit As String
Dim limit Data As Integer
Dim Cell2(1834) As Double
Dim alfa(300) As String
limit = "JD9"
limit Data = 261
Set Range = Sheets("Collector").Range("C3:" & limit)
For Each Cell In Range
If VC <= 1973 Then Cell2(VC) = Cell
VC = VC + 1
Next Cell
VC = 0
For Y = 0 To 10
For X = 1 To 26
alfa(VC) = B(Y) & B(X)
VC = VC + 1
Next X
Next Y
For X = 0 To limit Data
Sheets ("Factory #2"). Range("D14").Value = Cell2(X)
Sheets ("Factory #2").Range("D15"). Value = Cell2(X + 256)
Sheets ("Factory #2").Range("D16"). Value = Cell2(X + 512)
Sheets ("Factory #2").Range("D17"). Value = Cell2(X + 768)
Sheets ("Factory #2").Range("D18"). Value = Cell2(X + 1024)
Sheets ("Factory #2").Range("D19"). Value = Cell2(X + 1280)
Sheets ("Factory #2").Range("D20"). Value = Cell2(X + 1536)
Sheets("Collector").Range(alfa (X + 2) & 12).Value =
Sheets("Factory #2").Range("F20")
Next X
End Sub
```

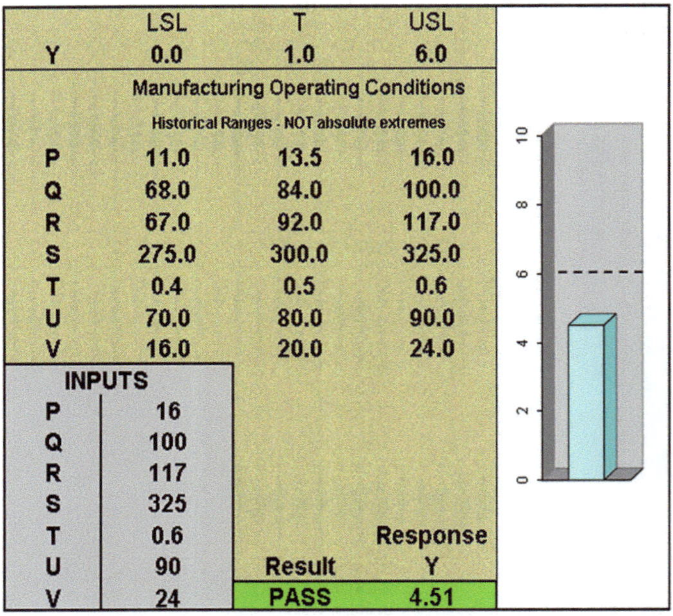

Fig. 4 The simulator illustrating the LSL, T, USL, CP, inputs, and response

Table 3 Example of the first five response values launched by the macro

Factors	Levels					
P	11	13.5	13.5	13.5	13.5	11
Q	68	84	84	68	84	68
R	67	67	92	67	92	67
S	300	275	300	275	275	275
T	0.5	0.4	0.5	0.5	0.5	0.4
U	80	70	80	70	80	70
V	20	16	20	20	20	20
Y	**2.50**	**2.36**	**2.36**	**2.23**	**2.15**	**2.07**

2.4 Statistical Analysis with R Software and Minitab®

The collected information is finally statistically tested using two of the most popular software packages for this purpose, *R software* and *Minitab®*. It will be observed that both statistical analyses are very similar. Three pieces of information are obtained as follows:

(a) *Analysis of variance (ANOVA).*
As said by Raman et al. [21], the analysis of variance (ANOVA) technique finds the significance control factors which help to improve parameters. In the case of this experiment, this analysis is developed with the mean of the factors.
(b) *Coded coefficient chart (CCC).*
The software packages have the capability to develop and deliver a chart with coded coefficients called *coded coefficient chart (CCC).* Coefficients are used to build the regression equation of coded units. This model could be built with all of the factors considered; nevertheless, the model that is commonly considered is the one that only has significant factors.
(c) *Regression model of main effects (RMME).*
The regression model of main effects (RMME) in coded units is a model that allows progressing with the analysis of improvement. Its form is

$$y = \beta_0 + \beta_1 X_1 + \cdots + \beta_k X_k.$$

For *Minitab*®, the responses are only copied back to the earlier-done *Minitab*® arrangement to run the statistical analysis. This analysis is used to compare and confirm results from *R software.*

When code is already charged to *R software*, the analysis is automatic. The next steps need to be followed, the first time that the software is run:

(a) Installation of package.
R software charges Excel through the package called *readxl* which is necessary to be previously installed through the code *install.packages("readxl").* Then, the package is charged using the code *library("readxl").*
(b) File importation.
In the environment section, there is an option called *Import Dataset.* The Excel file is chosen, and *R software* will store the file.
(c) File charging.
Once the file is storage, it is necessary to charge it or call it to the analysis environment through the code *str(nameofthefile).* The letters into parenthesis belong to the name that the user decided for the file. For this chapter, the name of the file is "*response*" so, the used code to charge the file of this analysis is *str(response).*

Definitely, optimization is one of the most important goals in organizations where target, minimum or maximum values are the objective. The decision in this stage lies on the selection of the tool that helps to optimize. Once the model is obtained, a convenient tool to examine the behavior of a response (to arrive to a region where optimization is achievable) is called Steepest Ascent or Descent Method (SADM). Myers, Montgomery, and Anderson-Cook [22] explain this method as a searcher for an optimization region through experimental design, model building and sequential experimentation. The decision to select the best value will be determined by the experimenter. A set of steps are needed to apply this method:

1. *RMME*

   After developing the experiment with ANOVA, CCC, a RMME is built. This linear model participates to star building the steepest path with a certain step size.
2. *Step size*

   Compute step size for natural and coded units for significant factors. This method basically involves sequential experimentation from one level in the factors to another, this dynamic builds a path. Therefore, a step size for this path is needed. It is the size of delta (the size from levels of one factor to levels of another factor). The values are obtained as follows:

   (a) *Natural value of first factor.*

   The *experimenter assigns* a natural step value to that factor with the highest absolute regression coefficient.

   (b) *Coded value of first factor.*

   That natural value is transformed to coded value considering the units of natural values that correspond to one coded using *common conversion*.

   (c) *Coded value of second factor.*

   To obtain the coded value of the factor with a lower coefficient, formula (1) is applied.

$$\Delta xj = \frac{bj}{\frac{bi}{\Delta xi}}; \ j = 1, 2, \ldots, k; i \neq j \tag{1}$$

   (d) *Natural value of second factor.*

   Now, to obtain the natural value of the factor with a lower coefficient, it is only necessary to *multiply* the result from previous step times the range (R) of its natural levels.

3. *Path building*

   Once increments of natural levels are set, the next step is to build the path. For this, it is necessary to consider the levels from nonsignificant factors that *Minitab®* delivered to be necessary to maximize the response. The only factors that will not follow this principle are those significant for the response variable.
4. *Increments*

   For those significant factors, the procedure mentions that it is necessary to start from the center points and follow the increments according to the step size previously done.

5. *New responses*
   Finally, the response value is obtained directly from the simulator. Fortunately, the Excel macro allows to run as many runs as the experimenter wishes.

After developing the procedure of SADM, next step is the exploration of a region where optimization is feasible. This procedure is shown in Sect. 3.1.

2.5 Human Intervention

Once the automation process modeling is finished, the statistical analysis offers important outcomes for decision-making.

The obtained information gives a useful framework to reach the objective of optimization. This information helps humans make certain important decisions to get the final aim.

On the one hand, the ANOVA allows to select the significant factors; it means the factors that affect the most the response variable. All of the variables somehow participate over the response; nevertheless, only a few of them participate in an important way. The decision in this stage lies on the selection of the significant factors according to the proper indicators such as the P-value.

On the other hand, the CCC explores the regression coefficients for each of the analyzed factors. The decision in this chart lies on the selection of the most important coefficients; it means those coefficients from significant factors.

Finally, the RMME is built. The goal is to submit this model into a tool, such as the path of SADM, to apply different strategies of improvement.

Human intervention takes place with the maximization of the presented outcome. The way to decide when a maximized $\boldsymbol{Y}$ is reached could be with the selection of the response before three consecutive drops in value. This procedure is well known to be empirical. Nevertheless, the literature has several rules to select the best response. They are called stopping rules (SR). For instance, Myers and Khuri [23] proposed a rule that assumes a normally distributed behavior in the observed response $\boldsymbol{Y(t)}$, a set of runs of a significance test using confidence intervals and an equation that defines a value to apply a decision test to stop at the best value in the path of improvement. Also, Miró-Quesada and Del Castillo [24] presented a Recursive Parabolic Rule (RPR) that assumes a quadratic behavior in the observed response $\boldsymbol{Y(t)}$ and estimates a set of parameters to determine a decision rule to be the one that points the best response out. Furthermore, Del Castillo [25] enhanced the RPR not only for quadratic responses but also for non-quadratic through a similar but more robust procedure.

For this chapter, the selection of the response before three consecutive drops in value is the way to empirically choose a maximized $\boldsymbol{Y}$.

3 Results

The results next shown are derived from the integration of the information flow. It includes the automation of stages with the intervention of user for decision-making using the outcome from the statistical analysis from statistical software packages. In each of the cases, the three pieces of information previously addressed are illustrated.

The outcome from *R software* is next explained. It has the capability to deliver the pieces of information mentioned related to the experiment explained. This software offers the capability to analyze the information in a free line-based code. Next the ANOVA, the CCC, and the RMME are presented.

(a) *ANOVA with R software.*

The first chart illustrated is the *ANOVA* shown in Fig. 4.

It includes the factors, the degrees of freedom (DF), the sums of squares (Sum Sq), the mean squares (Mean Sq), the *F*-value, and the *P*-value (***Pr*** (**>*F***)).

It is well known that a large value of F means the factor is significant, while a small value of P results in significance of the factor for the response value. The seven factors (P, Q, R, S, T, U, V) are presented. In this case, only two factors (Q and R) seem to be significant for the response variable due to its *P*-value which is lower than an alpha of 0.05. Factor *Q* presents a *P*-value of 0.006 and factor *R* has a *P*-value of 0.029. The rest of the factors are not significant; they present a *P*-value higher than 0.05. Figure 5 shows the code used to obtain the ANOVA, and Fig. 6 illustrates the ANOVA itself.

```
# ANOVA
anova<-aov(Y~Factor_P+Factor_Q+Factor_R+Factor_S+Factor_T+Factor_U+Factor_V,
data=response)
summary(anova)
```

Fig. 5 Code for ANOVA using *R* software

```
> # ANOVA
> anova<-aov(Y~Factor_P+Factor_Q+Factor_R+Factor_S+Factor_T+Factor_U+Factor_V,
+ data=response)
> summary(anova)
            Df Sum Sq Mean Sq F value  Pr(>F)
Factor_P     2   0.81  0.4056   1.078 0.34189
Factor_Q     1   2.85  2.8498   7.572 0.00636 **
Factor_R     1   1.81  1.8073   4.802 0.02933 *
Factor_S     1   0.34  0.3386   0.900 0.34378
Factor_T     1   0.03  0.0335   0.089 0.76556
Factor_U     1   0.78  0.7799   2.072 0.15123
Factor_V     1   0.08  0.0802   0.213 0.64483
Residuals  253  95.22  0.3764
---
Signif. codes:  0 '***' 0.001 '**' 0.01 '*' 0.05 '.' 0.1 ' ' 1
```

Fig. 6 ANOVA using *R* software

```
# Regression model with coded variables
coded.model<-lm(Y~xP+xQ+xR+xS+xT+xU+xV, data=response)
summary(coded.model)
```

Fig. 7 Code for the coded coefficient chart using *R* software

```
> # Regression model with coded variables
> coded.model<-lm(Y~xP+xQ+xR+xS+xT+xU+xV, data=response)
> summary(coded.model)

Call:
lm(formula = Y ~ xP + xQ + xR + xS + xT + xU + xV, data = response)

Residuals:
     Min       1Q   Median       3Q      Max
-2.21310 -0.35702  0.01053  0.38460  1.66714

Coefficients:
            Estimate Std. Error t value Pr(>|t|)
(Intercept)  4.22611    0.03793 111.419   <2e-16 ***
xP          -0.03348    0.03837  -0.872   0.3838
xQ          -0.10551    0.03837  -2.750   0.0064 **
xR           0.08402    0.03837   2.190   0.0295 *
xS          -0.03637    0.03837  -0.948   0.3442
xT           0.01145    0.03837   0.298   0.7657
xU           0.05520    0.03837   1.438   0.1515
xV          -0.01770    0.03837  -0.461   0.6451
---
Signif. codes:  0 '***' 0.001 '**' 0.01 '*' 0.05 '.' 0.1 ' ' 1

Residual standard error: 0.6139 on 254 degrees of freedom
Multiple R-squared:  0.0606,	Adjusted R-squared:  0.03471
F-statistic: 2.341 on 7 and 254 DF,  p-value: 0.02481
```

Fig. 8 Estimated coded coefficients using *R* software

(b) *CCC using R-software.*

The second chart delivered is the CCC shown in Fig. 8. Each of the factors illustrates its effect, coefficient, SE Coef (standard error of the coefficient), T-value, P-value, and the VIF (variance inflation factor). One of the most important pieces of information that is found here is related to the coefficient column. Each of those numbers represents an element of the RMME, and it is explained next. Figure 7 shows the code used to obtain the CCC.

(c) *RMME using R software.*

This regression Eq. (2) is shown next:

$$Y = 4.23 - 0.11Q + 0.08\text{R} \tag{2}$$

This equation in particular can have two possible interpretations or meanings:

Table 4 ANOVA of the experiment using Minitab®

Source	DF	Adj SS	Adj MS	*F*-value	*P*-value
P	1	0.287	0.28669	0.74	0.392
Q	1	2.85	2.84976	7.32	0.008
R	1	1.812	1.81166	4.65	0.033
S	1	0.34	0.34025	0.87	0.352
T	1	0.032	0.03193	0.08	0.775
U	1	0.782	0.78217	2.01	0.159
V	1	0.08	0.08002	0.21	0.651

Table 5 Coded coefficient chart using Minitab®

Term	Effect	Coef	SE Coef	*T*-value	*P*-value
Constant		4.2191	0.039	108.19	0
P	−0.067	−0.034	0.039	−0.86	0.392
Q	−0.211	−0.106	0.039	−2.71	0.008
R	0.1682	0.0841	0.039	2.16	0.033
S	−0.073	−0.037	0.039	−0.93	0.352
T	0.0223	0.0112	0.039	0.29	0.775
U	0.1106	0.0553	0.039	1.42	0.159
V	−0.035	−0.018	0.039	−0.45	0.651

1. It means that there is a reason of change of 11% in Y for every unit of change in *Q* if *R* remains constant.
2. It also means that there is a reason of change of 8% in Y for every unit of change in *R* if *Q* remains constant.

The outcome from *Minitab*® also contains the three pieces of information to be compared. ANOVA, CCC, and the RMME are presented.

(a) *ANOVA using Minitab*®.

The ANOVA using *Minitab*® is shown in Table 4. It illustrates the source or factors, the degrees of freedom (DF), the adjusted sums of squares (Adj SS), the adjusted mean squares (Adj MS), the *F*-value, and the *P*-value.

It is confirmed that the same two factors (Q and R) are significant for the response variable. Factor *Q* presents a *P*-value of 0.008 and factor *R* has a *P*-value of 0.033. The rest of the factors are not significant; they present a *P*-value higher than 0.05. All of the elements present one DF, their Adj SS goes in a range from 0.032 in T to 2.850 in Q, and their Adj MS is exactly the same as the Adj SS because it is only one DF.

(b) *CCC using Minitab*®.

The next chart illustrated is the CCC shown in Table 5. Each of the factors presents its effect, coefficient, SE Coef (standard error of the coefficient), *T*-value, *p*-value, and the VIF (variance inflation factor). The RMME is explained next.

(c) *RMME using Minitab®.*

The RMME or model (3) is extracted from the CCC. This equation is presented as follows:

$$Y = 4.22 - 0.11Q + 0.08\mathrm{R} \tag{3}$$

As it is shown, the model analyzed in *R software* is of course basically the same than the one analyzed by Minitab®. So the interpretation of the model can be considered to be exactly the same. Now, optimization is the duty.

3.1 Optimization Through SADM

As previously explained, the SADM helps to find a region where optimization is viable. A factorial experiment was developed in which effects were estimated; the resulted coefficients of significant factors of the linear model were considered to determine the improvement path of the SADM to optimize the response. *R software* also has the capability to automatically obtain a path of improvement given that the software already has the coefficients of the RMME. Considering the method of this chapter, this path can also be well integrated to the Excel macro to automatically obtain the responses of each experimental run that builds this path of improvement. The search of maximization of the response through this method is shown as follows:

1. *RMME*

   The resulted model is shown next in Eq. 4. This equation is used to build the path of improvement:

$$y = \mathbf{4.219 - 0.11Q + 0.08R} \tag{4}$$

2. *Step size*

   The procedure for the step size is shown next in Table 6. In this step, natural value of factor Q, coded value of factor Q, coded value of factor R, and natural value of factor *R* are shown. The conversion from natural to coded units of factor *Q* is shown in Table 7. As previously addressed, the way this table was obtained is not really related to a formula, but it is related to common conversion. If one coded unit values eighth natural units, then one natural unit values 0.1250 coded units. Table 8 shows a comparison between coded and natural units in factors *Q* and *R*. The range is the number of natural units needed to run from one coded unit to another.

Table 6 Computations needed to obtain the step size

	Step size	
	Q	R
Natural	1	−1.1364
Coded	0.1250	−0.0909

Table 7 Conversion from natural to coded units of factor *Q* using common conversion

Q	
Coded	Natural
1	8
0.1250	1

Table 8 Coded and natural values of factor *Q* and factor *R*

	Q	R	
Coded	Natural		
−1	68	67	
0	76	79.5	
1	84	92	
Range	**8**	**12.5**	**Step size**

3. *Path building*

   The SADM path is next shown in Table 9.

4. *Increments*

   The increments of levels for significant factors are shown in third and fourth columns of Table 9. It also shows constant levels for nonsignificant factors.

5. *New responses*

   Responses obtained from the simulator are also next shown in the last column of Table 9. They were obtained using the developed Excel macro in a second cycle of macro running.

Next, Fig. 9 shows the behavior of ***Y*** represented in a graph.

This path of improvement needed 73 runs to apply the mentioned empirical decision that states that the maximized ***Y*** is the response before three consecutive drops in values. As seen in Table 9, individual experiments 71, 72, and 73 show a drop in their responses consecutively. It fulfills the requirement of the empirical procedure; it means that iteration 70 presents the maximum *Y* with a value of **11.14.**

Definitely, all the factors are considered to get to the results; nevertheless, only significant factors have important participation in the maximization, minimization, or target values.

4 Discussion and Conclusion

Decision-making is a critical factor for organizations overall in this fourth industrial revolution. According to [26], this revolution is progressing exponentially, and it is reshaping schemes and resources. The correct usage of these resources is essential for productivity. Different tools to gain productivity and to reach optimization are

Table 9 Steepest ascent methodology building

Steps	P	Δ-Q	Δ-R	S	T	U	V	Y
0	13.5	76	79.5	275	0.4	70	20	**3.25**
1	13.5	77	78.36	275	0.4	70	20	**2.31**
2	13.5	78	77.23	275	0.4	70	20	**2.28**
3	13.5	79	76.09	275	0.4	70	20	**2.28**
4	13.5	80	74.95	275	0.4	70	20	**2.39**
5	13.5	81	73.82	275	0.4	70	20	**2.25**
6	13.5	82	72.68	275	0.4	70	20	**2.37**
7	13.5	83	71.55	275	0.4	70	20	**2.70**
8	13.5	84	70.41	275	0.4	70	20	**4.34**
9	13.5	85	69.27	275	0.4	70	20	**4.02**
10	13.5	86	68.14	275	0.4	70	20	**3.03**
11	13.5	87	67.00	275	0.4	70	20	**5.24**
12	13.5	88	65.86	275	0.4	70	20	**3.71**
13	13.5	89	64.73	275	0.4	70	20	**4.91**
14	13.5	90	63.59	275	0.4	70	20	**2.76**
15	13.5	91	62.45	275	0.4	70	20	**3.79**
16	13.5	92	61.32	275	0.4	70	20	**3.30**
17	13.5	93	60.18	275	0.4	70	20	**2.67**
18	13.5	94	59.05	275	0.4	70	20	**5.29**
19	13.5	95	57.91	275	0.4	70	20	**5.52**
20	13.5	96	56.77	275	0.4	70	20	**3.46**
21	13.5	97	55.64	275	0.4	70	20	**4.02**
22	13.5	98	54.50	275	0.4	70	20	**4.10**
23	13.5	99	53.36	275	0.4	70	20	**3.42**
24	13.5	100	52.23	275	0.4	70	20	**3.74**
25	13.5	101	51.09	275	0.4	70	20	**3.96**
26	13.5	102	49.95	275	0.4	70	20	**3.63**
27	13.5	103	48.82	275	0.4	70	20	**3.80**
28	13.5	104	47.68	275	0.4	70	20	**5.77**
29	13.5	105	46.55	275	0.4	70	20	**4.26**
30	13.5	106	45.41	275	0.4	70	20	**4.97**
31	13.5	107	44.27	275	0.4	70	20	**4.07**
32	13.5	108	43.14	275	0.4	70	20	**4.76**
33	13.5	109	42.00	275	0.4	70	20	**5.11**
34	13.5	110	40.86	275	0.4	70	20	**4.83**
35	13.5	111	39.73	275	0.4	70	20	**2.45**
36	13.5	112	38.59	275	0.4	70	20	**7.60**
37	13.5	113	37.45	275	0.4	70	20	**5.00**
38	13.5	114	36.32	275	0.4	70	20	**6.34**
39	13.5	115	35.18	275	0.4	70	20	**8.47**
40	13.5	116	34.05	275	0.4	70	20	**9.21**
41	13.5	117	32.91	275	0.4	70	20	**5.27**
42	13.5	118	31.77	275	0.4	70	20	**5.86**
43	13.5	119	30.64	275	0.4	70	20	**4.65**

(continued)

Table 9 (continued)

Steps	P	Δ-Q	Δ-R	S	T	U	V	Y
44	13.5	120	29.50	275	0.4	70	20	**2.91**
45	13.5	121	28.36	275	0.4	70	20	**5.20**
46	13.5	122	27.23	275	0.4	70	20	**8.15**
47	13.5	123	26.09	275	0.4	70	20	**5.89**
48	13.5	124	24.95	275	0.4	70	20	**6.86**
49	13.5	125	23.82	275	0.4	70	20	**6.03**
50	13.5	126	22.68	275	0.4	70	20	**6.13**
51	13.5	127	21.55	275	0.4	70	20	**8.02**
52	13.5	128	20.41	275	0.4	70	20	**5.52**
53	13.5	129	19.27	275	0.4	70	20	**3.82**
54	13.5	130	18.14	275	0.4	70	20	**7.76**
55	13.5	131	17.00	275	0.4	70	20	**3.89**
56	13.5	132	15.86	275	0.4	70	20	**8.48**
57	13.5	133	14.73	275	0.4	70	20	**6.01**
58	13.5	134	13.59	275	0.4	70	20	**4.39**
59	13.5	135	12.45	275	0.4	70	20	**6.49**
60	13.5	136	11.32	275	0.4	70	20	**4.34**
61	13.5	137	10.18	275	0.4	70	20	**3.45**
62	13.5	138	9.05	275	0.4	70	20	**8.95**
63	13.5	139	7.91	275	0.4	70	20	**5.74**
64	13.5	140	6.77	275	0.4	70	20	**8.37**
65	13.5	141	5.64	275	0.4	70	20	**10.31**
66	13.5	142	4.50	275	0.4	70	20	**7.55**
67	13.5	143	3.36	275	0.4	70	20	**4.34**
68	13.5	144	2.23	275	0.4	70	20	**6.34**
69	13.5	145	1.09	275	0.4	70	20	**6.63**
70	13.5	146	−0.05	275	0.4	70	20	**11.14**
71	13.5	147	−1.18	275	0.4	70	20	**7.99**
72	13.5	148	−2.32	275	0.4	70	20	**6.76**
73	13.5	149	−3.45	275	0.4	70	20	**5.86**

desirable, such as the response surface used, for example, by [27] as a statistical and mathematical technique to model data and perform optimization.

In general, response surface in addition to design experimentation for output analysis constitutes an initial stage to improve certain parameters. In the particular case of this manuscript, the improvement was shown in three scenarios: minimization, maximization, and a target value following the steepest ascent path. This strategy is widely used by different researchers such as [28–30] who studied multivariate responses for improvement purposes in certain processes.

However, this research could continue if a deeper analysis is run. Not only these two statistical software packages nor only these statistical resources could be applied. There are a huge number of tools and methods that are currently available in the literature to better understand and improve all kinds of processes.

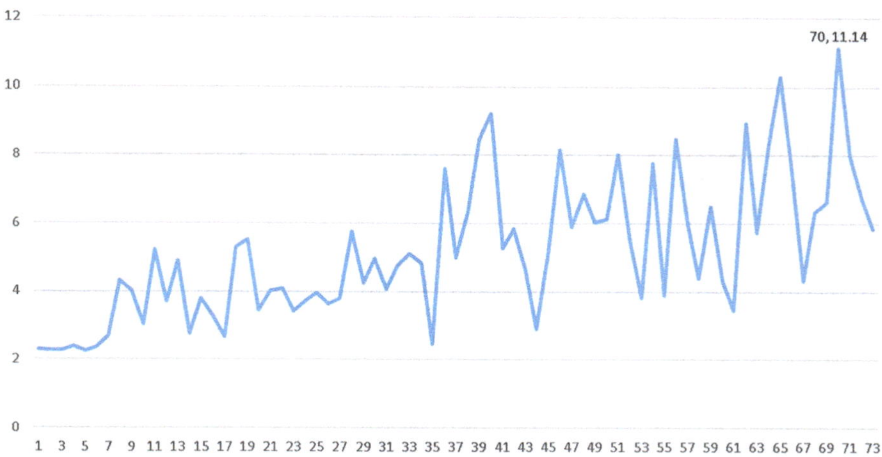

Fig. 9 Graph with the behavior of Y and the location of maximum point reached

The application of these statistical strategies can be applied to any process with controllable inputs and desirable outputs. Obtaining the input data is enough to develop the whole analysis. The strategies seen in this chapter were only examples of the endless possibilities when statistical analysis and process improvement are on stage for any organization in this age of Industry 4.0 where technology takes place through interconnectivity and smart automation. The only thing that the researcher needs is a technological process to improve, a code for the automation and interconnectivity of data, and a statistical software to analyze this data.

For the particular case here analyzed, the technological process consisted of seven factors (the inputs) that drop a response variable (the output). The interconnectivity and automation are related to the development of the code (or macro) for the data to be automatically collected from the technological process to the automated macro and finally the packages *R software* and *Minitab*® to analyze the data. All this is with the intention of decision-making.

It is important to note that this study is unique because the response variables obtained by the simulator were specially dropped for this particular experiment. This is why comparison with previous studies would not be feasible. It can be said that this specific analysis is certainly applicable to the special conditions where responses were exclusively obtained.

One of the limitations of this study is related to the stopping decision along the SADM path. The strategy here followed was the mentioned empirical decision to stop after three consecutive drops in the response. It is considered to be a limitation because this strategy does not have any mathematical support. Therefore, a better response could exist further on the path of improvement. A suitable recommendation is to develop a deeper analysis to apply existing stopping rules with mathematical support. The literature offers certain procedures such as the previously mentioned Myers and Khuri stopping rule [23], Miró-Quesada and Del Castillo [24] recursive parabolic rule, and Del Castillo [25] enhanced parabolic rule.

References

1. J. Trojanowska, A. Kolinski, D. Galusik, M.L.R. Varela, J. Machado, A methodology of improvement of manufacturing productivity through increasing operational efficiency of the production process, in *Advances in Manufacturing*, (Springer, 2018), pp. 23–32
2. M. Irfan, H. Zahoor, M. Abbas, Y. Ali, Determinants of labor productivity for building projects in Pakistan. Journal of Construction Engineering, Management & Innovation **3**(2), 85–100 (2020)
3. A. Nuvolari, Understanding successive industrial revolutions: A "development block" approach. Environmental Innovation and Societal **Transitions 32** (2018)
4. L. Melnyk, O. Kubatko, I. Dehtyarova, O. Matsenko, O. Rozhko, The effect of industrial revolutions on the transformation of social and economic systems. Probl. Perspect. Manag. **31**, 381–391 (2019)
5. E. Popkova, Y. Ragulina and A. Bogoviz, "Industry 4.0: Industrial revolution of the 21st century Springer,. Vols. Studies in Systems, Decision and Control, 2019
6. W.D. Leong, S.Y. Teng, B.S. How, S.L. Ngan, A.A. Rahman, C.P. Tan, S.G. Ponnambalam, H.L. Lam, Enhancing the adaptability: Lean and green strategy towards the industry revolution 4.0. J. Clean. Prod. **273** (2020)
7. D. Leech, *Intelligent Machine Technology and Productivity Growth* (The Social Value of New Technology, 2019), pp. 245–255
8. A. Khan, J. Keung, S. Hussain, M. Niazi, M. Tamimy, Understanding software process improvement in global software development. ACM SIGAPP Applied Computing Review **17**(2), 5–15 (2017)
9. S. Shim, S. Kim, Intervention meta-analysis: Application and practice using R software. Epidemiology and Health **41** (2019)
10. U. Grömping, R package DoE.Base for factorial experiments. J. Stat. Softw. **85**, 1–41 (2018)
11. M. Miranda-Ackerman, A. García-Lechuga, *An Overview of the Design of Experiment Workflow: Applications in Food Production Systems* (IGI Global, Baja California, 2020), p. 14
12. J. Jacyna, M. Kordalewska, M. Markuszewski, Design of Experiments in metabolomics-related studies: An overview. Science Direct **164**, 598–606 (2019)
13. M. Yolmeh, S.M. Jafari, Applications of response surface methodology in the food industry processes. Food Bioprocess Technol. **10**, 413–433 (2017)
14. S.J.S. Chelladurai, M.K.A. Pratip Ray, M. Upadhyaya, V. Narasimharaj, S. Gnanasekaran, Optimization of process parameters using response surface methodology: A review. Materials Today: Proceedings **37**(2), 1301–1304 (2021)
15. A. Law, A tutorial on design of experiments for simulation modeling. Proceedings - Winter Simulation Conference, 550–564 (2017)
16. T. Allen, Software overview and methods review: Minitab, in *Introduction to Engineering Statistics and Lean Six Sigma*, (Springer, 2018), pp. 575–600
17. H.I. Okagbue, P.E. Oguntunde, E.C.M. Obasi, E.M. Akhmetshin, Trends and usage pattern of SPSS and Minitab software in scientific research. J. Phys. Conf. Ser. **Conf. Ser. 1734**, 012017 (2021)
18. S. Lesik, *Applied Statistical Inference with MINITAB®* (Chapman and Hall/CRC, New York, 2018)
19. M. Akers, *Exploring, Analysing and Interpeting Data with Minitab 18: First Edition* (Compass Publishing, 2018)
20. T. Blackburn, The analyze phase with Minitab tools, in *Six Sigma*, (Springer, 2022), pp. 107–201
21. R.S. Raman, N. Kukreja, N. Singh, *Process Parameters Optimization through Taguchi and ANOVA Analysis - a Review* (IOP Publishing Ltd, 2021)
22. R.H. Myers, D.C. Montgomery, C.M. Anderson-Cook, *Response Surface Methodology: Process and Product Optimization Using Designed Experiments* (Wiley, 2016)

23. R. Myers, A. Khuri, A new procedure for steepest ascent. Communications in Statistics – Theory and Methods, 1359–1376 (1979)
24. G. Miró-Quesada, E. Del-Castillo, An enhanced recursive stopping rule for steepest ascent searches in response surface methodology. Communications in Statistics Simulation and Computation, 201–228 (2007)
25. E.D. Castillo, *Process Optimization: A Statistical Approach*, vol 476 (Springer, 2007)
26. M. Ghobakhloo, Industry 4.0, digitization, and opportunities for sustainability. J. Clean. Prod. **252** (2020)
27. Januardi, E. Widodo, A review of response surface methodology approach in supply chain management, in *Proceedings of the 3rd Asia Pacific Conference on Research in Industrial and Systems Engineering*, (2020), pp. 322–327
28. P. Maia, F. Alves-de-Almeida, V.-D.-C. Paes, H.-D.-F. Gomes, A. Paulo-de-Paiva, Multivariate steepest ascent method based on latent variables. Appl. Math. Model. **90**, 30–45 (2021)
29. D.-H. Lee, S.-H. Kim, J.-H. Byun, A method of steepest ascent for multiresponse surface optimization using a desirability function method. Qual. Reliab. Eng. Int. (2020)
30. Q. Zeng, W. Qiu, J. Liu, R. Xu, J. Shi, Y. Sun, A high dynamics algorithm based on steepest ascent method for GNSS receiver. Chin. J. Aeronaut. **34**, 177–186 (2021)

The Role of Industry 4.0 Technologies in the Energy Transition: Conceptual Design of Intelligent Battery Management System Based on Electrochemical Impedance Spectroscopy Analysis

W. J. Pech-Rodríguez, Enrique Rocha-Rangel, Eddie N. Armendáriz-Mireles, Gladis G. Suarez-Velázquez, and L. C. Ordóñez

Nomenclature

BMS	battery management system
EIS	electrochemical impedance spectroscopy
ESD	energy storage devices
ESS	energy storage systems
EV	electric vehicle
DC	direct current
GA	genetic algorithm
IoT	Internet of Things
I4.0	fourth industrial revolution
LIB	lithium-ion battery
ResNet	residual neural network
RUL	remaining useful life
SEM	scanning electronic microscopy
SoC	state of charge
SoH	state of health
TCN	temporal convolutional neural

W. J. Pech-Rodríguez (✉) · E. Rocha-Rangel · E. N. Armendáriz-Mireles · G. G. Suarez-Velázquez
Universidad Politécnica de Victoria, Departamento de maestría en Ingeniería, Parque Científico y Tecnológico de Tamaulipas, Ciudad Victoria, Tamaulipas, Mexico
e-mail: wpechr@upv.edu.mx

L. C. Ordóñez
Centro de Investigación Científica de Yucatán, Unidad de Energía Renovable, Yucatán, Mexico

L. C. Méndez-González et al. (eds.), *Innovation and Competitiveness in Industry 4.0 Based on Intelligent Systems*, EAI/Springer Innovations in Communication and Computing, https://doi.org/10.1007/978-3-031-29775-5_8

1 Introduction

Due to the gradual increase in energy demand and the aggravation of environmental issues, the scientific community has proposed the implementation of storage technologies as a clean alternative [1–3]. Technological progress must incorporate energy storage devices with high performance and longer life. The design of these systems involves a trade-off between several parameters, while advanced control strategies must be based on the instantaneous state of many indicators. Machine learning can drastically accelerate computations, monitor complex systems to improve prediction accuracy, and handle decisions based on given training data [4].

Currently, the widespread adoption of electric and hybrid vehicles has boosted the massive production of batteries. However, the high cost, limited chemical elements, durability, and relatively lower energy density are the main barriers associated with these electrochemical devices. Thus, additional strategies are needed to improve global efficiency and maintenance of storage energy devices. New materials and edge devices can help in the energy transition, but digitalization and networking can also be helpful to improve and potentiate storage devices. In this sense, unique approaches such as implementing an in-depth study of the balance of plants (considering the life cycle and sustainability) have been made. Besides the existing methodology, some authors use novel electrochemical solutions such as electrochemical impedance spectroscopy (EIS), accelerated durability test, and pseudo-random sequence signals to address the latter issues [5, 6]. Electrochemical techniques are helpful for batteries and cell performance evaluation since the nature of the chemical and electrochemical processes occurring in these devices can be determined. Both traditional and accelerated test methods are necessary to analyze and compare battery properties, obtain kinetic information, and predict the probability of failure.

EIS can not only provide precise kinetic details but can also be used to monitor changes in battery properties under use or storage conditions [7]. EIS applications include electric vehicle (EV) batteries, fuel cells, and solid-state batteries, such as polymer electrolyte batteries. Impedance can be used to monitor and control the battery's degradation while the car is being driven. Furthermore, from this technique, it is also possible to extract the diffusion coefficient of the intercalated species that could be further used to optimize battery materials and in degradation assessments.

The latter is possible due to advances in the electrochemical process. In fact, the rapid technological transformation has boosted the use of low-cost equipment. Yet, most of the researches conducted on battery diagnostic is obtained by standard bench instruments under laboratory conditions that are appropriate for online applications [8]. Recently, some authors have been adopting the advantages of the fourth industrial (I4.0) revolution by incorporating the Internet of Things (IoT), smart sensors, and big data into the monitoring of the state of health (SoH) and state of charge (SoC) in batteries [9, 10]. The former is changing the concept of battery management and the improvements that can be incorporated into this technology. Nonetheless, it is imperative to consider some technical difficulties of I4.0 (such as connectivity, cross-platform capability, data collection, and processing) before trying to implement their advantages in BMS. Because the current IoT architecture is mainly based on an open model, a universal consensus on IoT architecture needs to be adopted in the automotive sector.

This document aims to answer the research question: *To what extent can the adoption of I4.0 technologies and EIS measurements should be used to develop a practical framework for BMS?*

To answer the research question, we propose the implementation of I4.0 technologies' approaches and extend a study on how EIS can help determine the SoH and the RUL of batteries that can be used in electric transportation. In this regard, IoT, cloud computing, and GA are proposed and applied to solve multiobjective functions and obtain an intelligent BMS.

Concerns about environmental issues and global energy consumption motivate the study of the I4.0 adoption in BMS. The main goal of this research is to evaluate the versatility of EIS as a tool to estimate battery parameters by using the I4.0 approach. To this end, we used the battery Nyquist plot to compare the raw data with the simulated one. This study contributes to the existing I4.0 literature by identifying and highlighting their strengths for battery parameter estimation in vehicles and remarks on the necessity for guidelines and policies in the automotive sector.

Although some already BMS are developed, as you can see from the next section, most of them are based on voltage and electrical current measurements, and the SoH relies on these parameters. EIS has been widely applied to estimate battery parameters in off-line mode. Thus, we believe this work can contribute to the design of intelligent BMS based on online mode EIS measurements and I4.0 technology adoption.

The structure of this document is divided as follows: Section 2 presents related literature, Section 3 explains the experimental investigation, and Sect. 4 summarizes the obtained results.

2 Theoretical Background

2.1 Battery System and I4.0 Relationship

Recently, lithium-ion battery applications have gained significant attention due to their potential application in many electronic devices. A lithium-ion or Li-ion battery is a secondary battery in which lithium ions move from the anode electrode through the electrolyte to the cathode electrode during discharge and vice versa when charging. These batteries are characterized by their high energy density. Nevertheless, they represent a safety hazard since they contain flammable solvents and can lead to explosions and fires if damaged or incorrectly charged. On the other hand, Li-ion battery is prone to excessive heat, which can deteriorate the cells' battery life. Therefore, monitoring the battery parameters to extend its lifetime is crucial.

Additionally, I4.0 is characterized by the extended use of the Internet of Things (IoT) linked with radio frequency identification (RFI), sensors, and global positioning systems, among other devices. Thus, a promising approach was proposed by Dongliang et al. where the authors used RFI technology to acquire data from vehicles and battery compartments in charging stations [11]. The latter evidenced the impact of I4.0 in the battery management of electric vehicles. From a sustainable point of view, I4.0 could potentially boost the benefits of electric cars since it can help determine the SoH and the RUL of batteries to predict appropriate replacement. Furthermore, implementing I4.0 in electric vehicles could improve the understanding of the global efficiency of these technologies. Selecting batteries for secondary use or battery recovery and trace might be beneficial. For instance, the research group of Haijun implemented big data analysis for the correct battery recycling for electric vehicles [12]. Recent works proposed the implementation of digitalization, a challenging technology of I4.0 [13, 14]. These studies analyze virtual prototyping as a positive way to boost the production in real time of a Gigafactory. Thus, it might improve product reliability and quality. Moreover, the authors stated that digitalization plays a critical role in the sustainability of I4.0. By adopting this paradigm, enterprises can provide digital solutions in energy efficiency, flexibility, and material efficiency.

Furthermore, Tzu-Chia et al. indicated that hybrid energy storage devices are a promising tool to solve the problems of the energy power grid in terms of energy availability [15]. A dynamic and consistent strategy is essential while managing a hybrid energy storage system. Therefore, data obtained from energy storage devices during driving can be used to analyze the performance of the energy management systems. In this sense, providing an energy management system that can learn from the environment and the driving cycle is advantageous since this platform can send the necessary data to a control system for optimal management. In another work, Pham and coworkers studied the advantages of I4.0 in the circular economy of battery swapping. This group adopted a new business model by using embedded IoT and cloud computing technologies in the framework of I4.0. A digital platform achieved battery management to conduct battery maintenance and replacement. Also, the authors stated that I4.0 facilitates choosing the battery charging station location while providing information to the enterprise for stock battery numbers [16]. In this context, Wahab et al. explored the application of IoT in monitoring the performance of an electric vehicle battery through a battery monitoring system based on two main parts: the monitoring device and the user interface. The results showed that the proposed system detects degraded battery performance and sends notification to the user to take further action [17].

2.2 EIS and Battery Parameter Estimation

EIS is a powerful tool for battery life prediction, and the estimated parameters from measured impedance data have shown a good correlation between the health and aging of the battery [18]. Measuring EIS data and modeling is an emerging trend for battery management [19]. It is well known that electromobility applications need to have more than 70% of SoH to improve mileage and velocity; otherwise, the owner will have to replace it with a new one [20]. Thus, implementing EIS can reduce the cost of battery replacement and select battery packs or cells for second life use. The former could improve the economy for electric vehicles by streaming a new product revenue, addressing the environmental concern about the potential battery waste, and reducing by 56% the emissions of CO_2 when used for electrical generation, compared with the natural gas plant [21].

Shuangqi et al. reported a battery management method that employed a cyber-physical perspective to improve the reliability of electric vehicles [22]. Likewise, Lingjun et al. used big data analysis to estimate the SoH for lithium-ion battery packs [23]. They used real-world data from 700 electrical vehicles equipped with lithium nickel-manganese-cobalt oxide batteries and monitored them for 1 year. After all, the authors determined the aging battery time using the feedforward neural network and evaluated batteries' remaining useful life (RUL). One interesting point of this work was using a cloud platform to store the battery extracted data to estimate the vehicle's driving range. In this regard, a cloud battery management system that combines IoT and cloud computing technology can be used to monitor the battery parameters. Therefore, BMS software and IoT could be used to notify the manufacturer and customers about the system status [24].

The state of health (SoH) of lithium-ion batteries (LIBs) must be accurately determined to ensure safe operation. Mohamad et al. predicted that in the next decade, it would be possible to use electric cars in a safe, cost-effective, and environmentally friendly way but only if accurate and reliable predictions of the state parameters of their batteries were made [25]. Traditional SoH estimation techniques have been shown to have limits that can be overcome with data-driven methods. For example, temporal convolutional neural (TCN) network combines the advantages of residual neural networks (ResNet) with the computational efficiency of neural networks to produce an efficient and effective strategy. The simulation analysis showed that the proposed method had reduced the deployment cost. This algorithm can accurately estimate the SoH of a LIB with high precision during the battery lifetime.

Tran and coworkers stated that the current BMS is not flexible since it employs simple battery models which are not interconnected [26]. This research group proposes implementing I4.0 technology, specifically cloud computing for Li-ion batteries, to provide better monitoring and optimization solutions. In another research work, Yang et al. utilized a framework based on edge-edge cloud architecture for BMS systems to handle thermal management, cell balancing, and fault diagnosis [27]. Also they mentioned that BMS based on cloud computing could be used in battery energy storage systems. The authors highlighted the benefits that adopting EIS measurements can provide to multi-scale models in battery systems. Although they claimed this could be difficult, cloud computing and processes such as learning algorithms will play a key role in the near future. Similarly, Kahveci et al. consider that big data analytic solutions remain challenging in I4.0 digital transformation [28]. Thus, this group developed a case study where battery module assembly in electric vehicles was monitored and handled by the technologies of I4.0: IoT and big data analysis.

Generally speaking, EIS is a relatively simple experiment [29] that has been used to obtain physical properties. Nevertheless, the data interpretation of the EIS spectra can be confusing. Evidently, the battery performance depends on the charge-discharge cycles, the operating temperature, and the electrical current peaks. Furthermore, it is reported that the degradation mechanism decreases battery capacity and increases the equivalent resistance [30]. Thus, the SoH percentage of a battery can easily be obtained by determining the change in internal resistance [31–33]. The battery capacity and degradation can be determined by measuring the frequency response battery data (impedance at different frequencies) and using the concept of equivalent circuit models. A genetic algorithm optimizes multiobjective problems to perform a local search for nonlinear equations [34].

The already developed BMS estimates battery aging based on the capacity decrease and power fate. However, EIS has been proposed to identify the principal causes of battery degradation and quantify the aging mechanism over time by estimating the value of each electrical element in the proposed circuit model [35, 36]. On the other hand, EIS measurements can be conducted off-line or online mode through a digital platform with data storage and analysis [37]. In a recent study, Wimarshana et al. analyzed the EIS as a powerful tool for obtaining the physicochemical model of Li battery. Their research concluded that the behavior of each electrical element depends on the frequency range, and low frequency is the optimal condition for identifying the main electrode reactions [38].

2.3 Artificial Intelligence to Estimate EIS Battery Parameters

The main challenge for EIS analysis is the model solution and interpretation. The resolution of the circuit model results in nonlinear equations with multiple variables. In this sense, much effort has been made to develop a methodology that estimates EIS parameters easily. For example, Maghawry et al. used the concept of a hybrid genetic algorithm (GA) to solve the multiobjective function [39]. Indeed, GA is a population-based algorithm inspired by natural evolution that uses multiple candidates to solve a specific optimization problem [40]. The main advantage of GA relies on its versatility to find the value of one or more variables that satisfy the global optima [39]. The first step in the GA solution is initializing a random population composed of multiple chromosomes, selection, and mutation. Then, crossover and mutation operators are applied to produce a new individual to form part of a current population.

GA approach has already been implemented as part of I4.0 technologies. However, the manufacturing industry in India conducted novel research in the area of automotive battery recovery for secondary use [41]. To minimize the total supply chain cost, the authors adopted GA to solve the closed-loop supply chain model. Additionally, Mohamad et al. developed a prediction model for energy management in hybrid PV and battery systems using a deep learning architecture that can accurately estimate PV power production and load one step ahead [42]. The model was tested with dynamic electricity costs, concluding that the proposed strategy and prediction model can handle more than half of the annual load requirements compared to existing energy systems. The optimization problem is constrained by the produced CO_2 and the charge stored in a battery pack.

In 2020, Wei Li and coworkers built a model to predict battery capacity using an artificial neural network. The data was obtained by performing repeated charge and discharge cycles [43]. From the conceptual basis of I4.0, adopting this paradigm can be helpful for real-time monitoring of batteries and big data analysis for decision-making. Wireless communication permits the interaction of portable devices with external access points, and new solutions like data analysis and data storage can be introduced. Thus, with the help of I4.0 technologies and the deployment of low-cost wireless equipment, users can easily monitor the battery's health anywhere, and experts will have enough data to analyze its lifetime in mobile technology. Ngandjong et al. investigated the calendaring (compaction process in battery electrodes) impact of Li-ion electrodes [44]. For this purpose, they developed a computational model for battery manufacturing digital twin that incorporates a discrete element method calendaring model. Experimental validation was achieved through scanning electron microscopy (SEM) and EIS measurements. From the Nyquist plot, an electrode electronic resistance reduction of the calendaring electrodes was observed compared with the unmodified one. This result matches the electrode porosity and pore-size distribution trend.

3 Methodology

Figure 1 shows the foremost step in the BMS conceptual design. First, a literature review was conducted to gain insight into the aspect of current BMS and EIS applications. Then, to determine battery health, a proposal for intelligent BMS based on EIS measurements was achieved. In this sense, GA was chosen to assess the EIS parameters, and the obtained data was analyzed to identify the versatility of EIS and the main implications of the BMS in real life.

The raw EIS data analyzed in the present study was obtained from Zhang et al. [45]. This research group conducted several EIS measurements in a Eunicell LR2032 Li-ion battery composed of $LiCoO_2$/graphite under three different operational temperatures. Also, cyclability was achieved by charging and discharging the

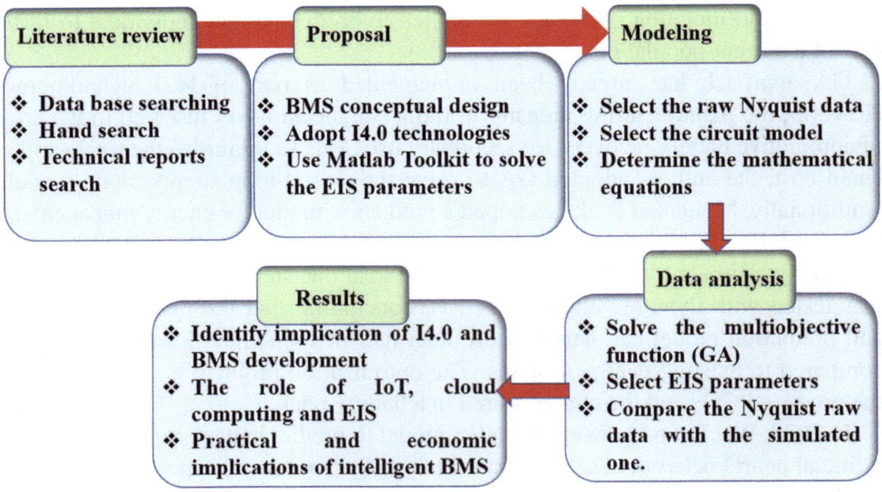

Fig. 1 Conceptual design of BMS

battery to 4.2 V and below 3 V, respectively. We employed the 25C01 sample (25 °C) to determine the EIS parameters and then assess the battery's aging considering the change in the calculated values. The Nyquist plot was fitted according to the circuit model proposed by Yang and coworkers and other research groups; see Fig. 2 [46, 47].

It is noteworthy that the complete circuit model for batteries considers an inductor at a high frequency [48]. However, since this work focuses on aging and SoH for batteries, we only consider the series resistance that occurs at high frequency. The diffusion mechanism (low frequency) was modeled bearing in mind a constant phase element. To reduce the circuit model, we considered frequencies below 12,516 Hz. Meanwhile, we adopted a two-series *Z*arc circuit composed of a constant phase element and resistance at a middle frequency, according to reference [49].

Parameter determination for the proposed electrical model was achieved by using the Matlab toolbox (community and technical college license) and the genetic algorithm for multiobjective functions. Simulations were performed in a 4 core, 16 GB, and 5 GHz Acer computers.

Figure 3 displays the main steps for battery parameter estimation. During the GA solution, a single point crossover operator was used to divide a pair of chromosomes and select a random site along the length of the string, which is further combined to generate a better offspring. The chosen fraction crossover was 0.65, while the Pareto set fraction was also maintained at the same value. Since the population size plays a critical role in searching for the optimal solution, we use a large population size, 9000. To improve the effectiveness of retrieval, which solves the multiobjective problem in reasonable computational time, the maximum number of generations was maintained at 3000.

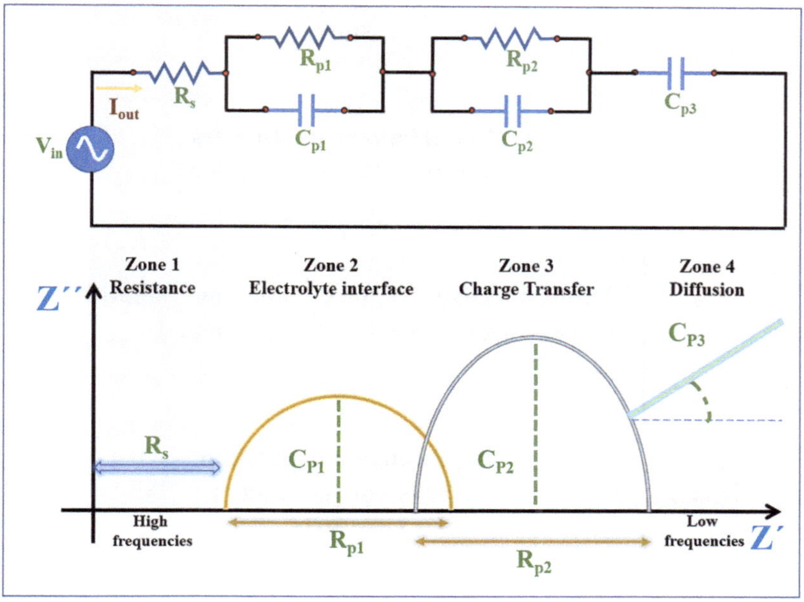

Fig. 2 Proposed circuit model for parameter estimation in batteries

As is well known, the solution of mathematical equations by GA is a statistical procedure that creates a subset of possible solutions. In this work, the diversity between each subset is tracked by implementing the crowding approach. In this work, a minimization process was used to obtain the local minima, and in this way, the nine unknown variables were calculated. Generally speaking, the number of individuals is adopted to estimate the minimizer. The nonlinear system equation for the proposed circuit model was transformed into a multiobjective optimization problem as follows:

$$\begin{cases} f_1\left(R_s, R_{P1}, C_{PE1}, \alpha_1, R_{P2}, C_{PE2}, \alpha_2, C_{PE3}, \alpha_3\right) = 0 \\ f_2\left(R_s, R_{P1}, C_{PE1}, \alpha_1, R_{P2}, C_{PE2}, \alpha_2, C_{PE3}, \alpha_3\right) = 0 \\ f_N\left(R_s, R_{P1}, C_{PE1}, \alpha_1, R_{P2}, C_{PE2}, \alpha_2, C_{PE3}, \alpha_3\right) = 0 \end{cases}$$

$$\begin{cases} F_1 = \min\left(abs\left(f_1\left(R_s, R_{P1}, C_{PE1}, \alpha_1, R_{P2}, C_{PE2}, \alpha_2, C_{PE3}, \alpha_3\right) = 0\right)\right) \\ F_2 = \min\left(abs\left(f_2\left(R_s, R_{P1}, C_{PE1}, \alpha_1, R_{P2}, C_{PE2}, \alpha_2, C_{PE3}, \alpha_3\right) = 0\right)\right) \\ F_N = \min\left(abs\left(f_N\left(R_s, R_{P1}, C_{PE1}, \alpha_1, R_{P2}, C_{PE2}, \alpha_2, C_{PE3}, \alpha_3\right) = 0\right)\right) \end{cases} \quad (1)$$

The above function aims to obtain the global optima, and constraints were unnecessary. Lower and upper bounds were defined to estimate the optimal values for R_s, R_{P1}, C_{PE1}, α_1, C_{PE2}, α_2, C_{PE3}, and α_3.

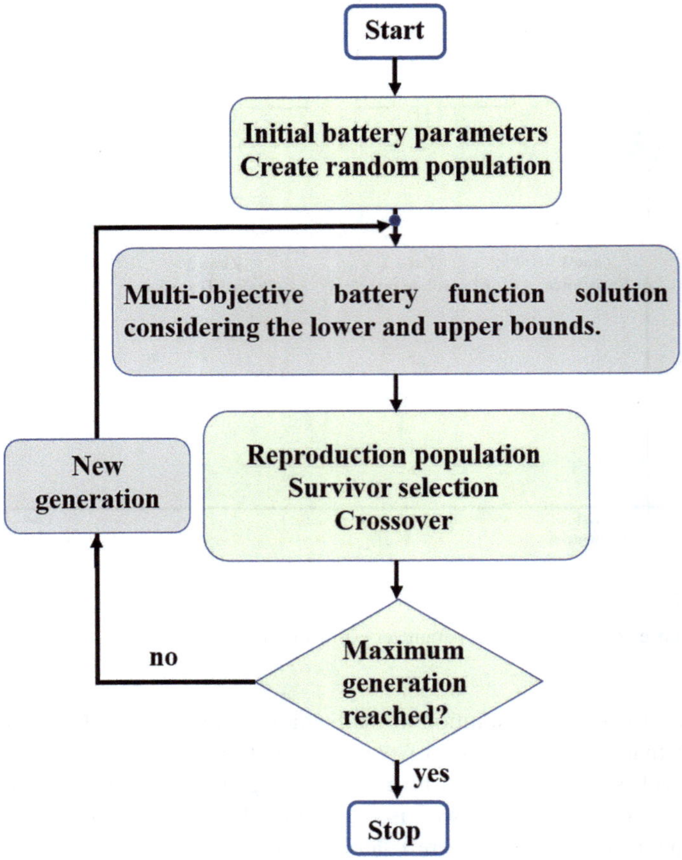

Fig. 3 Flowchart for the adopted GA for battery parameter estimation

4 Results

Determining the batteries' aging and SoH is mandatory to guarantee their reliability during operational conditions. Thus, we propose the implementation of I4.0 to conceptualize a BMS platform that can identify and quantify the main parameters of battery packs in mobile vehicles. In this sense, EIS online and off-line measurements are proposed. Furthermore, to estimate the parameters of the proposed circuit model through GA, we first determine the mathematical equation that describes the model's behavior. Indeed, the following equation defines the equivalent impedance (Z_{eq}) of the system; see Fig. 2:

$$Z_{\mathrm{eq}} = R_s + Z_{\mathrm{Zarc1}} + Z_{\mathrm{Zarc2}} + Z_{\mathrm{CPE}} \tag{2}$$

where R_s represents the series resistance, Z_{Zarc1} is the impedance of the parallel connection between the resistor and the constant phase element (the first semicircle), Z_{Zarc2} is the impedance of the second semicircle that occurs at mid-frequency, and, finally, Z_{CPE} is the impedance of the constant phase element, at low frequencies, that simulates the diffusion mechanisms in the Li-ion battery. This work adopts the implementation of the constant phase element (CPE) because the semicircles formed in the Nyquist plot have a deformation due to the physical properties of the materials that compose the battery, like porous electrodes [50].

The impedance of each Zarc element can be calculated as follows:

$$Z_{\text{CPE}} = \frac{1}{Qs^{\alpha}} \tag{3}$$

$$Z_{\text{Randles}} = \frac{R_p * Z_{\text{CPE}}}{R_p + Z_{\text{CPE}}} = \frac{\frac{R_p * e^{\frac{-j\pi\alpha}{2}}}{Q(w)^{\alpha}}}{R_P + \frac{1e^{\frac{-j\pi\alpha}{2}}}{Q(w)^{\alpha}}} = \frac{R_p * e^{\frac{-j\pi\alpha}{2}}}{\left(Q(w)^{\alpha} R_p + e^{\frac{-j\pi\alpha}{2}}\right)} \tag{4}$$

The above equation is modified after considering that

$$e^{\frac{-j\pi\alpha}{2}} = \frac{\cos\left(\frac{\pi\alpha}{2}\right)}{Q(w)^{\alpha}} - j\frac{\sin\left(\frac{\pi\alpha}{2}\right)}{Q(w)^{\alpha}} \tag{5}$$

then the resulting equation is multiplied by the complex conjugate and simplified to obtain the following equation:

$$Z_{\text{Randles}} = \frac{R_P\left(R_P * \frac{\cos\left(\frac{\pi\alpha}{2}\right)}{Q(w)^{\alpha}} + \left(\frac{\cos\left(\frac{\pi\alpha}{2}\right)}{Q(w)^{\alpha}}\right)^2 - j\frac{R_p \sin\left(\frac{\pi\alpha}{2}\right)}{Q(w)^{\alpha}} + \left(\frac{\sin\left(\frac{\pi\alpha}{2}\right)}{Q(w)^{\alpha}}\right)^2\right)}{R_p^2 + 2R_p * \frac{\cos\left(\frac{\pi\alpha}{2}\right)}{Q(w)^{\alpha}} + \left(\frac{\cos\left(\frac{\pi\alpha}{2}\right)}{Q(w)^{\alpha}}\right)^2 + \left(\frac{\sin\left(\frac{\pi\alpha}{2}\right)}{Q(w)^{\alpha}}\right)^2} \tag{6}$$

As can be observed, the latter equation can be reduced by applying trigonometric identities to obtain

$$Z_{\text{randles}} = \frac{R_p\left(Q(w)^{\alpha}\, R_P * \cos\left(\frac{\pi\alpha}{2}\right) + 1\right) - jR_p\left(Q(w)^{\alpha}\, R_P * \sin\left(\frac{\pi\alpha}{2}\right)\right)}{\left(Q(w)^{\alpha}\right)^2 R_p^2 + 2R_p Q(w)^{\alpha} * \cos\left(\frac{\pi\alpha}{2}\right) + 1} \tag{7}$$

To compute and solve the global minima of the equation and determine the parameters for the proposed fitting procedure, the obtained Z_{eq} impedance, Eq. 6, was separated into its real and imaginary parts. In this form, we obtain two equations at the same angular frequency:

$$\mathrm{Re}\left(Z_{eq}(w)\right) = R_s + \frac{R_{P_1}\left(Q_1(w)^{\alpha_1} R_{P_1} * \cos\left(\frac{\pi\alpha_1}{2}\right) + 1\right)}{\left(Q_1(w)^{\alpha_1}\right)^2 R_{P_1}^2 + 2R_{P1}Q_1(w)^{\alpha_1} * \cos\left(\frac{\pi\alpha_1}{2}\right) + 1}$$
$$+ \frac{R_{P_2}\left(Q_2(w)^{\alpha_2} R_{P_2} * \cos\left(\frac{\pi\alpha_2}{2}\right) + 1\right)}{\left(Q_2(w)^{\alpha_2}\right)^2 R_{P_2}^2 + 2R_{P2}Q_2(w)^{\alpha_2} * \cos\left(\frac{\pi\alpha_2}{2}\right) + 1} + \frac{\cos\left(\frac{\pi\alpha_1}{2}\right)}{Q_3(w)^{\alpha_3}} \quad (8)$$

$$\mathrm{Im}\left(Z_{eq}(w)\right) = \frac{-R_{P1}\left(Q_1(w)^{\alpha_1} R_{P_1} * \sin\left(\frac{\pi\alpha_1}{2}\right)\right)}{\left(Q_1(w)^{\alpha_1}\right)^2 R_{P_1}^2 + 2R_{P1}Q_1(w)^{\alpha_1} * \cos\left(\frac{\pi\alpha_1}{2}\right) + 1}$$
$$+ \frac{-R_{P2}\left(Q_2(w)^{\alpha_2} R_{P_2} * \sin\left(\frac{\pi\alpha_2}{2}\right)\right)}{\left(Q_2(w)^{\alpha_2}\right)^2 R_{P_2}^2 + 2R_{P2}Q_2(w)^{\alpha_2} * \cos\left(\frac{\pi\alpha_2}{2}\right) + 1} - \frac{\sin\left(\frac{\pi\alpha_1}{2}\right)}{Q_3(w)^{\alpha_3}} \quad (9)$$

It should be mentioned that the Zarc element cannot be converted from the frequency domain to the time domain without approximations [48]. Thus, both equations of the equivalent impedance for the proposed circuit model were solved using the GA in the Matlab toolkit. The low and upper limit values were selected considering the ranges reported in the literature.

Then Fig. 4 shows the raw Nyquist impedance plot and the model output computed by using the estimated parameters. The comparison was achieved to support the decision to use GA to determine the EIS parameters in batteries. As previously discussed, the inductor contribution was neglected because high frequencies have no practical relevance in batteries [48]. It is evident that the estimated parameters represent the experimental data with great precision. The Nyquist plot behavior correlates well with the graph presented in the supplementary information by the Zhang research group [45]. In addition, the author states that a change in the imaginary impedance is related to the battery cycle number, and this information is studied below (considering the 25th cycle). As can be observed, the semicircles at mid frequencies and the diffusion occurring at high frequency (range 1–0.1 Hz) were well fitted by the Zarc elements and the constant phase element, respectively. Therefore, from here we can notice that the proposed circuit model can give a good fit and the accuracy can be improved by refining the values of the obtained variables.

To study the effect of the charge and discharge processes, we also determined the parameters for the 25th cycle of the 25C01 sample; see Fig. 5. It can be observed that shape changes during the cycling process agree with the work developed by Deng et al. [47]. Furthermore, Zhang et al. conducted a study about the state of the 25C01 cell along cycle number. They observed that the Z_{im} shows a positive linear correlation with the cycle number. In other words, the value of Z_{im} increases with the cycle number.

Table 1 summarizes the calculated parameter values for the 25C01 sample at the first cycle and the 25th cycle. It can be seen that R_s, R_{p2}, and C_{PE2} computed values had a slight change after cyclability, and this can be used to identify degradation and assess the SoH and RUL in batteries. It is worth highlighting that the obtained

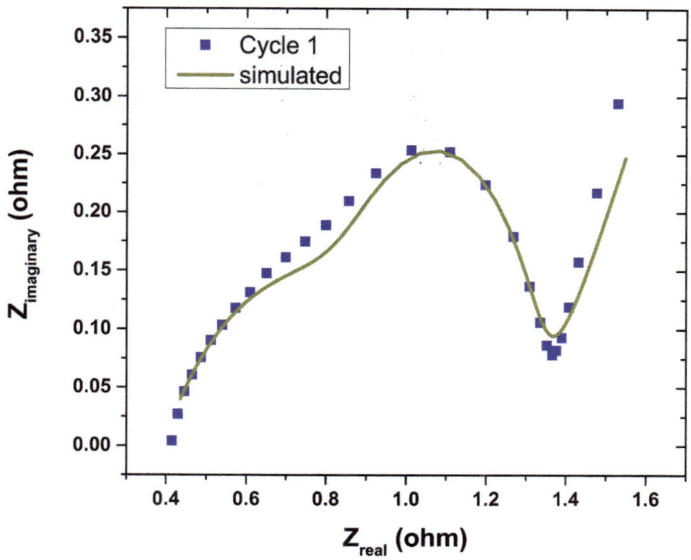

Fig. 4 Experimental and simulated Nyquist plot for the first cycle of battery

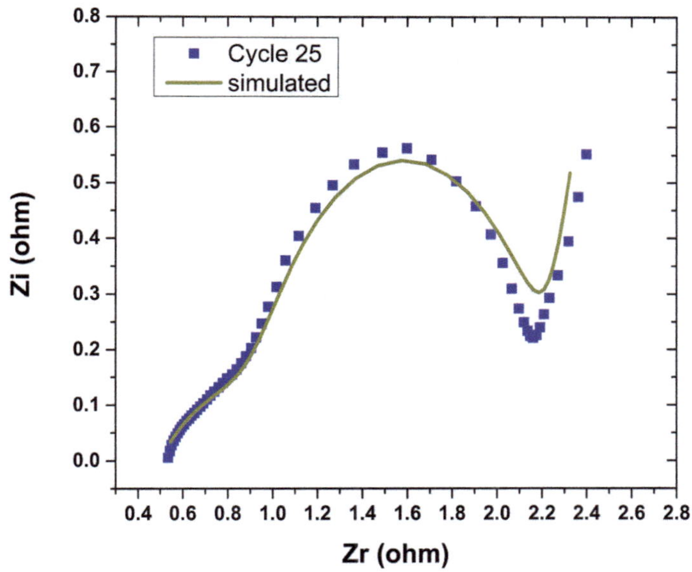

Fig. 5 Experimental and simulated Nyquist plot for the 25th cycle of battery

results are in good agreement with the literature. For example, Deng et al. studied the capacity fading of lithium batteries by analyzing EIS by Zview software and obtaining the same change trends in resistance [47]. They also observed that R_{p2} is the element that dramatically increases as a function of cycle number. In our case,

Table 1 Estimated parameter for the Nyquist plot with the adopted circuit model

Circuit	Rs	Rp1	CPE1	α1	RP2	CPE2	α2	CPE3	α3
A	0.391	0.35	0.068	1	0.6	0.05	0.518	8.408	0.548
B	0.51	0.492	0.035	0.588	1.257	0.092	0.855	6.28	0.72

the R*p2* value for the fresh battery was 0.6 ohms, and after the 25th cycle, this element increased its value to 1.257 ohms, a dramatic rise of 208% in the resistance value.

As mentioned by P. Hofmann, there is a current lack of information regarding battery aging for automotive applications [51]. Thus, EIS measurement and I4.0 technologies can boost the widespread application and management of batteries. The overall aging mechanism in batteries is fundamentally the contribution of two types of aging: calendar and cycle [52]. Calendar aging is characterized by the increase of the solid electrode interface layer. Table 1 shows that R_{P1} (zone 2, see Fig. 2) experiences a slight electrical resistance change after cyclability. The former indicates a solid electrolyte interface growth during the charge and discharge cycle. The increase in resistance means electrolyte decomposition resulting in a low battery capacity loss, which is more pronounced in the anode side [51, 53].

On the other hand, a well-marked change occurs in zone 3 that is ascribed to the second aging mechanism: cycling. This semicircle results from an electrochemical process where electron transfer is present. The increase in resistance indicates a change in materials' physical and chemical properties, such as phase transition and particle size. From Table 1, it can be noticed that R_{P2} changed its resistance after the cyclability. These phenomena cause an evident rise in ohmic resistance linked to a decrement in capacity or fading mechanism. The same behavior was observed by Lyu et al. where the shape of the EIS spectra changes with the cycle number [54]. They state that the semicircle arc in the middle frequency has a considerable increment related to the number of cycles. Besides, the Deng research group concluded that the charge transfer resistance in lithium/sulfur batteries plays a critical role in the capacity fading of this type of battery [47].

It has been reported that SoH can be determined in terms of the capacity of impedance variation. Haram and coworkers propose an equation to determine the SoH of the battery by considering the change in the internal resistance or impedance [20]. Thus, the SoH could be obtained by the following equation:

$$\mathrm{SoH}(Z) = \left(2 - \frac{Z_m}{Zn}\right) * 100 \tag{10}$$

where Z_m is the measured impedance and Z_n is the nominal impedance of a new battery. This group states that an increase of 200% impedance indicates battery replacement.

Furthermore, a complete study was developed by Pastor-Fernández, who compared the change in resistance under different SoC [35]. They found that the R_{ct} has the most significant difference compared with R_s an R_{P1}. Besides, they observed that loss of active material and loss of Li-ion were the most important aging mechanisms independent of the SoC.

Therefore, we can elucidate the battery's health by monitoring the change in resistance and determining the aging mechanism [35, 55]. Thus, from the experimental results, it can be concluded that EIS measurement is a versatile tool to estimate the parameters of batteries with reasonable accuracy. Furthermore, this work demonstrates GA's power to solve nonlinear multiobjective function that contains several variables.

Historically, most EIS equipment is highly complex, bulky, and unsuitable for online field measurements. The advent of I4.0 deploys the fabrication of low-cost portable devices. In this sense, Grossi and coworkers developed a device for in-the-field EIS measurements for liquid and semiliquid media [56]. They used a low-cost on-chip waveform generator that offers programmable frequencies to construct a versatile portable instrument for EIS measurements (about 100 USD). The prototype was tested by conducting concentration measurements in different saline solutions. On the other hand, Muñoz et al. proposed the construction of an eight-channel electrical impedance through a high-precision impedance converter (AD5933); the manipulation was achieved through Arduino Mega 2560 and four multiplexers [57]. The error in the measured impedance was less than 5%. In this regard, a programmable impedance emulator based on a low-cost digital finite impulse response filter has been proposed. Pseudo-random binary sequence (PRBS) has also been implemented to predict the SoH of energy storage systems. Surprisingly, PRBS results were obtained in 200 s, and the data were compared with laboratory instruments that took more than 2 h [58]. Generally, EIS can separate complex electrochemical processes into basic ones by considering different relaxation times in each electrical element [59]. This characteristic can be used to assess the SoH of the battery.

Figure 6 presents a conceptual proposal for the intelligent BMS based on EIS measurement. It should empathize that portable and fast response EIS equipment needs to be considered in this proposed BMS. In the first stage, battery information such as voltage, temperature, and EIS data is read in online or off-line mode. The obtained information can be used to assess the battery's health by conducting some big data analysis like the application of GA. Then, processed data could be saved in a cloud and sent to the local BMS (the front vehicle panel). Because the data is stored in a cloud system, customers can access it at any time to monitor the battery's status and information such as the accurate location mapping position, speed, and nearby service stations, among others.

Introducing I4.0 technology into the MBS can help explore and improve other philosophies, such as swapping. As is well known, batteries in electric vehicles need more than 70% of SoH to fulfil distance requirements. The stored data could help to trace batteries with a second life use that can be incorporated into the V2G vehicle to grid system [60].

On the other hand, two defining features characterize the IoT concept: firstly, automation, which involves direct communication between autonomous devices and machines without human intrusion. Secondly, connectivity facilitates access to

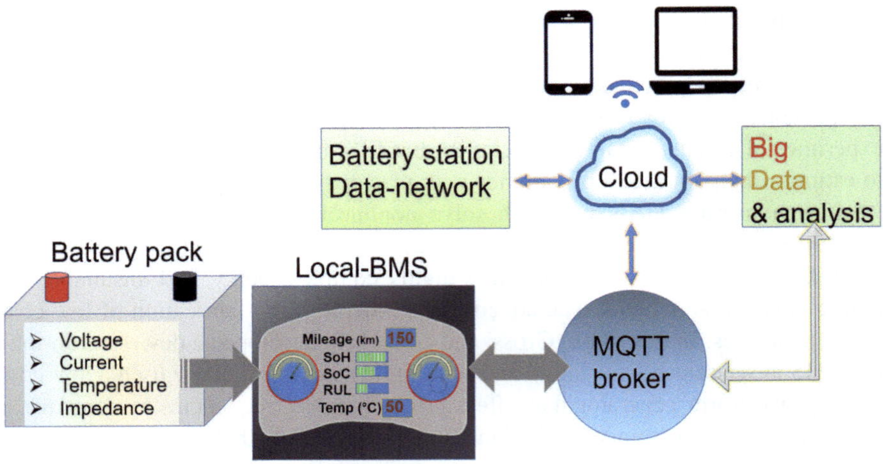

Fig. 6 Conceptual proposal for the BMS based on EIS measurement and GA analysis

diverse sources of information. Given these characteristics, the IoT must employ the I4.0 paradigm to ensure reliability between multiple devices. For example, automation requires artificial intelligence and big data analysis; meanwhile, connectivity is achieved by cloud computing and also wireless communication.

Previous BMS only monitored and detected the battery status and alerted the user via the front panel located inside the vehicle. Thus, the advancement in notification system design could be used to notify users and manufacturers about the status of the battery and take actions to improve their performance. For this reason, several battery monitoring systems using IoT technology have been proposed. System administrators can keep track of user details in a well-defined database. Among many other things, it can indicate to the user the status of the battery in terms of its working conditions, such as working voltages and currents, its temperature, the impedance of the system, and the nearest charging stations, and helps to reserve time slots. The idea of all this is to provide an online scheduling scheme to make it more convenient for all electric vehicle owners and users to monitor the battery status of their electric vehicles anytime and anywhere.

Finally, we want to mention that the present applied research can be an exciting topic for practitioners, and the conceptual framework could be further discussed as part of university projects. Also, the proposed model can be used to unlock international cooperation with leading organisms to create policy makers to overcome technical and financial obstacles.

5 Conclusions

This work introduces the implementation of I4.0 into assessing battery state of health by analyzing the EIS measurement through genetic algorithms. An equivalent circuit model was proposed to fit the EIS measured curve of the Li-ion battery, based on resistors and constant phase elements. The mathematical model was calculated, and the resulting Z_{eq} was expressed as two equations for the real and imaginary parts. The former equations were solved by using a genetic algorithm in the Matlab toolkit. The results indicate that I4.0 technologies can be adopted to estimate the SoH of batteries in an online and off-line mode. GA demonstrates versatility in solving multiobjective functions with nine unknown variables. On the other hand, the generated cloud data from the proposed BMS framework can be used to estimate battery efficiency and track batteries for second life use. Furthermore, because the BMS system can have a global positioning system, the users and enterprises can monitor the status of their vehicle fleet by using a portable application or a website. Finally, GA can also be used to generate data for battery maintenance and information about the stock of batteries in each battery service center.

Theoretical Contribution
This document originally contributes to the I4.0 literature by verifying the versatility of EIS in conjunction with IoT, cloud computing, and GA in determining the state of health of the battery in mobile applications. Although EIS has been extensively implemented to determine battery parameters, most measurements are achieved in a laboratory environment, in other words, in off-line mode. Hence, this study clearly shows the viability of applying EIS in BMS and opens the avenue for the widespread adoption of I4.0 paradigm.

Practical Implications
One difficulty in implementing a battery management system is the standardization of IoT devices in vehicles. Hence, a lot of work needs to be achieved to create manufacturing guidelines and policies for intelligent mobile vehicles. This will be a long journey from the current manufacturing paradigm. IoT potentializes opportunities in the mobile industry, but they need to be addressed by considering a well-defined structured methodology. In the present case, IoT is viewed as the cornerstone for the successful adoption of BMS because it permits the data flow from one point to another, and the installation of IoT infrastructure will allow for monitoring BMS parameters in real time and solve criticalities. The paradigm I4.0 should be adopted to create a collaborative platform where data, logistics, knowledge management, and so on can merge to promote economic, environmental, and social sustainability. It is worth mentioning that financial barriers are expected to the adoption of the proposed BMS. So governments should create economic incentives to persuade developers and customers. It is also very important to overcome different obstacles to the digitalization of processes, such as lack of digital culture, inadequate training staff, resistance to change, and reliability of digital security.

Limitations and Future Work

One limitation of the present work relates to the adopted GA. It should be emphasized that in this study, a Matlab toolkit was used to solve the multiobjective function. Although we obtain good accuracy in the obtained battery parameters, Matlab is a profit-seeking software which will increase the cost of developing practical BMS.

Another constraint of the present work relates to the proposed circuit model. As we know, the behavior of batteries depends on the type, material, capacity, model, and others. Thus, a circuit model database should be created together with an intelligent algorithm capable of selecting the better fit for the battery in question.

Future studies should consider technical aspects such as electronic vehicle circuits to carry out the EIS measurement in online mode to obtain voltage, current, and temperature from the battery system. In addition, we suggest an examination between companies and researchers to address main barriers to the intelligent BMS.

The future work of this research will be oriented toward adopting open-access software to develop specific genetic algorithms capable of minimizing error and interacting with a digital platform to assist the battery management system. Furthermore, it is necessary to look into the idea of using low frequency to develop portable EIS equipment.

Acknowledgments We gratefully acknowledge the time and infrastructure the Polytechnic University of Victoria provided.

Conflict of Interest The authors declare no competing interest.

References

1. Y.E. Durmus et al., Side by side battery technologies with lithium-ion based batteries. Adv. Energy Mater. **10**(24), 2000089 (2020). https://doi.org/10.1002/aenm.202000089
2. C. Song, Y. Shao, S. Song, S. Peng, F. Xiao, A novel electric bicycle battery monitoring system based on android client. J. Eng. **2017**, 2579084 (2017)
3. G. Zhao, X. Wang, M. Negnevitsky, Connecting battery technologies for electric vehicles from battery materials to management. iScience **25**(2), 103744 (2022)
4. T. Gao, W. Lu, Machine learning toward advanced energy storage devices and systems. iScience **24**(1), 101936 (2021)
5. J. Sihvo, D. Stroe, T. Messo, T. Roinila, Fast approach for battery impedance identification using pseudo-random sequence signals. IEEE Trans. Power Electron. **35**(3), 2548–2557 (2020)
6. R. Stadler, A. Maurer, Methods for durability testing and lifetime estimation of thermal Interface materials in batteries. Batteries **5**(1) (2019)
7. S. Hong, T.-C. Lo, Electrochemical impedance spectroscopy for battery Research and Development, in *Technical Report*, vol. 1, (Solartron Instruments, Farnborough, 1996) Available: https://www.ameteksi.com/-/media/ameteksi/download_links/documentations/library/solartonanalytical/electrochemistry/technical-report-31-hong-shih-eis-for-battery-research.pdf?la=en&revision=6723cd33-7ad6-4eea-81df-a71f4a292f48&hash=906EB05485B83685B9AF08E656BE806E

8. U.R. Koleti, T.Q. Dinh, J. Marco, A new on-line method for lithium plating detection in lithium-ion batteries. J. Power Sources **451**, 227798 (2020)
9. R. Lu, J. Lu, P. Liu, M. He, J. Liu, Design of the VRLA battery real-time monitoring system based on wireless communication. Sensors **20**(15) (2020)
10. Z. Wei, J. Zhao, H. He, G. Ding, H. Cui, L. Liu, Future smart battery and management: Advanced sensing from external to embedded multi-dimensional measurement. J. Power Sources **489**, 229462 (2021)
11. D.L. Gong, R.Y. Niu, J. Jin, H. Geng, RFID Technology application in electric Vehicle's battery charging stations. Adv. Mater. Res. **608–609**, 1545–1552 (2013)
12. H. Yu et al., Big-data-based power battery recycling for new energy vehicles: Information sharing platform and intelligent transportation optimization. IEEE Access **8**, 99605–99623 (2020)
13. E. Ayerbe, M. Berecibar, S. Clark, A.A. Franco, J. Ruhland, Digitalization of battery manufacturing: Current status, challenges, and opportunities. Adv. Energy Mater. **12**(17), 2102696 (2022). https://doi.org/10.1002/aenm.202102696
14. S. Thiede, Digital technologies, methods and tools towards sustainable manufacturing: Does industry 4.0 support to reach environmental targets? Procedia CIRP **98**, 1–6 (2021)
15. T.-C. Chen et al., Development of machine learning methods in hybrid energy storage systems in electric vehicles. Math. Probl. Eng. **2022**, 3693263 (2022)
16. T.T. Pham et al., Industry 4.0 to accelerate the circular economy: A case study of electric scooter sharing. Sustainability **11**(23) (2019)
17. M. Helmy et al., IoT-based battery monitoring system for electric vehicle. Int. J. Eng. Technol. **7**, 505–510 (2018)
18. A. Christensen, A. Adebusuyi, Using on-board electrochemical impedance spectroscopy in battery management systems, in *2013 World Electric Vehicle Symposium and Exhibition (EVS27)*, (2013), pp. 1–7
19. W. Li, M. Rentemeister, J. Badeda, D. Jöst, D. Schulte, D.U. Sauer, Digital twin for battery systems: Cloud battery management system with online state-of-charge and state-of-health estimation. J. Energy Storage **30**, 101557 (2020)
20. M.H.S.M. Haram, J.W. Lee, G. Ramasamy, E.E. Ngu, S.P. Thiagarajah, Y.H. Lee, Feasibility of utilising second life EV batteries: Applications, lifespan, economics, environmental impact, assessment, and challenges. Alex. Eng. J. **60**(5), 4517–4536 (2021)
21. L. Ahmadi, A. Yip, M. Fowler, S.B. Young, R.A. Fraser, Environmental feasibility of re-use of electric vehicle batteries. Sustain. Energy Technol. Assess. **6**, 64–74 (2014)
22. S. Li, P. Zhao, Big data driven vehicle battery management method: A novel cyber-physical system perspective. J. Energy Storage **33**, 102064 (2021)
23. L. Song, K. Zhang, T. Liang, X. Han, Y. Zhang, Intelligent state of health estimation for lithium-ion battery pack based on big data analysis. J. Energy Storage **32**, 101836 (2020)
24. A. Gairola, IoT based battery management system in electric vehicles, in *Bacancy Systems*, ed. by B. Technology, vol. 2022, (Bacancy Technology, Sweden, 2022)
25. M.R.A. Refaai, J.N. Bharothu, T.V.V.P. Kumar, C. Srinivas, M. Sudhakar, A. Bhowmick, Battery energy forecasting in electric vehicle using deep residual neural network. Int. J. Photoenergy **2022**, 5959443 (2022)
26. M.-K. Tran, S. Panchal, T.D. Khang, K. Panchal, R. Fraser, M. Fowler, Concept review of a cloud-based smart battery management system for lithium-ion batteries: Feasibility, logistics, and functionality. Batteries **8**(2) (2022)
27. S. Yang et al., Implementation for a cloud battery management system based on the CHAIN framework. Energy and AI **5**, 100088 (2021)
28. S. Kahveci, B. Alkan, M.A.H. Ahmad, B. Ahmad, R. Harrison, An end-to-end big data analytics platform for IoT-enabled smart factories: A case study of battery module assembly system for electric vehicles. J. Manuf. Syst. **63**, 214–223 (2022)
29. F. Ciucci, Modeling electrochemical impedance spectroscopy. Curr. Opin. Electrochem. **13**, 132–139 (2019)

30. B. Agudelo, W. Ospina, E.M. Zamboni, G. Spagnuolo, *Identification of Battery Circuit Model from EIS Data* (Presented at the JCGE - Congrès des Jeunes Chercheurs en Génie Electrique, Saint Pierre d'Oléron, France, 2019–06, 2019) Available: https://hal.archives-ouvertes.fr/hal-02915697
31. L. Lu, X. Han, J. Li, J. Hua, M. Ouyang, A review on the key issues for lithium-ion battery management in electric vehicles. J. Power Sources **226**, 272–288 (2013)
32. D. Andre, C. Appel, T. Soczka-Guth, D.U. Sauer, Advanced mathematical methods of SOC and SOH estimation for lithium-ion batteries. J. Power Sources **224**, 20–27 (2013)
33. U. Tröltzsch, O. Kanoun, H.-R. Tränkler, Characterizing aging effects of lithium ion batteries by impedance spectroscopy. Electrochim. Acta **51**(8), 1664–1672 (2006)
34. M.A. El-Shorbagy, A.M. El-Refaey, Hybridization of grasshopper optimization algorithm with genetic algorithm for solving system of non-linear equations. IEEE Access **8**, 220944–220961 (2020)
35. C. Pastor-Fernández, W.D. Widanage, J. Marco, M.Á. Gama-Valdez, G.H. Chouchelamane, Identification and quantification of ageing mechanisms in Lithium-ion batteries using the EIS technique, in *2016 IEEE Transportation Electrification Conference and Expo (ITEC)*, (2016), pp. 1–6
36. X. Han, M. Ouyang, L. Lu, J. Li, Y. Zheng, Z. Li, A comparative study of commercial lithium ion battery cycle life in electrical vehicle: Aging mechanism identification. J. Power Sources **251**, 38–54 (2014)
37. D.A. Howey, P.D. Mitcheson, V. Yufit, G.J. Offer, N.P. Brandon, Online measurement of battery impedance using motor controller excitation. IEEE Trans. Veh. Technol. **63**(6), 2557–2566 (2014)
38. B. Wimarshana, I. Bin-Mat-Arishad, A. Fly, Parameter sensitivity analysis of a physico-chemical lithium-ion battery model with combined discharge voltage and electrochemical impedance data. J. Power Sources **527**, 231125 (2022)
39. A. Maghawry, R. Hodhod, Y. Omar, M. Kholief, An approach for optimizing multiobjective problems using hybrid genetic algorithms. Soft. Comput. **25**(1), 389–405 (2021)
40. S. Katoch, S.S. Chauhan, V. Kumar, A review on genetic algorithm: Past, present, and future. Multimed. Tools Appl. **80**(5), 8091–8126 (2021)
41. G. Kannan, P. Sasikumar, K. Devika, A genetic algorithm approach for solving a closed loop supply chain model: A case of battery recycling. Appl. Math. Model. **34**(3), 655–670 (2010)
42. M.R.A. Refaai, S.N.R. Vonteddu, P.K. Nunna, P.S. Kumar, C. Anbu, M. Markos, Energy management prediction in hybrid PV-battery systems using deep learning architecture. Int. J. Photoenergy **2022**, 6844853 (2022)
43. W. Li, L. Gao, A. Garg, A framework based on big data for intelligent monitoring of battery packs. IOP Conf. Series Earth Environ. Sci. **463**(1), 012158 (2020)
44. A.C. Ngandjong et al., Investigating electrode calendering and its impact on electrochemical performance by means of a new discrete element method model: Towards a digital twin of Li-ion battery manufacturing. J. Power Sources **485**, 229320 (2021)
45. Y. Zhang, Q. Tang, Y. Zhang, J. Wang, U. Stimming, A.A. Lee, Identifying degradation patterns of lithium ion batteries from impedance spectroscopy using machine learning. Nat. Commun. **11**(1), 1706 (2020)
46. Q. Yang, J. Xu, B. Cao, X. Li, A simplified fractional order impedance model and parameter identification method for lithium-ion batteries. PLoS One **12**, e0172424 (2017)
47. Z. Deng, Z. Zhang, Y. Lai, J. Liu, J. Li, Y. Liu, Electrochemical impedance spectroscopy study of a lithium/Sulfur battery: Modeling and analysis of capacity fading. J. Electrochem. Soc. **160**(4), A553–A558 (2013)
48. D. Andre, M. Meiler, K. Steiner, H. Walz, T. Soczka-Guth, D.U. Sauer, Characterization of high-power lithium-ion batteries by electrochemical impedance spectroscopy. II: Modelling. J. Power Sources **196**(12), 5349–5356 (2011)
49. D. Andre, M. Meiler, K. Steiner, C. Wimmer, T. Soczka-Guth, D.U. Sauer, Characterization of high-power lithium-ion batteries by electrochemical impedance spectroscopy. I. Experimental investigation. J. Power Sources **196**(12), 5334–5341 (2011)

50. M.R. Shoar Abouzari, F. Berkemeier, G. Schmitz, D. Wilmer, On the physical interpretation of constant phase elements. Solid State Ionics **180**(14), 922–927 (2009)
51. D. Geringer, P. Hofmann, J. Girard, E. Trunner, W. Knefel, Aging investigations and consideration for automotive high power lithium-ion batteries in a 48 V mild hybrid operating strategy. Automotive and Engine Technology **6**(3), 219–234 (2021)
52. E. Redondo-Iglesias, P. Venet, S. Pelissier, Calendar and cycling ageing combination of batteries in electric vehicles. Microelectron. Reliab. **88-90**, 1212–1215 (2018)
53. W. Choi, H.-C. Shin, J.M. Kim, J.-Y. Choi, W.-S. Yoon, Modeling and applications of Electrochemical Impedance Spectroscopy (EIS) for Lithium-ion batteries. J. Electrochem. Sci. Technol **11**(1), 1–13 (2020)
54. C. Lyu, T. Zhang, W. Luo, G. Wei, B. Ma, L. Wang, SOH estimation of Lithium-ion batteries based on fast time domain impedance spectroscopy, in *2019 14th IEEE Conference on Industrial Electronics and Applications (ICIEA)*, (2019), pp. 2142–2147
55. N. Somakettarin, A. Pichetjamroen, A study on modeling of effective series resistance for Lithium-ion batteries under life cycle consideration. IOP Conf. Series Earth Environ. Sci. **322**(1), 012008 (2019)
56. M. Grossi, C. Parolin, B. Vitali, B. Riccò, Electrical Impedance Spectroscopy (EIS) characterization of saline solutions with a low-cost portable measurement system. Eng. Sci. Technol. Int. J. **22**(1), 102–108 (2019)
57. J.D. Muñoz, V.H. Mosquera, C.F. Rengifo, A low-cost, portable, two-dimensional bioimpedance distribution estimation system based on the AD5933 impedance converter. HardwareX **11**, e00274 (2022)
58. D.A. Stone, M.P. Foster, E. Ballantyne, H. Price, A low cost, rapid impedance measurement technique suitable for Li-ion health diagnosis in battery energy storage systems. Int. J. Smart Grid Clean Energy **9** (2020)
59. M. Gaberšček, Understanding Li-based battery materials via electrochemical impedance spectroscopy. Nat. Commun. **12**(1), 6513 (2021)
60. J. Hu, X. Xu, D. Cao, T. Zheng, and W. Hao, *Research and Application of Electric Vehicle Battery Swapping Technology*. Presented at the 2021 2nd International Conference on Artificial Intelligence and Information Systems, Chongqing, China (2021). Available: https://doi.org/10.1145/3469213.3470421

Performance Analysis of Eight-Channel WDM Optical Network with Different Optical Amplifiers for Industry 4.0

Anuj Kumar Gupta, Raju Sharma, Digvijay Pandey, Vinay Kumar Nassa, Binay Kumar Pandey, A. Shaji George, and Pankaj Dadheech

1 Introduction

Wavelength division multiplexing is a technology that increases bandwidth by transmitting various optical signals over a single optical fiber. It works on the principle as frequency division multiplexing. In WDM, various signals are transmitted by using different frequencies known as carrier, and signals at different wavelengths are independent of each other. WDM used wavelengths 1300 nm and 1550 nm at which optical fibers have very low signal loss [1, 9]. In optical communication, information is sent by simple pulses of light. To represent 1 s and 0 s, light was

A. K. Gupta
Chandigarh Group of Colleges, Mohali, Punjab, India

R. Sharma
Baba Banda Singh Bahadur Engineering College, Fatehgarh Sahib, Punjab, India

D. Pandey (✉)
Department of Technical Education, Kanpur, Uttar Pradesh, India

V. K. Nassa
Department Rajarambapu Institute of Technology, Sangli, India

B. K. Pandey
Department of Information Technology, College of Technology, Govind Ballabh Pant University of Agriculture and Technology, Pantnagar, Uttarakhand, India

A. S. George
Department of Information and Communication Technology, Crown University Int'l. Chartered Inc. (CUICI), Santa Cruz, South America, Argentina

P. Dadheech
Department of Computer Science & Engineering (NBA Accredited), Swami Keshvanand Institute of Technology, Management & Gramothan (SKIT), Jaipur, Rajasthan, India
e-mail: pankaj@skit.ac.in

L. C. Méndez-González et al. (eds.), *Innovation and Competitiveness in Industry 4.0 Based on Intelligent Systems*, EAI/Springer Innovations in Communication and Computing, https://doi.org/10.1007/978-3-031-29775-5_9

flashed on and off, respectively. The actual light could be of any wavelength in the range from 670 nm to 1550 nm. WDM is a technique in optical communication that uses multiple light wavelengths to send information over the single fiber [2, 3, 11]. There are two types of WDM network in optical communication:

Coarse WDM (CWDM): CWDM is defined by the WDM systems that operated with eight channels per fiber with spacing between the channels 20 nm to provide room for wavelength drift when lasers heat up and cool down during its working. As compared to DWDM, energy consumption is less in CWDM, and it is compact and cost effective. It is used for short distance communication.

Dense WDM (DWDM): In DWDM, the number of channels is much larger as compared to the CWDM. It has 40 channels at 100GHz spacing and 80 with 50 GHz spacing. Due to this large amount of data, it can be sent through a single fiber link. In other words, DWDM is a system with more than eight active wavelengths per fiber. It is also finding its applications in cloud data centers for their IaaS services.

1.1 Advantages of WDM Optical Networks [4]

1.1.1 Large Capacity

Optical fiber provides large transmission bandwidth but still its utilization rate is low. To increase the utilization, WDM technology is used that increases the transmission capacity and makes full use of large bandwidth provided by optical fiber. DWDM networks use up to 96 channels that can be multiplexed into a composite optical signal that send through a single optical fiber pair. If networks multiplex 96 channels of 100GBPS onto one fiber, the capacity of the network will increase from 100Gbps to 9.6Tbps. WDM technique increases the capacity of the network.

1.1.2 Transparent Transmission of Data

Optical networks are of layer 1; they perform multiplexing and demultiplexing of signals according to their wavelengths of light. Because it works on the wavelength, they are transparent to the transmission of the data regardless of the electrical modulation mode and signal's data rate [4, 10]. CWDM and DWDM techniques used in WDM technology take data from different light sources and put together on a single fiber link with each signal carried on its separate wavelength at the same time. Channels used in CWDM and DWDM are physically separated and do not interfere with each other. Different data formats are used in each channel due to the properties of light, and each can be transmitted at different data rates. It allows channels to hold varying services such as Video, SONET, and Gigabit Ethernet over a single fiber pair.

1.1.3 Simplified Operations

WDM networks operate in very simple ways nowadays. To improve the operational efficiency, they offer integrated management solutions. To provide simple, fast, and dynamic provisioning of network connections, DWDM networks are combined with ROADM technology. It gives the ability to the organizations to provide high bandwidth services in days rather than months.

1.1.4 Flexibility

Optical networks use multiple topology schemes such as P2P, Linear, and Ring for deployment. DWDM ROADM platforms use mesh topologies for deployment and can be implemented with various schemes to provide reliability and protection mechanisms.

1.1.5 Overcome Distance Limitations

Optical networks are capable to transmit large capacity data over long distances with DWDM systems. It supports transmission link beyond 120 km and thousands of kilometers with the use of optical amplifiers and signal regeneration.

1.1.6 Maximize Dark Fiber Utilization

Unlimited transmission capacity is the primary advantage of the WDM optical networks in both technical and economic perspectives.

1.1.7 Scalability

DWDM and CWDM systems that are easily expanded because of single fiber can be divided into many channels. When the demand changes, its capacity can be increased by either adding additional equipment or by increasing the number of wavelengths used on the fiber without expensive upgrades.

Figure 1 shows the WDM transmission system. At transmitter site, multiple optical light sources are transmitters emitting light at different wavelengths. Individually, each transmitter sends its different wavelength signals, and these signals are multiplexed by a wavelength multiplexer (WDM mux). The multiplexed signals are then transmitted to the receiver by using single fiber line. At receiver section, a demultiplexer is used to demultiplex the signal and transmitted to multiple receivers. At the receiver side, this signal is demultiplexed by the wavelength demultiplexer and sent to multiple receivers.

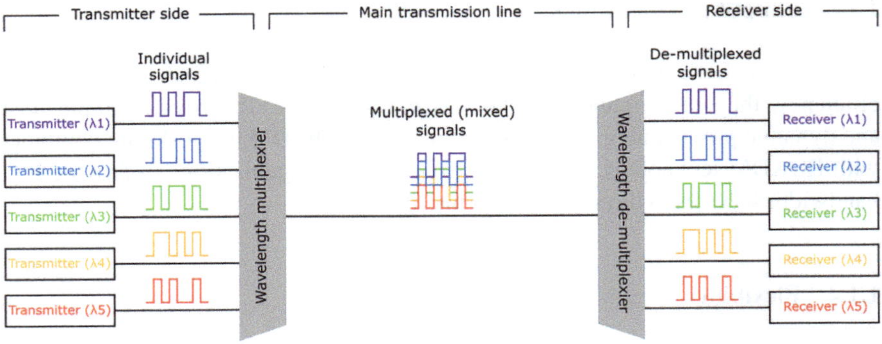

Fig. 1 Schematic of WDM transmission system

2 Optical Amplifiers

Optical amplifiers allow the transmission of many data over distances from 100 km to longer distances in optical communication networks [5]. Along with telecommunication systems, they provide the data capacity that are required for current and future communication networks. Optical amplifiers are used as a repeater circuit in optical networks to provide long-distance communication. Optical amplifiers amplify an optical signal directly without converting the signal from optical to electrical. An optical communication link consists of a transmitter, a fiber cable, and a receiver. Optical fiber cable provides less absorption and energy consumption to the propagating signals. Still the media have a limit in distance of 140 km at which the signals can propagate before producing disturbances [1]. Optical amplifiers are used to regenerate the optical signals at every 80–140 km electronically to provide the long-distance communication. Optical amplifiers are of four types [2, 3]: fiber amplifies, Raman amplifiers (RA), erbium-doped fiber amplifiers, and solid-state optical amplifiers.

2.1 Raman Amplifiers (RA)

A Raman amplifier is an optical amplifier that is based on the stimulated Raman scattering (SRS) phenomenon. In Raman amplifier, signal is amplified due to this phenomenon [5]. Raman amplifier is divided into two types: distributed types (DRA) and lumped types (LRA). IT amplify the signal band that is not covered by EDFA. Distributed Raman amplifier has high gain as compared to lumped Raman amplifier (Fig. 2).

When strong excitation light enters the optical fiber, it causes excited emission based on SRS. The light signal is amplified. Wavelength of amplified signal is approximately 100 nm that is greater than the excitation light wavelength. By the

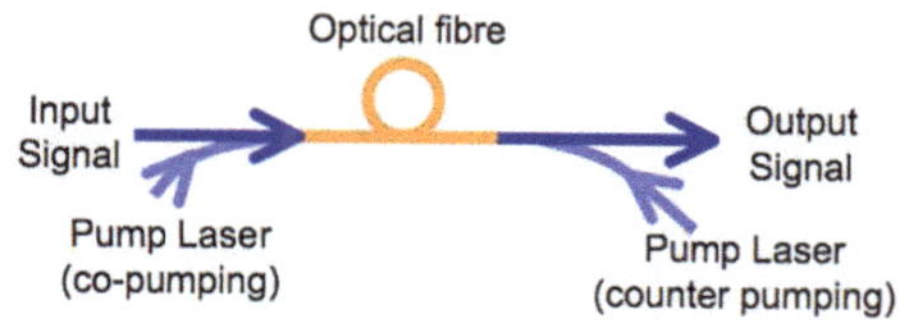

Fig. 2 Configuration of Raman amplifier

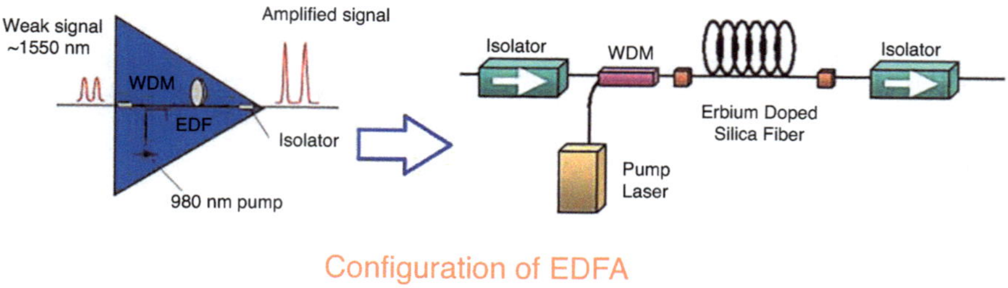

Fig. 3 Configuration of EDFA

wavelength of the excitation light, the wavelength of the amplified signal can be determined. There are certain advantages and disadvantages of Raman amplifier. The advantages are: amplification of variable wavelength is possible; with installed SM fiber, Raman amplifier is compatible; broadband operation is possible; the structure of the amplifier is simple; and it has low noise and gain wavelength flexibility. The disadvantages are: Raman amplifier needs high pump power, noise is a major issue, and gain control is also needed in the amplifier.

2.2 *Erbium-Doped Fiber Amplifiers*

Erbium-doped fiber amplifier (EDFA) is made of erbium-doped fiber (EDF), optical couplers, pump light source, optical isolators, optical filters, and other components [6]. They are widely used amplifiers in optical communications (Fig. 3).

Optical pump laser of selective wavelength is used to excite the erbium ions from where they can amplify light and reach high energy states and stay at this state (980 nm or 1450 nm). When they return to low-energy state called the geo-energy state. There are various advantages and disadvantages of EDFA. The advantages are: EDFA has gain in excess of 40–50 dB, high pump power utilization (>50%), and low noise figure 3–5 dB suitable for long haul applications, and EDFA provides high gain with small cross talk. EDFA amplify wide wavelength band in the range of 1530 nm–1560 nm by gain-flattening optical fibers, and they provide high-energy conversion efficiency. The disadvantages are: size of EDFA is large, not possible to integrate EDFA with other semiconductor devices, EDFA has fixed gain range and gain up flatness, and it needs high power pump consumption.

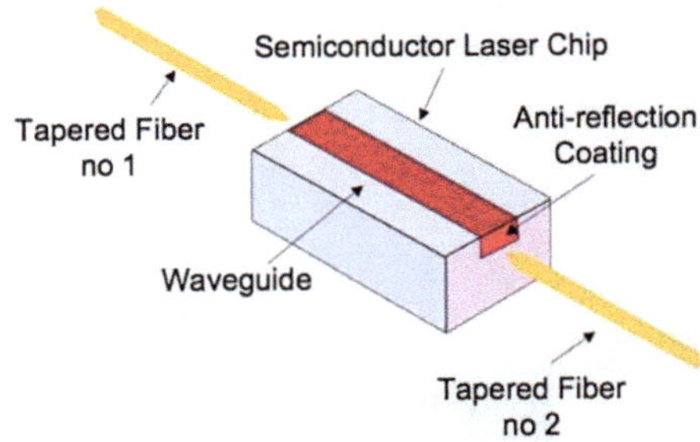

Fig. 4 Geometry of EDFA

Table 1 Comparison of optical amplifiers [8]

Characteristic	SOA	Raman	EDFA
Gain (dB)	>30	>25	>40
Bandwidth (3 dB0	60	Pump dependent	30–60
Wavelength (nm)	1280–1650	1280–1650	1530–1560
Noise figure (dB)	8	5	5
Pump power	<400 nm	>30 dBm	25 dBm
Size	Compact	Bulk module	Rack mounted
Cost factor	Low	High	Medium
Max. Saturation (dBm)	18	0.75*pump	22
Switchable	Yes	No	No
Polarization sensitivity	Yes	No	No

2.3 Semiconductor Optical Amplifiers

Semiconductor optical amplifier uses a semiconductor element to provide gain medium. It does not require any external pump laser because it pumped electronically [7]. Running cost of SOA amplifier is low as compared to EDFA amplifier, and it builds in compact size (Fig. 4).

The advantages of SOA are: the size of SOA amplifier is compact, it is electronically pumped, it is economically cheaper than EDFA, it can be integrated with semiconductor devices like lasers and modulators, and it needs low power laser for its operation. The disadvantages are: it has lower gain, has high noise, and has linearity with fast transient time (Table 1).

3 Industry 4.0

Industry 4.0 is basically the digital transformation of production and manufacturing industries with the help of modern smart technology and industrial value change control. It represents an organizational stage where Internet of Things (IoT) and machine-to-machine communication at large scale are integrated to increase the automation and improve self-monitoring and communication [12, 13]. Industry 4.0 enables the production of smart machines which help in the analysis and diagnosis issues without the use of any human intervention. It is the intelligent networking of industrial process and machines with the help of communication and information technology. The basis of Industry 4.0 is the cyber physical systems which use embedded software systems in the modern control systems which can be connected and addressed through IoT. In this way, there is a networking of products which enables new ways of value creation, production, and real-time optimization. It is characterized by creating a bridge between the physical world and the digital world through cyber physical systems that are enabled by IoT. It is the manufacturing transformation that is information intensive in a connected environment of cybersecurity, augmented reality, big data, people, autonomous robots, additive manufacturing, simulation, system integration, cloud computing, and IoT. IoT indeed plays a key role in Industry 4.0 scope from its IoT gateways to IoT stack components to IoT platforms, devices, and much more [14]. Yet it is not just the course of IoT; big data, cloud computing, data analysis, artificial intelligence, network technologies, mobile and data communication, enterprise resource planning (ERP), manufacturing executive systems (MES), servers and actuators, programmable logic controllers (PLA), data exchange models, and edge computing (which store and compute power at the edge of network) all play a key role [15–17]. There are certain goals and design principles which are integral to Industry 4.0. The first one is interconnection. It is the ability of people, sensors, devices, and machines to connect with each other and communicate through IoT. The second one is information transparency which provides comprehensive information to the operators that allows them to make decisions and collect large amount of information and data from the manufacturing process points which identify the key areas to improve and increase the functionality. The third one is technical assistance which is the technological facility provided by the systems that assist humans in problem-solving and decision-making. The fourth one is decentralized decisions, which allow the cyber physical systems to make their own decisions and autonomously prefer the task [18–22] (Fig. 5).

3.1 Optical Networks for Industry 4.0

The exact potential of Industry 4.0 is unleashed by some fundamental prerequisites such as providing real-time communication between robots, humans, products, and factory logistics. The telecom network has many challenging requirements, such as large capacity, dense connectivity, various mobility needs, and low latency, which are met by 5G network and its upcoming evolutions. Network slicing and distributed edge cloud are the concepts that serve profiles of different traffics in same network infrastructure [23]. Optical networks also play a very important role in fulfilling the requirements of latency, bandwidth, and radio access network (RAN) reliability. Vertical industries like logistics systems (warehouses, maritime ports, airports, etc.)

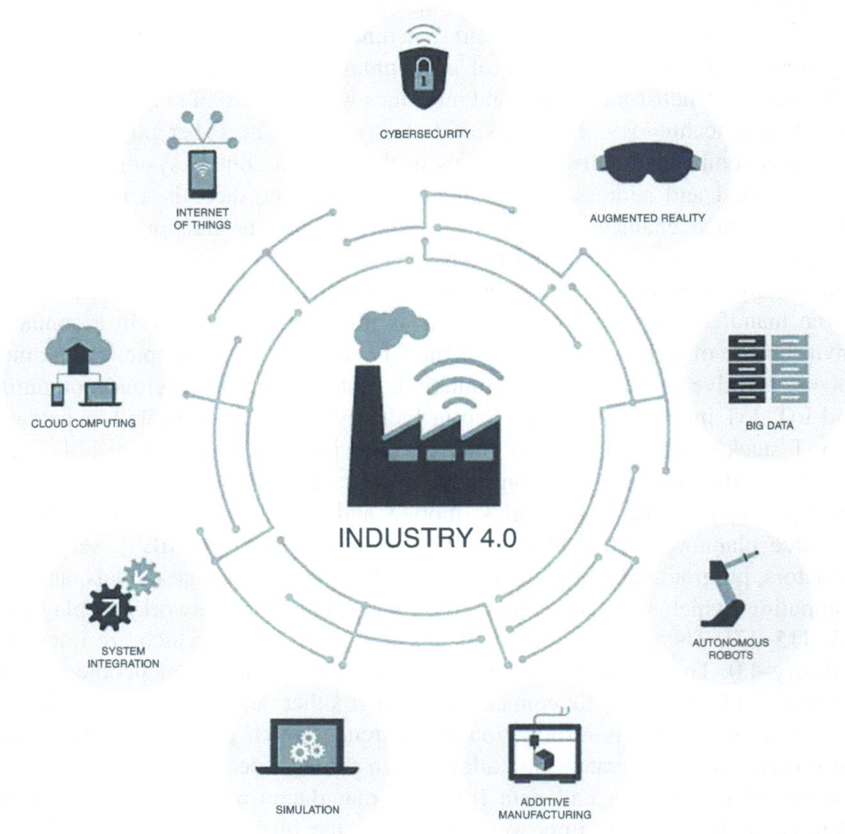

Fig. 5 Digital transformation of Industry 4.0

and manufacturing plants will require telecom architectures of specific form which rely on the different forms of optical network. The most important driving force in the evolution of cellular radio networks is Industry 4.0, which spans all the industrial sectors. The transport network is very crucial to ensure the performance level of network which are required to serve various vertical services. Therefore, a reliable, intelligent, and automated coordination is required between the transport networks and the radio to ensure robustness and meet the requirements of operators in multiple use cases. The industrial applications have so much wide range that they mandate the transport segment to cope with various challenges such as switching or routing capacity, huge transmission, high flexibility, low energy consumption, and latency control which enables the network to react with the changes automatically [24–26]. Optical network is the basis to realize the transport infrastructure for Industry 4.0.

4 Simulation Model

The model and its simulation are done in Optisystem environment. Figures 6, 7, and 8 show the eight-channel WDM transmission system with EDFA, SOA, and Raman amplifier, respectively. In transmitter section to generate data sequence, user-defined bit sequence generator is used to generate a data rate of 2.5Gbps, for encoding the data sequence generated by sequence generator, NRZ pulse generator is used, CW laser sources are used to generate light signals at wavelengths of 193.1–193.8THz. For modulation, Mach-Zehnder modulator is used to modulate the intensity light in accordance with the sequence generated by the NRZ pulse generator. WDM multiplexer is used to combine the signals from modulators and sent this signal through optical fiber to the WDM demultiplexer used at receiver side.

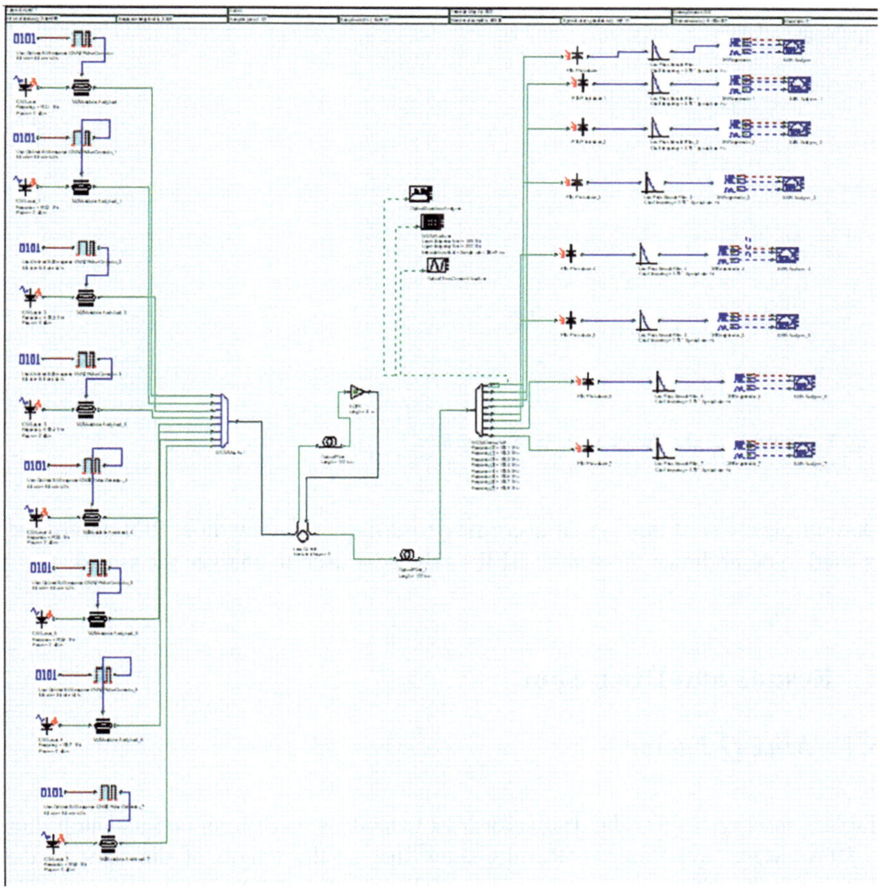

Fig. 6 WDM optical network with EDFA amplifier

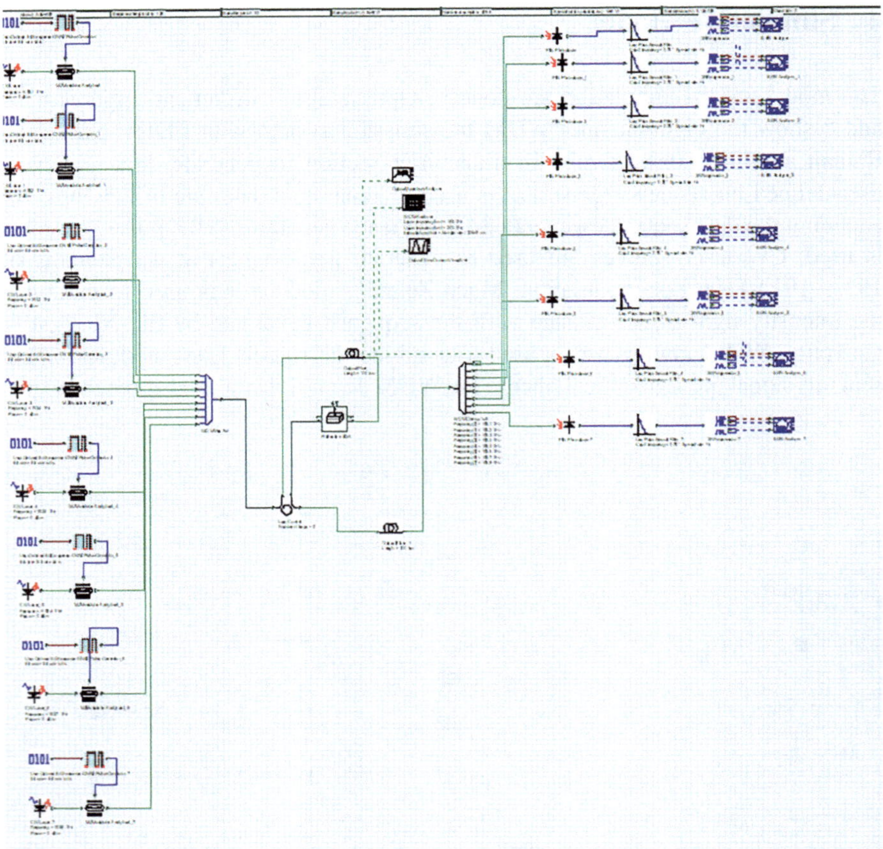

Fig. 7 WDM optical network with SOA amplifier

Demultiplexer split this signal according to different bandwidths. PIN photodiode is used to demodulate the signal. BER analyzer is used to analyze the signal.

5 Results and Discussion

5.1 Max Q Factor

Table 2 and Fig. 9 show the results for Max Q factor with different optical amplifiers (EDFA, SOA, and Raman). Results show that as the length of the optical fiber increases, Max Q factor decreases. It is also clear that WDM optical network with SOA in dynamic mode performs better than EDFA and Raman. Difference between EDFA and SOA dynamic mode is small, but the difference between these two and

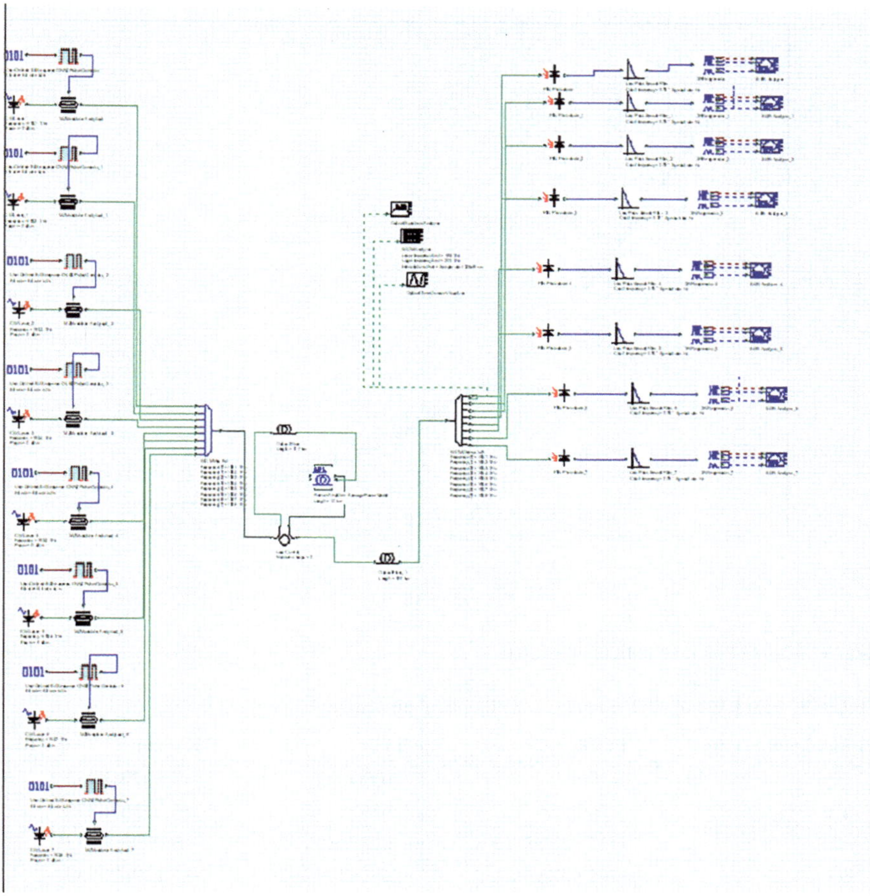

Fig. 8 WDM optical network with Raman amplifier

Table 2 Max Q factor of optical amplifiers

Fiber length (km)	EDFA	SOA	Raman
20	97.81817606	102.815224	5.388826444
40	47.13568789	48.59735539	4.776323049
60	22.44312522	24.23911871	4.656323049
80	9.666773211	11.11503119	4.331206361
100	4.052278547	4.579309392	2.648966201

Raman amplifier is large. From results, it is concluded that Max Q factor decreases with increase in transmission length and WDM optical network with EDFA and SOA amplifiers performed better than Raman amplifier.

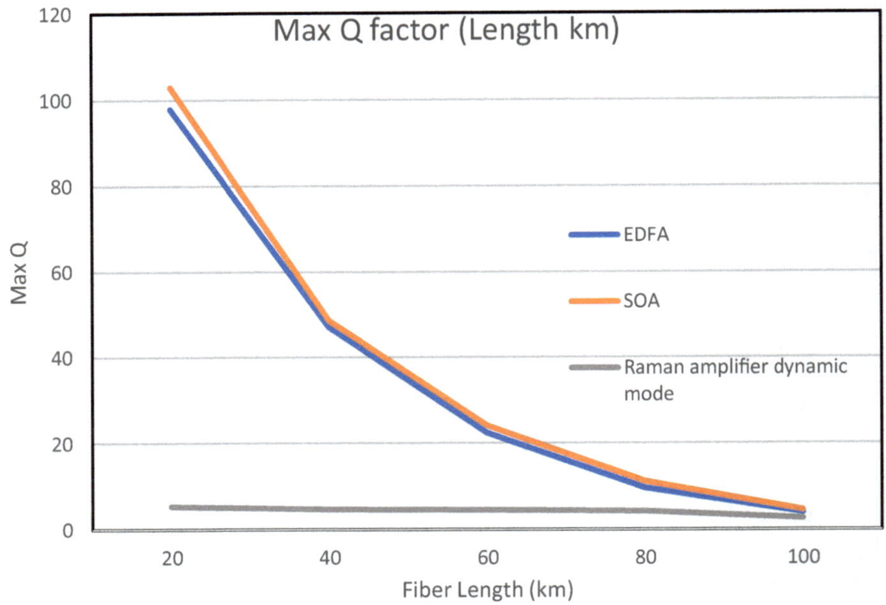

Fig. 9 Max Q factor of optical amplifiers

Table 3 Min BER of WDM network with optical amplifiers

Fiber length (km)	EDFA	SOA	Raman
20	−1000	−1000	−7.507918955
40	−1000	−1000	−6.091696077
60	−111.1267745	−129.3665792	−6.000696077
80	−21.68288936	−28.27694405	−5.130808753
100	−4.59901314	−5.632854923	−2.419596066

5.2 *Min BER*

Table 3 and Fig. 10 show the results for Min BER of WDM network with different optical amplifiers (EDFA, SOA, and Raman). Results show that as the length of the optical fiber increases, Min BER increases. It is also clear that WDM optical network with Raman amplifier performs better than EDFA and SOA dynamic mode. Difference between EDFA and SOA dynamic mode is small, but the difference between these two and Raman amplifier is large. From results, it is concluded that Min BER increases with increase in transmission length and WDM optical network with Raman amplifier performed better than EDFA and SOA dynamic mode amplifier.

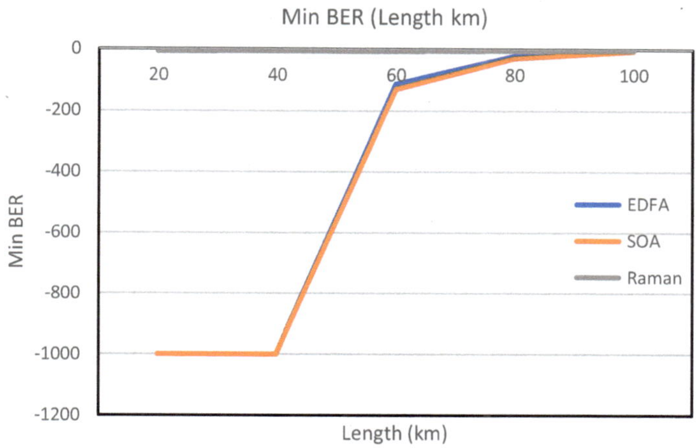

Fig. 10 Min BER of WDM network with optical amplifiers

Table 4 Eye Height of WDM network with optical amplifiers

Fiber length (km)	EDFA	SOA	Raman
20	0.000384895	0.00038575	0.000127596
40	0.000147647	0.00014727	4.09E-05
60	5.45E-05	5.45994E-05	5.09295E-05
80	1.72E-05	1.83383E-05	6.06E-06
100	2.45E-06	1.58E-06	−1.09E-06

5.3 *Eye Height*

Table 4 and Fig. 11 show the results for Eye Height of WDM network with different optical amplifiers (EDFA, SOA, and Raman). Results show that as the length of the optical fiber increases, Eye Height decreases; it happens because of distortion, which increases with the transmission distance. It is also clear that WDM optical network EDFA and SOA dynamic mode performed better than Raman amplifier. Difference between EDFA and SOA dynamic mode is small, but the difference between these two and Raman amplifier is large. From results, it is concluded that Eye Height decreases with increase in transmission length and WDM optical network with EDFA and SOA dynamic mode performed better than Raman amplifier.

5.4 *OSNR*

Table 5 and Fig. 12 show the results for OSNR of WDM network with different optical amplifiers (EDFA, SOA, and Raman). Results show that as the length of

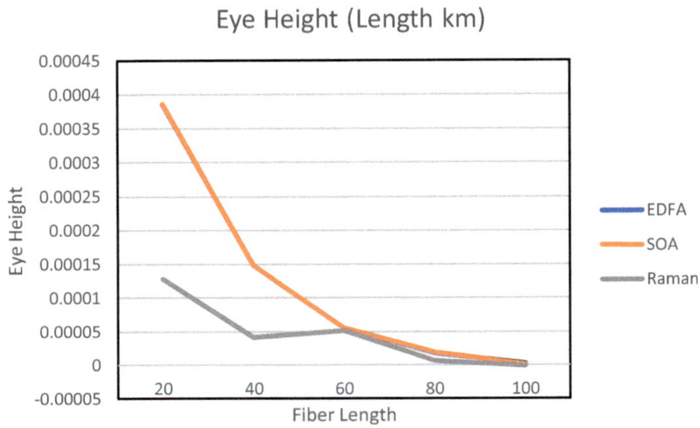

Fig. 11 Eye Height of WDM network with optical amplifiers

Table 5 OSNR of WDM network with optical amplifiers

Fiber length (km)	EDFA	SOA	Raman
20	90.10602332	89.9954796	92.63648793
40	89.3284952	89.36544861	88.63825459
60	85.33019765	85.37592164	84.62516434
80	81.32826416	81.3618346	80.64718494
100	77.32920641	77.36651639	76.63980469

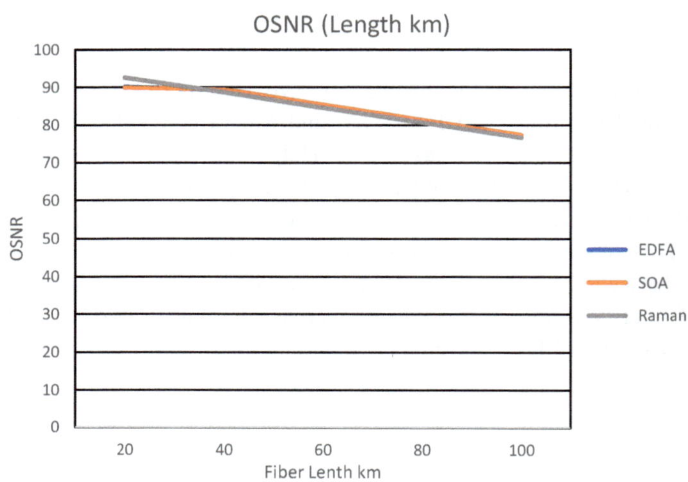

Fig. 12 OSNR of WDM network with optical amplifiers

the optical fiber increases, eye OSNR decreases. It is also clear that WDM optical network EDFA and SOA dynamic mode performed better than Raman amplifier. At 20 km, OSNR of WDM optical network with Raman amplifier is more as compared to EDFA and SOA, but when the fiber length increases, EDFA and SOA performed better than Raman. From results, it is concluded that OSNR decreases with increase in transmission length and WDM optical network with EDFA and SOA dynamic mode performed better than Raman amplifier.

6 Conclusion

For Industry 4.0, optical networks also play a crucial role in the form of latency, bandwidth, and radio access network (RAN) reliability for industries like maritime ports and airports. Manufacturing plants will also require telecom architectures of specific form which rely on the different forms of optical network. Industry 4.0 with optical networks incorporates better and faster networks. This chapter analyzes the performance of eight-channel WDM optical network with different optical amplifiers (EDFA, SOA dynamic mode, and Raman amplifier) in software environment. This chapter presents three configurations of WDM optical networks. The first configuration is with EDFA amplifier, the second configuration is with SOA amplifier, and the third configuration is with Raman amplifier. These networks are simulated using Optisystem 18 by varying optical fiber length from 20 km to 180 km. Three configurations are compared using Max Q factor, Min BER, Eye Height, and OSNR parameters. After simulation, it is seen that Max Q factor, Eye Height, and OSNR decrease with increase in optical fiber length. Network with EDFA and SOA amplifiers performed better than Raman amplifier. Min BER increases with increase in optical fiber length. Network with Raman amplifier performed better than EDFA and SOA amplifiers.

References

1. Finisar White Paper, *Introduction to Optical Amplifiers* (Finisar Corporation, Sunnyvale, CA, 2012)
2. M. PreethiAnushiya, V. Sasirekha, S. Sumitha, A.T. Banu, M., S. Geerthana, Performance analysis of 8-channel & 16-channel optical Fiber using WDM system. Int. J. Eng. Rese. Technol. (IJERT) Iconnect **5**(13) (2017, 2017)
3. S. Bhalaik, A. Sharma, R. Kumar, N. Sharma, Performance Modeling and analysis of WDM optical networks under wavelength continuity constraint using MILP. Recent Adv. Electr. Electron. Eng. **13**(2) (2020). https://doi.org/10.2174/2352096512666190214105927
4. M. Sharma, D. Pandey, D. Khosla, S. Goyal, B.K. Pandey, A.K. Gupta, Design of a GaN-based Flip Chip Light Emitting Diode (FC-LED) with au Bumps & Thermal Analysis with different sizes and adhesive materials for performance considerations. SILICON, 1–12 (2021)
5. M.S. Bhamrah, A. Atieh, SOA/EDFA/RAMAN optical amplifier for DWDM systems at the edge of L & U wavelength bands. Opt. Fiber Technol. **52**, 101971 (2019)

6. N.N.H. Saris, A. Hamzah, S. Ambran, Investigation on gain improvement of Erbium Doped Fiber Amplifier (EDFA) by using dual pumped double pass scheme. J. Adv. Res. Appl. Sci. Eng. Technol. **7**(1) (2017)
7. S.P. Kaur, M. Sharma, Radially optimized zone-divided energy-aware wireless sensor networks (WSN) protocol using BA (bat algorithm). IETE J. Res. **61**(2), 170–179 (2015)
8. T. Ivaniga, P. Ivaniga, Comparison of the optical amplifiers EDFA and SOA based on the BER and Q-factor in C-band. Adv. Opt. Technol. (2017)
9. R. Sharma, H.S. Ryait, A.K. Gupta, Study of Zigbee protocol using OPNET. Int. J. Comput. Appl. **1**, 23–29 (2016)
10. R. Sharma, H.S. Ryait, A.K. Gupta, Wireless body area network a review. Res. Cell: Int. J. Eng. Sci. **17**, 494–499 (2016)
11. R. Sharma, H.S. Ryait, A.K. Gupta, Analysing the effect of posture mobility and sink node placement on the performance of routing protocols in WBAN. Indian J. Sci. Technol. **9**, 40 (2016)
12. M. Sharma, S. Singh, S.G. DishantKhosla, A. Gupta, Waveguide diplexer: Design and analysis for 5G communication, in *2018 Fifth International Conference on Parallel, Distributed and Grid Computing (PDGC)*, (IEEE, 2018), pp. 586–590
13. C. Zhang, Y. Chen, A review of research relevant to the emerging industry trends: Industry 4.0, IoT, blockchain, and business analytics. Jo. Ind. Integr. Management **5**(01), 165–180 (2020)
14. S. Aheleroff, Y.L. XunXu, M. Aristizabal, J.P. Velásquez, B. Joa, Y. Valencia, IoT-enabled smart appliances under industry 4.0: A case study. Adv. Eng. Inform. **43**, 101043 (2020)
15. M. Sharma, B. Sharma, A.K. Gupta, B.S. Singla, Design of 7 GHz microstrip patch antenna for satellite IoT-and IoE-based devices, in *The International Conference on Recent Innovations in Computing*, (Springer, Singapore, 2020), pp. 627–637
16. L.S. Dalenogare, G.B. Benitez, N.F. Ayala, A.G. Frank, The expected contribution of industry 4.0 technologies for industrial performance. Int. J. Prod. Econ. **204**, 383–394 (2018)
17. A.G. Frank, L.S. Dalenogare, N.F. Ayala, Industry 4.0 technologies: Implementation patterns in manufacturing companies. Int. J. Prod. Econ. **210**, 15–26 (2019)
18. M. Sharma, H. Singh, Contactless methods for respiration monitoring and design of SIW-LWA for real-time respiratory rate monitoring. IETE J. Res., 1–11 (2022)
19. A.K. Gupta, M. Sharma, A. Sharma, VikasMenon., A study on SARS-CoV-2 (COVID-19) and machine learning based approach to detect COVID-19 through X-ray images. Int. J. Image Graph., 2140010 (2020)
20. M. Ghobakhloo, Industry 4.0, digitization, and opportunities for sustainability. J. Clean. Prod. **252**, 119869 (2020)
21. M. Sharma, D. Khosla, D. Pandey, S. Goyal, A.K. Gupta, B.K. Pandey, Design of a GaN-based Flip Chip Light Emitting Diode (FC-LED) with au Bumps & Thermal Analysis with different sizes and adhesive materials for performance considerations. SILICON, 1–12 (2021)
22. R. Sabella, P. Iovanna, G. Bottari, F. Cavaliere, Optical transport for industry 4.0. J. Opt. Commun. Netw. **12**(8), 264–276 (2020)
23. M. Sharma, H. Singh, Substrate integrated waveguide based leaky wave antenna for high frequency applications and IoT. Int. J. Sens. Wirel. Commun. Control **11**(1), 5–13 (2021)
24. M. Wollschlaeger, T. Sauter, J. Jasperneite, The future of industrial communication: Automation networks in the era of the internet of things and industry 4.0. IEEE Ind. Electron. Mag. **11**(1), 17–27 (2017)
25. M. Sharma, D. Pandey, P. Palta, et al., Design and power dissipation consideration of PFAL CMOS V/S conventional CMOS based 2:1 multiplexer and full adder. SILICON (2021). https://doi.org/10.1007/s12633-021-01221-1
26. M. Faheem, S.B.H. Shah, R.A. Butt, B. Raza, M. Anwar, M.W. Ashraf, M.A. Ngadi, V.C. Gungor, Smart grid communication and information technologies in the perspective of industry 4.0: Opportunities and challenges. Comput. Sci. Rev. **30**, 1–30 (2018)

Part III
Soft Computing Application in the Industry 4.0

Traffic Signs Configuration with a Geo-simulation Approach

Ariadna C. Moreno Román and Mailyn Moreno Espino

1 Introduction

Reduction of traffic congestion and accidents is a goal pursued by governments around the world. To ensure so, it is necessary, among other factors, to achieve a correct configuration and synchronization of the signals located on the roads [1].

In 2015, the traffic circle of the Ciudad Deportiva, Havana, Cuba, suffered a case of poor signs configuration that caused a collapse in the roads for over 25 min, with vehicles stopped on the railroad level crossing and obstructing the access to a clinical-surgical hospital. A proper analysis prior to this signal installation would have helped to avoid economic losses and material costs resulting from a counterproductive signal installation [2, 3]. This situation serves as motivation to start researching on traffic modeling to reduce congestion and accidents.

Although signal configuration is the main element to reduce congestion and accidents, there is another group of factors that influence traffic and need to be modeled. Vehicles, human characteristics, and relations between traffic actors may be potential causes of undesirable traffic situations [4]. The decisions of drivers and pedestrians on the road are difficult to model because they respond to various personal and psychological elements [5]. Agent-based simulation is widely used to mimic scenarios where social behavior is present. It will serve to model events such as obedience to signals, inter-vehicle relationships, and response to weather or other external events [6].

Additionally, weather and area conditions also affect the performance of signals and influence human behaviors [7]. These data in models must be reliable to ensure

A. C. M. Román · M. M. Espino (✉)
Facultad de Ingeniería Informática, Universidad Tecnológica de La Habana "José Antonio Echeverría", La Habana, Cuba
e-mail: my@ceis.cujae.edu.cu; amoreno@ceis.cujae.edu.cu

L. C. Méndez-González et al. (eds.), *Innovation and Competitiveness in Industry 4.0 Based on Intelligent Systems*, EAI/Springer Innovations in Communication and Computing, https://doi.org/10.1007/978-3-031-29775-5_10

the model fit. For this purpose, data from Geographic Information Systems (GIS) and weather forecasting systems can be consumed.

The research question of this study is: What is the impact of the integration of factors on the performance of traffic signaling? There is wide research on the optimization of signals to reduce negative effects on the road, but no proposal has been found in the literature that considers the whole set of factors influencing traffic [6].

The aim of this research is to present an agent-based simulation model for evaluating traffic events. The model uses real spatial information and includes parameters of the road, the weather, and the individuals that travel on it as well as other external factors to the simulation environment. The main contribution of this study is to achieve an integration between techniques, parameters, and factors that have been used by previous studies to analyze traffic in order to correctly evaluate the configurations of the signals located on the road. The proposed model is flexible and extensible and allows integration with external systems that will use the data generated by the simulation to optimize the configuration of signals, extract patterns, or notify the corresponding traffic authorities to support decision-making.

The document is structured as follows: Sect. 2 presents a literature review on models for traffic accidents and congestion. Section 3 describes the research methodology of the study. Section 4 explains what a correct signal configuration consists of and outlines the consequences of not performing this task correctly. Section 5 introduces intelligent agents as a solution for traffic-related analysis and compares agent-based simulators developed for this purpose. Section 6 presents the simulation model proposed in this work, each of its modules and its possible integration with external systems, and it is evaluated using a design of experiments. A discussion of the results obtained with the study is presented in Sect. 7. Finally, Sects. 8 and 9 present the conclusions and recommendations reached with the development of this study.

2 Related Studies

There are several models proposed for the analysis, simulation, and configuration of road signs [6]. They vary in the artificial intelligence techniques used, the parameters considered, the objective they aim to optimize, and the factors they consider influential in their model.

In [8] a simulation model is proposed that includes vehicle, tram, and pedestrian interactions on roads, to be used for decision-making on road management, signal control, and public traffic. It is initially based in an area of Japan and uses real maps obtained from a GIS.

In [9] several models called car-following models that represent the vehicular interaction in traffic are presented, with the aim of analyzing and understanding the causes of traffic congestion and bottlenecks. The simulation models used are based on the city of Chennai. They analyze among other performance measures the flow concentration and the speed concentration of the road. Other examples of vehicle interactions are presented in [10, 11], where a mathematical model of traffic flow dynamics is presented for reducing congestion in large cities.

In [12] a simulation model is developed that aims to analyze the behavior of drivers on the road, based on a rural region of several cities in Iraq. In addition, an S-Paramics model is used to make graphs and statistical analysis with the data collected from the simulation. In [4] the influence of human factors in traffic and causing accidents is analyzed. This study also includes analysis of human conditions and "psychologist issues."

In [13] the main objective is to identify the causes of traffic congestion from a micro-level traffic simulation model. The environmental consequences of these situations are also studied. The model was generated and calibrated from Croatian cities.

The influence of traffic demand, climate factors, and pavement structure in pavement conditions over time is analyzed in [7]. They obtained a new classification of climate zones based on weather influence and pavement deterioration. These parameters directly determine traffic flow behavior.

In [14] the conflicting areas of the road networks of a medium-sized city are analyzed with the aim of reducing air pollution in the area. The simulations carried out in the area followed a discrete event approach and showed that by optimizing traffic performance, the emission of pollutants can be reduced.

The studies described above and others found in the literature focus on some of the factors described, but do not analyze them as a whole. Although some studies include the analysis of drivers and pedestrians, the simulation is not always agent-based, which is recommended because intelligent agents have greater autonomy to imitate human behavior. It is important to highlight the integration that exists in previous studies between simulations and other systems, such as statistical models and GIS.

3 Methodology

This study is considered a developmental research that aims to relate the influential factors on traffic behavior in a single model. This relationship has been partially described in previous studies, but not in an integrated manner or with the possibility of including external systems in the model.

The data used in the study are considered quantitative, since they refer to records of accidents, congestion, and traffic events. They are also classified as secondary, since they were collected by the Dirección Nacional de Tránsito de Cuba. In this data set, there are both experimental and descriptive data.

Prior to analysis, the collected data were prepared. The data set was checked for missing data and outliers.

The model is built from the data obtained, and a design of experiments is performed using the case study of Ciudad Deportiva, which was the motivation for the research. For the experiment we used Minitab software [15] and a factorial design that allowed us to study all the treatments among the data [16].

The results of the simulation are compared with the real data collected.

4 Traffic Signs Configuration

Traffic signs configuration has become a prioritized task in traffic planning and management in most developed and developing countries [17]. Signs configuration is understood as the selection of the times that make up the traffic light cycle and the location of vertical signs when a traffic light is not necessary in a given region. The installation of signs that optimize traffic flow contributes to the proper functioning of cities and minimizes unfortunate situations such as traffic congestion or accidents [18].

The implementation of poor signs configurations affects road users and governments in several dimensions. Among the most well-known damages are the loss of human lives in traffic accidents and increased fuel consumption [18]. However, in practice, there are many more implications of counterproductive sign placement. The following are some of the most concerning consequences of poor configuration.

4.1 Traffic Congestion

Traffic congestion is a problem that affects every country in the world [19]. They cause delays in drivers' arrival at their destination, inconvenience, economic losses for drivers, and environmental pollution [1, 20]. Figure 1 shows a relation of most common causes and consequences of congestion.

Several authors have dedicated their research to measure the impact of vehicle congestion through speed reduction and time spent in congestion, among other metrics. There are mitigation proposals for traffic congestion situations that are generally applied in developed countries [21].

A typical characteristic of traffic jams is vehicles moving at very low speeds and with frequent stops. This means that engines work much harder and, therefore, generate more polluting gases into the atmosphere, pollutes the environment, and contributes to the increasingly alarming global warming. Vehicles older than 10 years consume more energy, produce more pollution, and are noisier. On the other hand, the concentration of vehicle claxons simultaneously generates noise pollution in the area where congestion occurs [17].

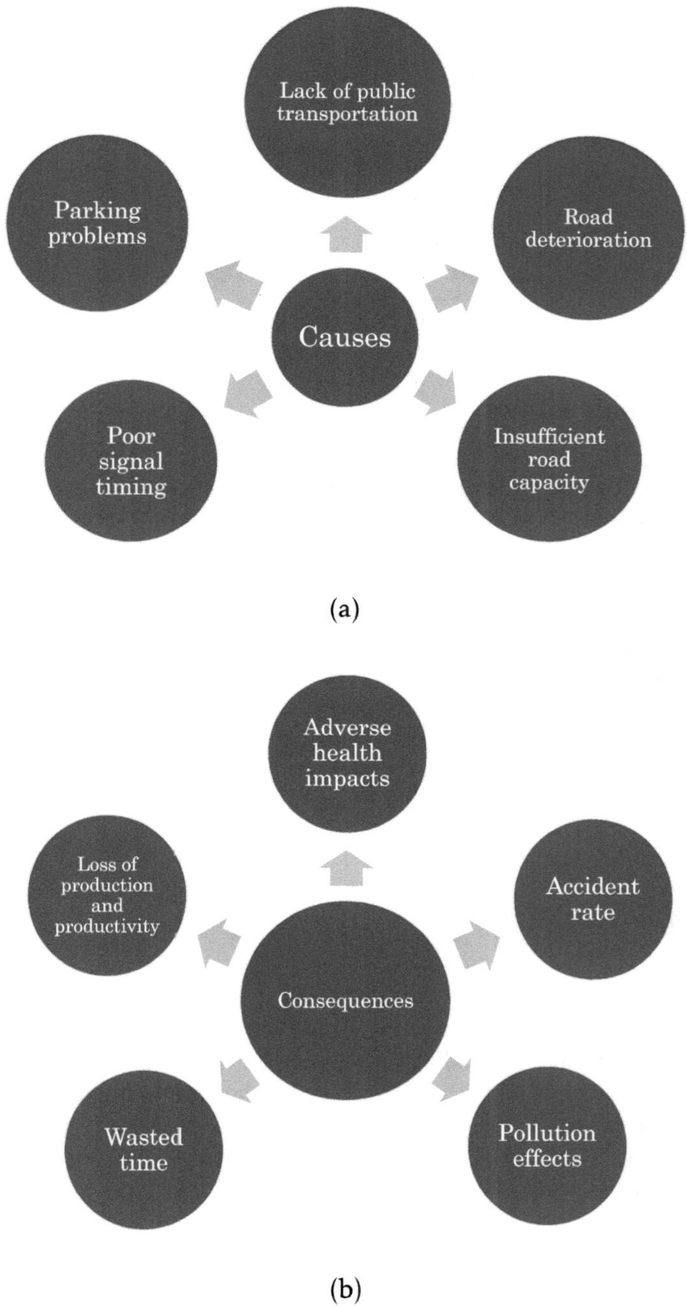

Fig. 1 Traffic congestion's influencing factors. (**a**) Causes. (**b**) Consequences

Table 1 Costs associated with traffic congestion

City	Total (millions of dollars)	Per capita (units of dollars)
Sao Paulo	2100	190
Buenos Aires	1700	250
Mexico City	1200	175
Santiago	1050	400
Rio de Janeiro	950	225
Bogota	650	350
Lima	600	200
Montevideo	350	475

4.2 Economic Losses

Poor signs configurations generate costs for individuals and cities. In 2020, in the case of US cities, for example, the total direct cost has been estimated at US$ 151.2 billion. In the European Union, the estimate for Germany is US$ 37.3 billion, followed by France and the United Kingdom (US$ 25.4 billion each) [19]. Table 1 shows a summary of the costs related to traffic congestion in some Latin American cities in the year 2019.

These costs are associated with excessive fuel prices proportional to the duration of congestion. Other factors that increases direct costs are the materials and labor used in the installation of signs that are counterproductive and must be removed. Finally, in the case of accidents, costs associated with the value of vehicles involved and medical care for injuries or fatalities are reported [22].

4.3 Traffic Accidents

The most serious consequence associated with poor signs configuration is traffic accidents. Traffic accidents are the eighth leading cause of death in the world for people of all ages and the leading cause of death for children and young people between 5 and 29 years old. These statistics are three times higher in low-income countries than in middle- and high-income countries. In 2018, it was recorded an estimated of 1.35 million deaths per year due to road traffic accidents, with an average of one road traffic death every 24 s globally [23].

Many accidents occurring annually are related to drivers' disobedience of traffic signs. Other reasons are speeding, overconfidence, and dangerous operations [24]. These data can be corroborated thanks to the statistics of the World Health Organization [22].

Figure 2 is a report by causes of accidents in Latin America, showing the most important in relation to the amounts for the years 2018 and 2019. Given the

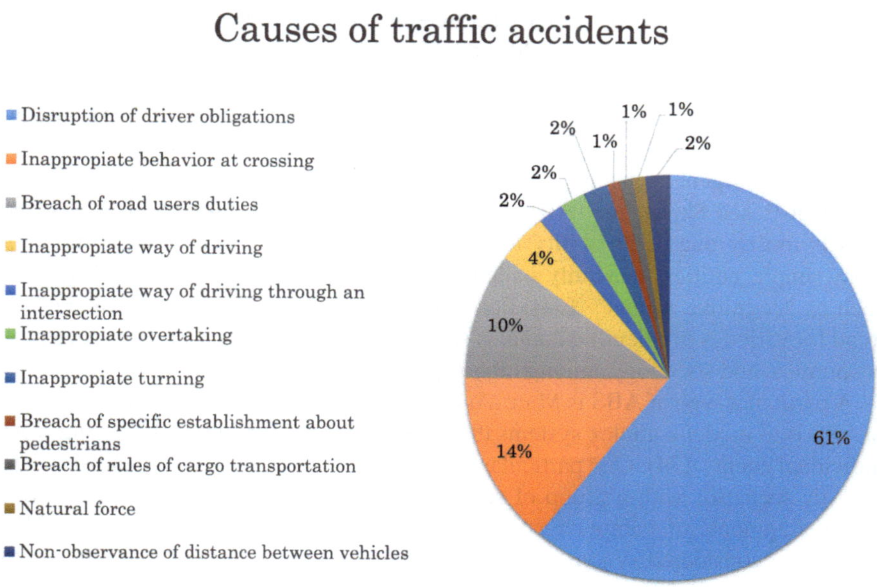

Fig. 2 Causes of accidents in Latin America

consequences related to the poor signs configuration, it is necessary to evaluate them before the installation stage [19].

5 Agent-Based Simulation and Its Application for Traffic Problems

The implementation of a process or system without first having a certain degree of confidence in how it will work is not a good practice. This is caused by the risk of crashes, insufficiencies, and errors. Simulation techniques have long been used to mimic the performance of a system or artifact to be developed in order to estimate what would be its actual performance. Simulation of a process can help to identify relevant problems and quantitatively evaluate alternative solutions [25].

There are several types of computational simulation; among them, it is not possible to determine which one is the most convenient to use for all types of problems [25, 26]. Depending on the characteristics of a model, the type of simulation to be applied can provide favorable or unfavorable results.

One area in which simulation can provide advantages is traffic engineering [6]. The simulation of road networks is of great importance since, with these, traffic behavior can be analyzed for candidate configurations in traffic lights and other controllers, thus obtaining statistical data that support decision-making.

Following this objective, several traffic simulators have been developed around the world. Each of them uses a different type of simulation and pursues different objectives. Examples of these simulators are TRANsYT [27], SimMobility [28], OpenTraffic [29], SimTraffic [30], SUMO [11], and TrafficWare [31], among many others [6, 32, 33]. From the first years of the twenty-first century, an approach based on intelligent agents begins to appear in these simulations.

Agent-Based Simulation (ABS) is currently used in a growing number of areas. This is mostly due to its ability to deal with a wide variety of models, ranging from simple entities, generally called "reactive agents," to more complex ones, such as "cognitive agents" [34]. For more than two decades, its use has been a trend for varieties of scientific domains such as biology, physics, chemistry, ecology, economics, and sociology, among others [6, 35].

A particular type of ABS is Multi-Agent-Based Simulation (MABS). The goal of MABS is to model complex systems that adopt a bottom-up approach starting from individual agents [34]. One particular approach of MABS is to model and simulate realistic scenarios with a group of self-governing agents, either as simple entities within fragments of computer codes or as considerably intelligent objects. This is possibly considered synonymous with human problem-solving capabilities with infinite states, beliefs, trusts, decisions, actions, and responses. Acquiring adequate knowledge of the system to build an appropriate conceptual and logical model is one of the most challenging tasks of simulation [36].

With this MABS concept, several traffic simulators emerge to model not only static behaviors of the entities but also autonomous and intelligent behaviors.

5.1 Agent-Based Traffic Simulators

MATISSE is a large-scale agent-based simulation platform written in Java [37]. The simulator has been released by UTD MAVS (University of Texas at Dallas) for non-commercial use under GPLv3. Within this set of fully agent-based simulation models, it is criticized the lack of core agent mechanisms such as sensing and diverse communication types. MATISSE specializes in the simulation of scenarios related to traffic safety.

POLARIS is an open-source agent-based software framework written in C++. The motivation behind POLARIS was to combine different traffic-related modelling aspects into a single framework that otherwise requires a number of separate standalone software applications. POLARIS focuses on large-scale transportation scenarios and has been used to analyze energy consumption of vehicles in the city of Detroit comparing scenarios that include current and future vehicle technologies [38].

MATSim is an agent-based software framework implemented in Java and licensed under GPLv2. The framework has a general focus and is designed for the simulation of large-scale transportation scenarios. Hence, a particular effort was made for efficient computational processing and parallelization. MATSim has been used in particular to simulate energy demand planning in transportation [33].

AgentPolis is a fully agent-based software framework written in Java and licensed under GPLv3 [39]. The creators noted that existing simulation approaches fail to implement the ability to model ad hoc interactions among the entities of the transport system as well as the spontaneous decision behavior that is required for this form of interaction. However, current mobility services rely on frequent, ad hoc interactions between various entities of the transport system. Hence, AgentPolis focuses particularly on the simulation of interaction-rich transport systems.

The advantage that these new simulators offer over non-agent-based simulators is the possibility of modeling complex traffic situations, which do not operate statically during a given time interval. The application of agent technology is particularly suitable to investigate road traffic from the individual perspective, as it allows for modelling of individuals with intelligent and autonomous behavior [6].

Despite the breadth of behaviors and scenarios covered by existing platforms, there is a need for a model that can be integrated into a comprehensive configuration system, considering all the factors that affect the development and behavior of road users.

Following this objective, agent-based simulation models have been created that consider not only the autonomous behavior of people on the road but also the factors that can influence this behavior.

6 Simulation Proposed Model

The proposal presented in this chapter brings together a set of human, technical, and external factors that positively or negatively determine the behavior of road users on a real road section. The simulation model, the agents used, the factors considered, and finally a case study where the simulation is applied are presented.

The model integrates several areas of artificial intelligence in order to mimic the behavior of road users and provide a result that supports decision-making.

The model receives parameters from various sources, shown in Fig. 3, to obtain a final evaluation of the simulation. Historical statistics of the area, the geographic data of the map, the signs, and the weather conditions are necessary to execute the simulation model.

6.1 Geo-spatial Information

To guarantee the veracity of the data obtained by the proposed model, it was decided to use real road maps. The spatial data supporting the model are obtained from a GIS.

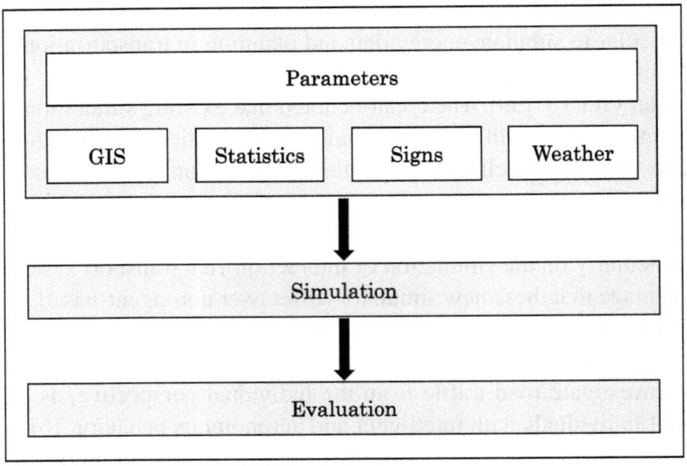

Fig. 3 Proposed model parameters

Table 2 Data extracted from GIS

Attribute	Map equivalent	Description
Lines	Street	Space where vehicles drive in the simulation, they have a section that corresponds to the sidewalk for pedestrians
Points	Corners, signals	Places where the signs are placed or can be placed
Polygons	Buildings	Buildings where people leave or enter and stop interacting in the simulation
Direction	Direction of the street	Direction vehicles can travel on the street
Lanes	Lanes of the street	Number of lanes on a street determines the number of vehicles that can travel in parallel
Speed	Maximum speed allowed	Maximum speed vehicles can travel without committing infractions
Arrivals	Corners with input streets only	Corners where vehicles start their way through the simulation

GIS are integrated and organized systems of hardware, software, and geographic data that allow to analyze geographically referenced information and also to represent it. In this case, it is used the OpenStreetMap GIS [40], which is a free and interactive platform with a large set of updated spatial data on streets, landmarks, buildings, and geographic elements around the world [41, 42].

Numerous traffic control-related researches use GIS to obtain spatial data of the regions and road sections they wish to simulate. Each proposal extracts from the maps the information that is relevant to their objectives [42]. For the proposed model, the data extracted from the GIS are listed in Table 2.

Traffic analysis systems that use real spatial information have advantages over those that simulate pre-determined or hypothetical scenarios. By studying traffic performance using the actual characteristics of the intersection where the signs will eventually be placed, the results are guaranteed to be more faithful to reality. In addition, the use of GIS allows simulations and analysis for any map available on the Internet [43].

6.2 *Factors That Influence Traffic Safety*

The behavior of vehicles on the road is conditioned by several factors that are divided into three categories: human, vehicle, and external [10]. Characteristics and technical condition of vehicles are determining factors when driving [44], since the speed and correct execution of operations depend on them, including, in the case of accidents, the impact they would have. Drivers are affected by temperature, increasing their speed if they feel uncomfortable and impatient to reach their destination, as well as other psychological factors or situations that may affect their driving [7]. The state of the pavement is another key factor for vehicles to moderate their speed, in order to take care of the technical condition of the vehicle [7]. Finally, the road can be interrupted by pedestrians or animals obstructing traffic, endangering their lives as well as the lives of the drivers.

6.2.1 Physical Condition of Road Pavement

The pavement is the layer made up of one or more materials that is placed on the natural or level ground to increase its resistance and serve for the circulation of people or vehicles [45]. As the main part of the road, the physical condition of the pavement is decisive in the behavior of the vehicles that travel on it. Pavement deterioration, especially the presence of potholes, generates increasing capacity restrictions and increases congestion. Rainfall accumulated on roadways reduces road capacity and also generates congestion [7]. On the other hand, worn roads become dangerous even if they do not have potholes, since rain causes vehicles to slide and thus collide.

6.2.2 Environmental Sensation

There are adverse characteristics related to road conditions, which can increase the driver's skill level, as well as the risk in different actions. Besides the geometric characteristics of the traffic area, there are the climatic conditions, which, from moderate increases in temperature, strong winds, and rain, can alter normal driving and easily cause the occurrence of an accident on the road [46].

With rain, visibility conditions decrease, and windshields fog internally and make it difficult to detect vehicles and people on the road. Drivers' visibility can also be affected by humidity and fog. Likewise, tires lose grip, and the wheel tends to skid on water with little contact with the pavement [7].

Faced with high temperatures, which are reflected even warmer inside the vehicles, drivers manifest behaviors of desperation to reach their destination, disobeying the traffic signs in their way, and as expected also increase their speed, thus being more prone to cause or suffer accidents.

The time of the day is also decisive. In the morning and afternoon, drivers tend to be more attentive to what is happening on the road; however, at night and in the early morning, as the flow of vehicles is lower, drivers relax and stop paying attention to the signs and thus are potential accident causers.

6.2.3 Animals in the Road

Unfortunately, there are a not inconsiderable number of animals living on the streets, trying to survive the cold, rain, hunger, and other miseries. These animals are part of the roadway and often interrupt the movement of vehicles. Sometimes, carelessness or not having enough time to maneuver the vehicles ends up crashing into them, leaving them injured and in the worst case dead. By drivers, it is also a danger because when trying to save these animals, they realize abrupt and hurried turns that can end in a collapse or an accident. It is a common phenomenon today and deserves to be considered in all traffic studies [47].

6.2.4 Vehicles Overtaking Lanes

One of the most common behaviors adopted by drivers is to overtake vehicles in front of them. This behavior may be conditioned by the factors mentioned above, as well as by the simple fact that the driver wants to reach his destination early. The driver of a vehicle that is going to overtake another vehicle is obliged to check that he can carry out the maneuver without interfering with other vehicles circulating and, without risk of accident, carry out the overtaking in the left lane and gradually and safely join the lane in which he was circulating, whenever he does not force the driver of the vehicle overtaken to modify his direction or speed. The driver of the vehicle overtaken is obliged to not to increase the speed nor to carry out maneuvers that impede or hinder the overtaking and provide sufficient space for the overtaking vehicle to rejoin the lane or lane in which it is traveling [48].

6.2.5 Vehicle Characteristics

The characteristics of each vehicle have a lot to say about the probability of collision. A study conducted at Monash University [49] establishes a variation in the danger of vehicles according to their color and time of day, starting with the danger of white vehicles. Table 3 presents the relationship between traffic accidents and the different car colors defined for Australia. The unit used is the danger of the white color. For example, during the morning, silver (1.10) is 10% more dangerous

Table 3 Collision probability by color and time of the day

Accident probability by day moments (white: 1.00)			
Color	Morning	Sunset	Night
White	1.01	1	1.05
Orange	0.09999	1.21	0.77
Fuchsia	1.07	1	0.65
Khaki	0.93	1.16	0.97
Yellow	1	0.88	1
Golden	0.98	1.04	1.1
Pink	1.19	0.66	1.06
Red	1.07	1.02	1.1
Blue	1.07	0.82	1.09
Violet	1.11	1.01	1.1
Silver	1.1	1.15	1.08
Green	1.04	1.03	1.04
Brown	1.05	1.12	0.98
Gray	1.1	1.25	1.07
Black	1.12	1.47	0.98

than white, at sunset (1.15) is 15% more dangerous, and at night (1.08) is 8% more dangerous.

In addition to the color, the weight of the vehicle, its body type, and the use it is given are also determining factors in its behavior on the road, and the impact they have in an accident is different.

Generally, larger vehicles travel at a lower speed on the road, as they are more likely to overturn and suffer accidents. In terms of damage and impact, an accident involving a trailer will be much more dangerous than an accident between two-passenger-carrying vehicles. Drivers must be able to assume the responsibility of driving such a large vehicle [44].

Another issue to consider is the technical condition of the vehicles. Mechanical aspects such as brake failure, steering failure, suspension failure, lack of maintenance, overloading, improper modifications, or oversizing of the vehicle can cause damage on the road [50]. The responsibility of drivers and vehicle owners in making periodic revisions to their vehicles helps to a correct circulation and avoids accidents.

6.2.6 Human Factor on the Road

Human error is one of the most frequent causes of traffic accidents and many times it is avoidable. Overconfidence; distraction; drug and alcohol consumption; speeding; driving while tired or fatigued; using electronic, navigation, or radio devices; performing dangerous maneuvers; and improper overtaking are some of the reasons that cause accidents based on the human factor [4]. Pedestrians, as road users, are also sometimes to blame for accidents, either by crossing signalized intersections without obeying, being distracted, and running in front of vehicles, among other actions that endanger their lives. In addition, the state of mind also influences this behavior. In general, people's expertise on the road, whether as drivers or pedestrians, plays a very important role. Knowing the traffic laws and the intersections through which you are traveling always helps you to get along better [5].

6.3 *Different Agents' Behaviors*

The simulation model is agent-based and includes the autonomous and intelligent behaviors of each actor present on the road, from the users to the previously mentioned conditions that are time-varying and affect people's actions. Figure 4 shows the influence between agents presented below which also serves as the conceptual framework for the study.

6.3.1 Agent Traffic Light

Traffic light agents follow the operation of these devices on the public road. From the beginning of the simulation, it has configured its times, and these are modifiable during the simulation. At the end of the time of a light, it proceeds to turn on the indicator of the next light.

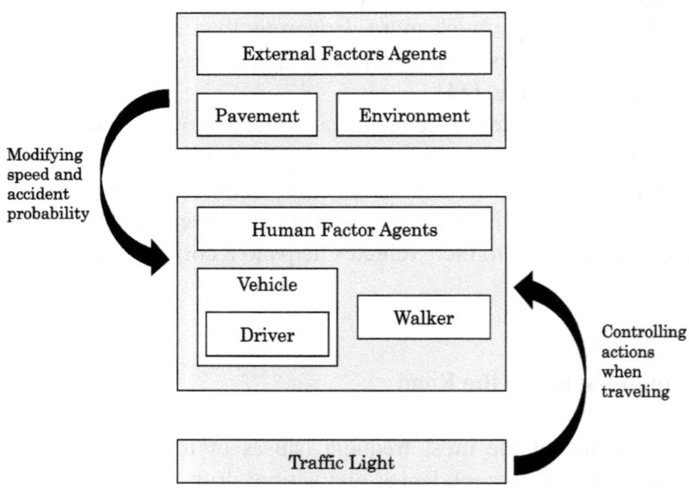

Fig. 4 Relations of simulation agents

Traffic lights have different phases, one for each access street. At the same time, for each access street, there will be a group of exit streets, where vehicles can continue to move on if they are stopped at the traffic light.

6.3.2 Agent Vehicle

It is one of the main entities of the simulation. These are autonomous agents with a defined route from the beginning of their trajectory, created using Dijkstra's algorithm [51] to find the shortest path from the arrival node of the vehicle to one of the terminal nodes of the simulation environment.

While driving, the vehicles check that the street they are driving on has some sign, either traffic lights or vertical or horizontal signs, and store it to decide what action to take when they reach the sign line. Vehicles are driven by drivers, and the characteristics of the driver, added to the size and color of the vehicle, define through probabilistic events the actions performed by the vehicle.

Vehicles may collide, overtake, pick up, or drop off passengers, break or repair themselves on the road, and commit any violation.

6.3.3 Agent Driver

Driver agents are created to consider the characteristics of the road users in the simulation. Drivers are created randomly based on values for age, gender, experience on the road, mood at the time of the simulation, and a level of distraction that may influence the behavior of the driving vehicle. The membership of the drivers in each of the age and experience sets will determine their membership also in the sets corresponding to a "Risk Level" which is used to establish the violation and accident probabilities of those drivers.

6.3.4 Agent Walker

They move through the road network making decisions such as the speed of travel, when to cross the street, and whether a vehicle may collide with them. Pedestrians follow a set route when they are created. They can be generated from a vehicle, and they can have simple disabilities that indicate whether a pedestrian has vision, motor, or hearing deficiencies, influencing the pedestrian's perception of risk.

There is a probability that a pedestrian will faint on the road, due to lack of energy. For every step a pedestrian takes, his or her energy decreases until it reaches the value where that pedestrian can no longer walk.

Finally, relationships between pedestrians were added, each person having a list of possible individuals that may be of interest to him, and a new state called chatting was created, which the pedestrian can enter if he finds in his path a pedestrian that is in his list of people of interest.

Each pedestrian has an independent perception of risk in a situation, which is given by gender, decreasing if the pedestrian is male; age, increasing the perception for older adults, since they overestimate their own safety, and decreasing for young people, who tend to underestimate it; years of experience as a driver, which provide better judgment when assessing risk situations; energy or mood, a variable that gradually decreases with each step a pedestrian takes, which improves risk perception if the pedestrian is in a good mood or decreases otherwise; the recklessness that may characterize them, regardless of their other traits; and the individual's disabilities.

6.3.5 Agent Pavement

Pavement agent behavior is based on improving or worsening its capability for vehicles to circulate. An initial state is configured before starting the simulation, and this will change during the course of it.

When drivers are about to drive on a road, the first step is to check its state, and from then on they modify their speed in order to protect the vehicle from damage and breakage.

6.3.6 Agent Environment

The environmental conditions considered for the simulation are temperature, humidity, moment of the day, and rainfall. There are two ways to obtain the weather data: to insert them manually or to consume them from the Internet.

In case the user executes the simulation online, the ambient agent will obtain the current weather forecast using the OpenWeatherMap API [52]. The update of the observed variables is displayed by the environment at each step of the simulation, and the vehicles are notified accordingly to moderate their speed considering the environmental state at all times.

6.4 Integration with Intelligent Traffic System

After analyzing the simulation model, the behavior of its agents, the parameters it receives, and the factors it considers, it is necessary to explain how it could be integrated with other devices and systems in order to constitute an intelligent traffic system.

Intelligent Transportation Systems (ITS) is a combination of leading-edge information and communication technologies used in transportation and traffic management systems to improve the safety, efficiency, and sustainability of transportation networks, to reduce traffic congestion, and to enhance drivers' experiences [53].

6.4.1 Internet of Things for Traffic Signaling

The Internet of Things (IoT) is a collection of sensors and software that has a purpose to connect and exchange data over the Internet. IoT plays an important role whenever there is technology and Internet as it collects data with minimal human intervention. Nowadays, IoT has low-power, low-cost sensor technology, which means it is very affordable and reliable [54].

A traffic monitoring system has many applications like traffic congestion, accident detection, vehicle identification/detection, automatic vehicle guidance, smart signaling, forensics, traffic density, safe pedestrian movement, emergency vehicles transit, etc. The system can be designed by using traffic detector technologies, vehicular sensor networks, or probe vehicles [55]. The number of devices which connect to the Internet is increasing. IoT can add some level of digital intelligence to devices, enabling them to communicate and also to merge the digital and physical world. IoT has massive possibilities for developing new sensible applications in almost every discipline [56]. A traffic monitoring system is one of the applications of IoT.

In [57] is developed a vehicle detection scheme known as vehicle detector based on EMD-HT (empirical mode decomposition-null-hypothesis interval threshold) and multi-channel GLRT (multi-channel generalized likelihood ratio test) detector. The research of [54] was organized to determine the risk factors and the nature of accident and identify the causes that lead to fatality, non-fatality of injuries, and damage to property classifications of road traffic accidents. The results are used as input to formulate and adopt IoT application in reducing road traffic accidents. In [58] the main objective is to detect accidents in real time and minimize the response time of medical help. For accident avoidance, tire pressure is measured, whereas in accident detection it is implemented with the help of node MCU. The proposed system is useful in reducing vehicular accidents, and pollution monitoring will help to know the environmental status.

6.4.2 Possible Integration for Simulation Model Proposed

The integration proposal is based on ITS consuming information from IoT devices installed on the streets [59], which could be motion sensors, photo or video cameras, vehicle-to-vehicle infrastructures, or any other existing method of capturing information on the road.

The captured information is provided as a parameter to the simulation model, and the evaluation obtained will be returned to the ITS.

An intermediate step in this connection will be the integration of another module that handles the optimization of the model, so that the simulation is executed several times, until the optimization module decides that it is returning the optimal solution for the intersection concerned.

Once the simulation evaluation is returned and a solution for signaling is obtained, the ITS can autonomously modify the configuration of the existing signs,

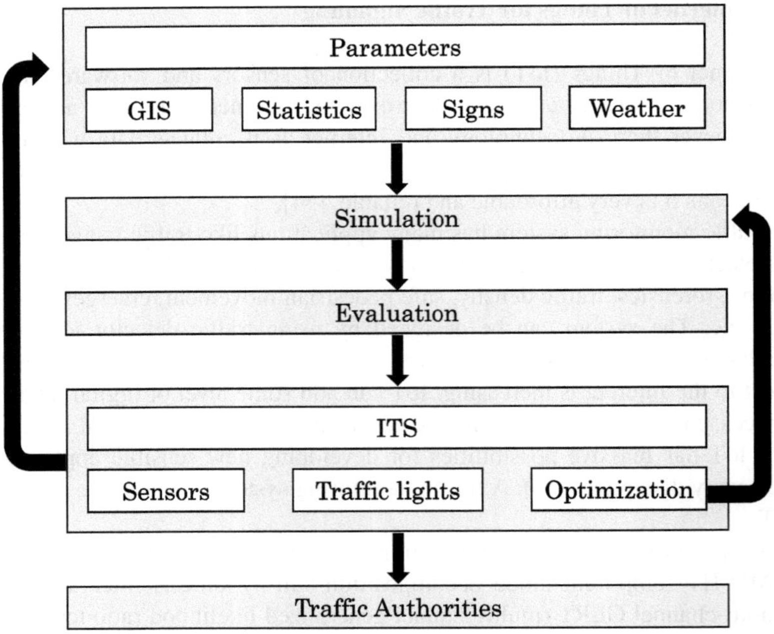

Fig. 5 Future integration with intelligent traffic system

if they are intelligent signs and have a remote configuration option available. Otherwise, if this procedure is not possible, the ITS will have the task of notifying the traffic authorities of the configuration to be performed so that they can proceed with its installation.

These relations and information flows are illustrated in Fig. 5.

6.5 Case Study Ciudad Deportiva, Havana

The instance to be analyzed corresponds to the Ciudad Deportiva traffic circle, in Havana, Cuba. In 2015, a signaling project was carried out in this place, which turned out to be unsuccessful. The study intends to demonstrate that a correct evaluation of the project, before its installation, would have shown that it would not be efficient.

Figure 6 shows the map section to be simulated, and the traffic lights placed in the project, one for each access to the traffic circle. In the road, vehicles will circulate counterclockwise, and all types of vehicles are allowed to pass through the traffic lights.

For the validation of the proposal, a design of experiments (DoE) is carried out. This is defined as "a sequence of steps taken in advance to ensure that the

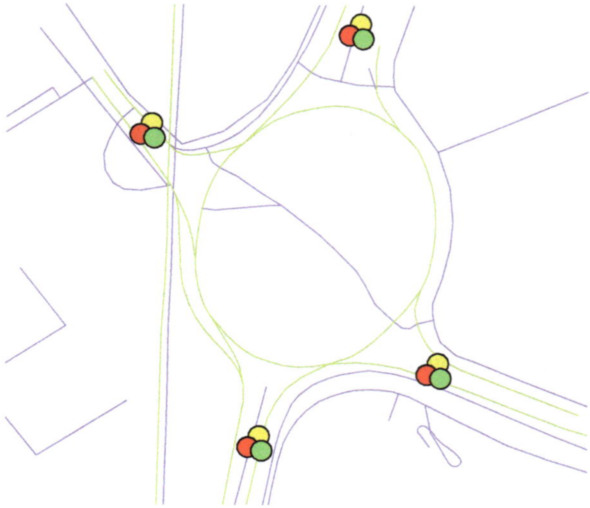

Fig. 6 Simulated region and traffic lights

Table 4 Controllable factors

ID	Factor	Level		Unit
		Low (−1)	High (1)	
A	Traffic light presence	No	Yes	–
B	Vehicle average speed	40	60	Km/h
C	Vehicle arrival amount	1	3	u/s

appropriate data will be obtained in a way that allows an objective analysis leading to valid deductions regarding the problem posed" [16]. The DoE developed is oriented to the case: "Traffic-light Ciudad Deportiva." For this purpose, the DoE guidelines are followed.

6.5.1 Planning Phase

This first phase begins with the recognition and formulation of the problem. We want to perform a simulation in the Ciudad Deportiva, so that vehicles circulate through this area, including the traffic circle, having to wait as little time as possible. The performance is the average waiting time, in minutes, of the vehicles after a 12-h simulation. This time is affected in each vehicle when it is waiting for the traffic light, obstructed by an accident or queued, or waiting by another vehicle which is also waiting. The objective is to minimize this time. The factors identified are divided into two groups, controllable and non-controllable. The first group, together with the levels, is described in Table 4. Other factors that cannot be controlled (second group) were also identified, but are no less important, since they have a considerable influence on performance. These are the number of accidents, number of violations, number of violators, and number of vehicles reaching their destination.

6.5.2 Design Phase

In this phase, the design resolution to be applied is chosen. The number of controllable factors is three, which implies that the total number of treatments (T) is eight according to the following equation:

$$T = L^F \tag{1}$$

where L is the number of levels and F is the number of factors.

This quantity is acceptable (considerably small in terms of executions), so it is decided to apply a full factorial resolution. The most important benefit obtained with this design resolution is that it avoids confounding or aliasing, which would be present if fewer executions were performed than the total number of treatments [16].

6.5.3 Conduction Phase

In order to carry out the conduction phase, the eight executions with the corresponding treatments are performed. The Minitab statistical software [15] was used as a support tool. Table 5 shows the results in terms of performance of each experiment.

6.5.4 Analysis Phase

Minitab, after inserting the data, performs data processing, which results in a set of graphs that will make the final analysis possible. The graph in Fig. 7 provides the main influential factors for the performance of the problem.

It shows that the slope of factor A is considerably larger than the slopes of factors B and C. This suggests that the most influential main effect on the performance of the problem is the main factor A (traffic light presence).

The Pareto plot in Fig. 8 shows a clear comparison between the effect of each factor or interaction. It confirms that factor A (traffic light presence) is the influential factor and adds that there is no interaction of factors with a considerable influence.

Table 5 Results of treatments

No.	A. Traffic Light	B. Speed	C. Arrival	Time
1	1	1	1	31
2	1	1	−1	22
3	1	−1	1	37
4	1	−1	−1	28
5	−1	1	1	13
6	−1	1	−1	9
7	−1	−1	1	12
8	−1	−1	−1	10

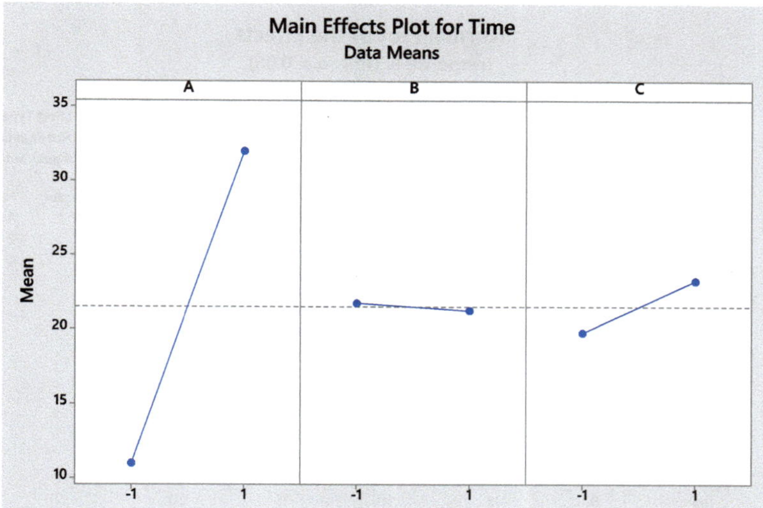

Fig. 7 Main influential factors

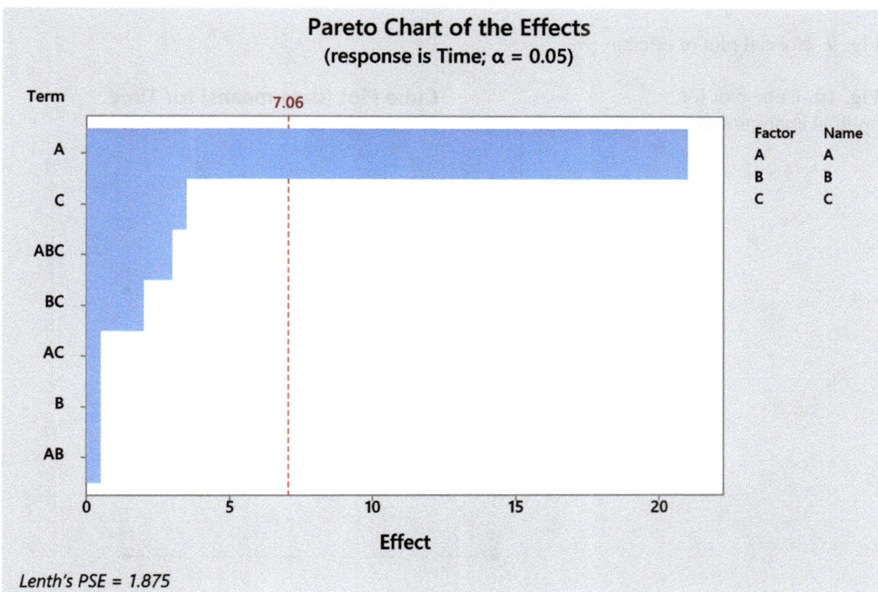

Fig. 8 Pareto plot

To be more certain that factor A (traffic light presence) is the only influential factor, Minitab provides the graph illustrated in Fig. 9. This graph shows that, after plotting the points of each effect, the point of factor A is the only one that is considerably distant from the plotted line.

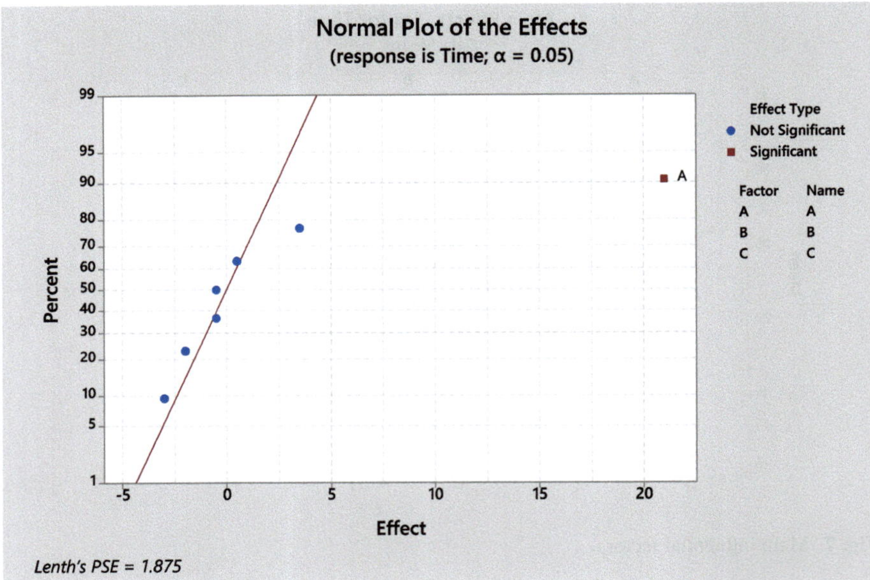

Fig. 9 Normal plot of effects

Fig. 10 Cube plot for optimal configuration

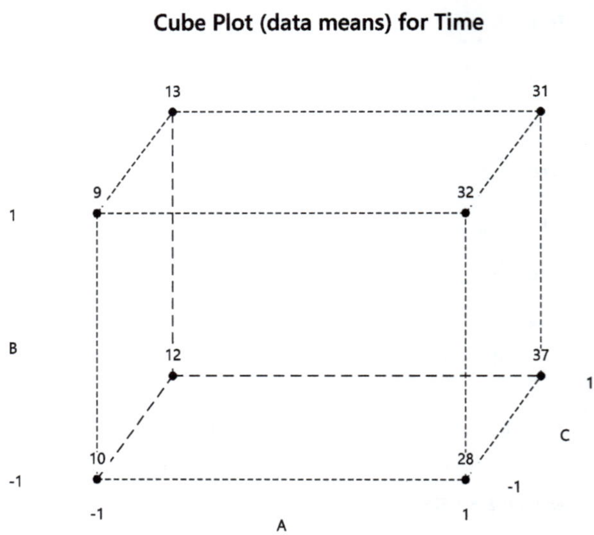

To know which is the optimal configuration for this DoE, Minitab generates a cube plot. Figure 10 denotes each vertex of the cube with the performance result. The design objective is to minimize throughput; therefore, when looking for the minimum, it can be seen that the shortest average waiting time for vehicles is 9 min. Then, the coordinates of this vertex indicate the treatment that achieved the best performance.

Table 6 Recommended configuration

ID	Factor	Level	Value	Unit
A	Traffic light presence	Low (−1)	No	–
B	Vehicle average speed	Alto (1)	60	Km/h
C	Vehicle arrival amount	Low (−1)	1	u/s

After the application of the DoE, it can be concluded that the choice of a full factorial design avoids confounding or aliasing. Compared to vehicle arrival and speed, the presence of traffic lights at the Ciudad Deportiva traffic circle is the most influential factor. For the Ciudad Deportiva DoE, there is no interaction of factors with an influential effect. All treatments where traffic lights were present at the Ciudad Deportiva traffic circle resulted in significantly longer vehicle waiting times than those without traffic lights. The configuration achieving the best performance is shown in Table 6, with an average waiting time of 9 min.

7 Discussion

The presented simulation model allows the representation of traffic considering a wide range of external factors such as weather, pavement condition, presence of animals and obstacles on the road, and technical conditions of the vehicles. In addition, it allows the analysis of human behavior based on people's characteristics such as age, gender, disabilities, or their particular level of stress. The model can obtain its parameters from external systems or can be added manually by the traffic analyst, which makes it flexible and adaptable.

In previous studies on the subject, the authors focus their attention on a particular type of parameters. In [4] they studied the human factors that most influence traffic accidents and showed that the most influential is driver distraction, although they may include other factors such as age, gender, and familiarity with the road. These factors are all considered in this model.

In [46] the simulations performed demonstrate the impact that weather conditions have on the behavior of the vehicular flow. They also incorporate some behaviors of the vehicles as part of the simulation and the interactions between them. These behaviors are also included in the model presented.

On the other hand, the authors of [7] present an integration between several external parameters such as pavement and weather. The importance of these and their influence on traffic is demonstrated. However, it is not possible to have a real approach to the road if human behavior is not observed.

From the previous studies and models, the factors that have been analyzed separately are obtained to achieve the integration presented in this research. This new model, besides including most of the factors that interact in traffic, is also easily integrated with other systems, so that the data and solutions generated can be used automatically. This guarantees that the application of the results will be immediate; otherwise they could be obsolete at the time of installation.

7.1 Limitation of the Study

In this study only instances pertaining to Cuba have been used, because the simulation is adapted to Cuban traffic laws. For the simulation, 1 year's accident and traffic jam data were used, when more data should be used to ensure the fit. The authors have Spanish as their native language, which influences the reading of related literature and the writing of the results. Finally, measures related to the frequency and flow of vehicles on the streets were sometimes the result of observation by the authors.

8 Conclusions

The correct traffic signs configuration ensures an adequate performance of road networks and minimizes negative effects such as accidents, traffic jams, environmental pollution, and associated economic losses. The developed case study demonstrated that pre-installation analysis of configurations can help and predict their performance, thus saving costs on deficient configurations.

The proposed simulation model constitutes an approach to intelligent traffic management, where several artificial intelligence techniques, Internet of Things devices, and statistical data recorded by traffic authorities will be integrated. This integration will be useful to facilitate signaling in high-demand road networks and to reduce the workload of the staff that today is dedicated to perform these tasks manually.

The advantage of developing this system in a modular design will allow its adaptation and implementation in any region or country, enabling easy modification of parameters such as traffic laws, signaling rules, or characteristics of the volume of vehicles in each region. It is extremely important to execute these simulations and systems with real data, as foreseen in this proposal, in order to guarantee that the results obtained are faithful to the road reality.

9 Recommendations

The simulation model obtained is capable of imitating human behavior due to the use of intelligent agents; however, it is advisable to carry out experiments where the influence of each of the characteristics of the people is evidenced. In this way, it would be possible to reach conclusions about which age group is more vulnerable in traffic, how disabled users affect the flow of vehicles, and how gender influences the speed of travel of drivers, among other analyses.

Moreover, it is necessary to study new instances, including medium and large cities, to verify that the model supports a high vehicle flow for the simulation.

To validate the simulation model, further experiments and comparison with the results obtained in previous studies, using performance measures such as the waiting time of vehicles on the road, are also recommended.

As part of the integration with external systems, it is also proposed to analyze the data generated by the simulation using data mining techniques, in order to identify the causes and most influential factors in traffic events.

References

1. M. Krzysztof, The importance of automatic traffic lights time algorithms to reduce the negative impact of transport on the urban environment. Transp. Res. Proc. **16**, 329–342. https://doi.org/10.1016/j.trpro.2016.11.032
2. D.N. de Tránsito, A partir del viernes 9 funcionarán semáforos en la Rotonda de la Ciudad Deportiva. Cubadebate **10**(8) (2015). http://cubadebate.cu/noticias/2015/10/08/apartirdelviernes9funcionaransemaforosenlaRotonda
3. O.F. Reinaldo, Rotondas, semáforos y comunicación por descongestionar. Cubadebate **12**(23) (2015). http://cubadebate.cu/opinion/2015/12/23/rotondas-semaforos-y-comunicacion
4. K. Bucsuházy, E. Matuchová, R. Zuvala, P. Moravcová, M. Kostíková, R. Mikulec, Human factors contributing to the road traffic accident occurrence. Transp. Res. Proc. **45**, 555–561 (2020). https://doi.org/10.1016/j.trpro.2020.03.057
5. K. Chebanyuk, O. Prasolenko, D. Burko, A. Galkin, O. Lobashov, A. Shevchenko, D.S. Usami et al., Pedestrians influence on the traffic flow parameters and road safety indicators at the pedestrian crossing. Transp. Res. Proc. **45**, 858–865 (2020). https://doi.org/10.1016/j.trpro.2020.02.083
6. J. Nguyen, S.T. Powers, N. Urquhart, T. Farrenkopf, M. Guckert, An overview of agent-based traffic simulators. Transp. Res. Interdiscip. Perspect. **12**, 100486 (2021). https://doi.org/10.1016/j.trip.2021.100486
7. D. Llopis-Castelló, T. García-Segura, L. Montalbán-Domingo, A. Sanz-Benlloch, E. Pellicer, Influence of pavement structure, traffic, and weather on urban flexible pavement deterioration. Sustainability **12**(22), 9717 (2020). https://doi.org/10.3390/su12229717
8. H. Fujii, H. Uchida,T. Yamada, S. Yoshimura, *Mixed Traffic Simulation of Cars and Pedestrians for Transportation Policy Assessment* (Springer, Cham, 2020), pp. 199–222. https://doi.org/10.1007/978-3-030-50450-2_9
9. G. Asaithambi, V. Kanagaraj, K.K. Srinivasan, R. Sivanandan, Study of traffic flow characteristics using different vehicle-following models under mixed traffic conditions. Transp. Lett. **10**(2), 92–103 (2018). https://doi.org/10.1080/19427867.2016.1190887
10. O. Ozerova, S. Lytvynenko, R. Sushchenko, Y. Zapara, P. Ovchar, Y. Lavrushchenko, Factors influencing the modelling of transport flow dynamics in cities. Comptes Rendus de l'Academie Bulgare des Sciences **75**(2), 259–265 (2022). https://doi.org/10.7546/CRABS.2022.02.11
11. N.C. Sarkar, A. Bhaskar, Z. Zheng, M.P. Miska, Microscopic modelling of area-based heterogeneous traffic flow: area selection and vehicle movement. Transp. Res. C: Emerg. Technol. **111**, 373–396 (2020). https://doi.org/10.1016/j.trc.2019.12.013
12. H.A. Al-Jameel, A.J. Kadhim, Rural traffic characteristics using field data and the developed simulation model, *IOP Conf. Series: Materials Science and Engineering*, vol. 888(012058) (2020). https://doi.org/10.1088/1757-899X/888/1/012058
13. N. Grubisic, T. Krljan, L. Maglić, S. Vilke, The microsimulation model for assessing the impact of inbound traffic flows for container terminals located near city centers. Sustainability **12**(22) (2020). https://doi.org/10.3390/su12229478

14. V. Mavrin, K. Magdin, V. Shepelev, I. Danilov, Reduction of environmental impact from road transport using analysis and simulation methods. Transp. Res. Proc. **50**, 451–457 (2020). https://doi.org/10.1016/j.trpro.2020.10.053. XIV International Conference on Organization and Traffic Safety Management in Large Cities (OTS-2020)
15. P.G. Mathews, in *Design of Experiments with MINITAB*. ANSI/ ISO 9000 Series Standards (American Society for Quality, Quality Press, Milwaukee, 2005). https://www.academia.edu/23892705/
16. V.L. Anderson, R.A. McLean, *Design of Experiments: A Realistic Approach*. Statistics: A Series of Textbooks and Monographs (CRC Press, Boca Raton, 2018). https://doi.org/10.1201/9781315141039
17. P. Sangaradasse, S. Eswari, Importance of traffic and transportation plan in the context of land use planning for cities—a review. Int. J. Appl. Eng. Res. **14**(9), 2275–2281 (2019). https://www.ripublication.com/ijaer19/ijaerv14n9_33.pdf
18. T. Afrin, N. Yodo, A survey of road traffic congestion measures towards a sustainable and resilient transportation system. Sustainability **12**(11), 4660 (2020). https://doi.org/10.3390/su12114660
19. A. Calatayud, S.S. González, F.B. Maya, F. Giraldez, J.M. Márquez, *Congestión urbana en América Latina y el Caribe: características, costos y mitigación*. Tech. rep., Banco Interamericano de Desarrollo: División de Transporte (2021). https://publications.iadb.org/es/congestion-urbana-en-america-latina-y-el-caribe
20. F. Ibili, A.O. Owolabi, W. Ackaahc, A.B. Massaquoi, Statistical modelling for urban roads traffic noise levels. Sci. Afr. **15**(e01131) (2022). https://doi.org/10.1016/j.sciaf.2022.e01131
21. I. Karakikes, E. Nathanail, M. Savrasovs, Techniques for smart urban logisticssolutions' simulation: a systematic review. RelStat **68**, 551–561 (2019). https://doi.org/10.1007/978-3-030-12450-2_53
22. World Health Organization, *Global Status Report on Road Safety* (Geneva, 2018). https://www.who.int/publications-detail-redirect/9789241565684
23. M. Touahmia, Identification of risk factors influencing road traffic accidents. Eng. Technol. Appl. Sci. Res. **8**(1), 2417–2421 (2018). https://doi.org/10.48084/etasr.1615
24. V. Harantová, S. Kubíková, L. Rumanovský, Traffic accident occurrence, its prediction and causes, in *Development of Transport by Telematics*, ed. by J. Mikulski (Springer, Cham, 2019), pp. 123–136. https://doi.org/10.1007/978-3-030-27547-1_10
25. F. Hillier, G. Lieberman, *Introduction to Operations Research*, 11th edn. (McGraw-Hill, 2020). https://www.abebooks.com/Introduction-Operations-Research-11th
26. A.M. Law, W.D. Kelton, *Simulation Modelling and Analysis*, 5th edn. (McGraw-Hill, New York, 2015). http://www.averill-law.com/simulation-book/
27. D.I. Robertson, TRANSYT: a traffic network study tool. Highway Safety Literature (1969). https://trid.trb.org/view/115048
28. M. Adnan, F.C. Pereira, C.M.L. Azevedo, K. Basak, M. Lovric, S. Raveau, Y. Zhu et al., Simmobility: a multi-scale integrated agent-based simulation platform, in *95th Annual Meeting of the Transportation Research Board Forthcoming in Transportation Research Record* (The National Academies of Sciences, Engineering, and Medicine, Washington, DC, 2016). http://eprints.soton.ac.uk/id/eprint/390938
29. M. Miska, E. Santos, E. Chung, H. Prendinger, Opentraffic-an open source platform for traffic simulation, in *Australasian Transport Research Forum*, ed. by ARRB, vol. 34(12) (Citeseer, 2011). https://trid.trb.org/view/1138239
30. K. Shaaban, I. Kim, Comparison of SimTraffic and VISSIM microscopic traffic simulation tools in modeling roundabouts. Proc. Comput. Sci. **52**, 43–50 (2015). https://doi.org/10.1016/j.procs.2015.05.016
31. G. Kotusevski, K.A. Hawick, A review of traffic simulation software. Res. Lett. Inform. Math. Sci. **13**, 35–54 (2009). https://mro.massey.ac.nz/handle/10179/4506
32. D. Krajzewicz, Traffic simulation with SUMO-simulation of urban mobility, in *Fundamentals of Traffic Simulation* (Springer, Berlin, 2010), pp. 269–293. https://doi.org/10.1007/978-1-4419-6142-6_7

33. K.W. Axhausen, A. Horni, K. Nagel, *The Multi-Agent Transport Simulation MATSim* (Ubiquity Press, 2016). https://doi.org/10.5334/baw
34. M. Wooldridge, *An Introduction to MultiAgente Systems*, 2nd edn. (Wiley, London, 2009). https://www.wiley.com/en-be/p-9780470519462
35. A. Jimenez, P. Cardenas, A. Canales, F. Jimenez, A. Portacio, A survey on intelligent agents and multi-agents for irrigation scheduling. Comput. Electron. Agr. **176**(2020), 105474 (2020). https://doi.org/10.1016/j.compag.2020.105474
36. S. Abar, G.K. Theodoropoulos, P. Lemarinier, Agent based modelling and simulation tools: a review of the state-of-art software. Comput. Sci. Rev. **24**, 13–33 (2017). https://doi.org/10.1016/j.cosrev.2017.03.001
37. B. Torabi, M. Al-Zinati, R. Wenkstern, MATISSE 3.0: a large-scale multi-agent simulation system for intelligent transportation systems, in *Advances in Practical Applications of Agents, Multi-Agents Systems and Complexity: The PAAMS Collection* (Springer, Berlin, 2018), pp. 357–360. https://doi.org/10.1007/978-3-319-94580-4_38
38. J. Auld, M. Hope, H. Ley, V. Sokolov, B. Xu, K. Zhang, POLARIS: agent-based modeling framework development and implementation for integrated traveldemand and network and operations simulations. Transp. Res. C **64**, 101–116 (2016). https://doi.org/10.1016/j.trc.2015.07.017
39. M. Jakob, Z. Moler, Modular framework for simulation modelling of interaction-rich transport systems, in *16th International IEEE Conference on IntelligentTransportation Systems*, ed. by IEEE (2013), pp. 2152–2159. https://doi.org/10.1109/itsc.2013.6728547
40. P. Mooney, M. Minghini et al., A review of OpenStreetMap data. Mapping and the Citizen Sensor (2017), pp. 37–59. https://doi.org/10.5334/bbf.c
41. S.M. Quintana, E.O. Pérez, L.M. Fernández, A.C.V. Criado, *Libr SIG: Aprendiendo a manejar los SIG en la gestión ambiental.* (Libr SIG, 2008). https://oa.upm.es/2080/
42. E.S. Ansola, *Servicios basados en localización para una infraestructura de datos espaciales.* Master's thesis, Instituto Superior Politécnico, José Antonio Echeverría, CUJAE (2015). https://repositorio.cujae.edu.cu/items/07b6d05c-f7f5-441a-8d63-0e324edb6910
43. A.M. Román, M.M. Espino, D.B. Fraga, Herramienta de simulación para evaluar configuraciones semafóricas. Revista Cubana de Transformación Digital **2**(1), 102–114 (2021). https://rctd.uic.cu/rctd/article/download/100/40
44. C. Gao, J. Xu, X. Jia, Y. Dong, H. Ru, Influence of large vehicles on the speed of expressway traffic flow. Adv. Civil Eng. **2020** (2020). https://doi.org/10.1155/2020/2454106
45. C. Morris, J.J. Yang, A machine learning model pipeline for detecting wet pavement condition from live scenes of traffic cameras. Mach. Learn. Appl. **5**, 100070 (2021). https://doi.org/10.1016/j.mlwa.2021.100070
46. C. Chen, X. Zhao, H. Liu, G. Ren, Y. Zhang, X. Liu, Assessing the influence of adverse weather on traffic flow characteristics using a driving simulator and VISSIM. Sustainability **11**(3), 830 (2019). https://doi.org/10.3390/su11030830
47. J.E.L. de Ramos, P. Mareschal, G.A. Closs, A.R. Maqueda, S.E. Galeano, M.D.P. de Florenciano, L.D. Cristaldo et al., Reducción de animales en las calles de la ciudad de Concepción dentro de un marco ético, equitativo y transparente. Investigaciones y Estudios-UNA **13**(1), 29–35 (2022). https://doi.org/10.47133/IEUNA22103a
48. C. Mo, Y. Li, L. Zheng, Simulation and analysis on overtaking safety assistance system based on vehicle-to-vehicle communication. Autom. Innovation **1**(2), 158–166 (2018). https://doi.org/10.1007/s42154-018-0017-9
49. S. Newstead, *An Investigation into the Relationship Between Vehicle Colour and Crash risk*, vol. 1(263) (Monash University. Accident Research Center, 2007). https://www.monash.edu/muarc/archive/our-publications/reports/muarc263
50. I. Chernyaev., E. Oleshchenko, I. Danilov, Methods for continuous monitoring of compliance of vehicles' technical condition with safety requirements during operation. Transp. Res. Proc. **50**, 77–85 (2020). https://doi.org/10.1016/j.trpro.2020.10.010.
51. A. Sedeño-Noda, M. Colebrook, A biobjective Dijkstra algorithm. Eur. J. Oper. Res. **276**(1), 106–118 (2019). https://doi.org/10.1016/j.ejor.2019.01.007

52. C. Dewi, R.C. Chen, Integrating real-time weather forecasts data using OpenWeatherMap and Twitter. Int. J. Inform. Technol. Bus. **1**(2), 48–52 (2019). https://ejournal.uksw.edu/ijiteb/article/view/2302
53. N.O. Alsrehin, A.F. Klaib, A. Magableh, Intelligent transportation and control systemsusing data mining and machine learningtechniques: a comprehensive study. IEEE Access **7**, 49830–49857 (2019). https://doi.org/10.1109/ACCESS.2019.2909114
54. R.M. Paradina, M.I. Noroña, Applications and challenges of adopting Internet of Things (IoT) in reducing road traffic accidents, in International, I.S. [ed.] *Second Asia Pacific International Conference on Industrial Engineering and Operations Management* (Surakarta, Indonesia, 2021), pp. 1210–1230. http://ieomsociety.org/proceedings/2021indonesia/204.pdf
55. A. Prasad, P. Chawda, Power management factors and techniques for IoT design devices, in *2018 19th International Symposium on Quality Electronic Design (ISQED)* (2018), pp. 364–369. https://doi.org/10.1109/ISQED.2018.8357314
56. P.P. Ray, A survey on Internet of Things architectures. J. King Saud Univ. Comput. Inform. Sci. **30**(3), 291–319 (2018). https://doi.org/10.1016/j.jksuci.2016.10.003
57. J. Li., Y. Xiang, J. Fang, W. Wang, Y. Pi, Research on multiple sensors vehicle detection with EMD-based denoising. IEEE Internet Things J. **6**(4), 6262–6270 (2019). https://doi.org/10.1109/JIOT.2018.2890541
58. M.A. Rakhonde, S.A. Khoje, R.D. Komati, Vehicle collision detection and avoidance with pollution monitoring system using IoT, in *2018 IEEE Global Conference on Wireless Computing and Networking (GCWCN)* (2018), pp. 75–79. https://doi.org/10.1109/GCWCN.2018.8668622
59. A. Sharif., J.P. Li, M.A. Saleem, Internet of things enabled vehicular and ad hoc networks for smart city traffic monitoring and controlling: a review. Int. J. Adv. Netw. Appl. **10**(3), 3833–3842 (2018). https://doi.org/10.35444/IJANA.2018.10031

Emotional Diagnosis for Employees Within the Framework of Industry 4.0: A Case Study in Ciudad Juarez

Florencio Abraham Roldan-Castellanos, Ivan Juan Carlos Pérez Olguín, Luis Carlos Méndez-González, and Luis Ricardo Vidal-Portilla

1 Introduction

Technology, society, economics, and many other aspects transform synergistically over time. Furthermore, few changes have such a profound impact on the preceding categories as industrial revolutions. The current industrial revolution has demonstrated how different modes of production and services can and will affect society on all levels, from the mass paradigm shift to an individual's adaptability. Due to the sheer importance of Industry 4.0, it is merely logical that there will be a plethora of consequences and ramifications of its implementation. Technology in all its areas of application has proven to develop along with the demand for entertainment, manufacturing production, war effort, etc. [1]. These demands often catalyze such technological innovations. As a result, the technology that underpins Industry 4.0 behaves similarly, initially changing and evolving in response to industrial requirements before progressing in response to social and legislative demands [2].

Based on the previous idea, it is frequently stated that new technological advancements provide a common social innovation order on the working and educational framework, which is a refocus of working skills [3]. Many demands on workers and students are shifting to learning and managing digital technologies instead of more traditional ones. Abilities like advanced programming and data analysis are starting to outshine the classic techniques and tacit knowledge required by manufacturing, service, and digital companies. Given that the workload in a traditional industry can already cause a hefty amount of acute stress (a typical non-long-lasting type of stress) within the workers, when adding such acute stress

F. A. Roldan-Castellanos (✉) · I. J. C. Pérez Olguín · L. C. Méndez-González
L. R. Vidal-Portilla
Universidad Autonoma de Ciudad Juarez, Ciudad Juarez, Chihuahua, Mexico
e-mail: florencio.roldan@uacj.mx; ivan.perez@uacj.mx; luis.mendez@uacj.mx; lvidal@uacj.mx

L. C. Méndez-González et al. (eds.), *Innovation and Competitiveness in Industry 4.0 Based on Intelligent Systems*, EAI/Springer Innovations in Communication and Computing, https://doi.org/10.1007/978-3-031-29775-5_11

to a constant connection setting, it is highly probable that chronic mental stress (a condition of continuous stress considered as clinical condition) will develop. In addition to such a stressful life setting, relaxation proves to be difficult in a mass-connected environment because of the difficulty of unplugging. As a result, some research suggests that employees not used to constant employment of digital technology (e.g., employees doing hand work or repetitive low-flexible actions) encounter an increased difficulty in adopting new Industry 4.0 measures, directly or indirectly affecting their emotional health either by increasing anger, frustration, boredom, and awareness or even by overstimulation [4].

Companies that attempt to deploy smart factory technologies want to reduce the risks associated with planning, determine the effects of the new setup on workers, avoid needing to redesign equipment, optimize resource use, decrease waste, and improve performance and adaptability [4]. As a consequence, it is reasonable to think that workers' safety and healthcare represent fields in which Industry 4.0 is fulfilling an increasing role [5]. Added to that trend, the digitalization of medical records, like diseases and medical checks, delivers better healthcare for workers and directly improves operational personnel's safety [6]. However, the enormity and complexity of these datasets present great challenges for analysis and subsequent applications in a practical clinical environment within an industrial framework [7, 8]. As a result, most of the data records or measures generate diagnosis indirectly (by a professional) and can represent the difference between a reliable and fast identifications from a slow and wrong one. Hence, several approaches for clinical detection in industrial environments have been delivered on many occasions [9–11], although most of them are centered on a one-diagnosis range [12, 13], meaning that if they focus on a sole condition which can be detected by an abnormal lecture of biometric data (as arrhythmias or elevated pressure), measuring emotions, for example, has become important in many countries since the creation of laws that promote the mental well-being of workers, such as the NOM-035 [14] in Mexico or OHSAS 18001 and CSA Z1000-06 in other nations. The majority of these laws are primarily concerned with workplace stress, requiring the containment and reduction of stressful stimuli for workers. The fundamental problem with this legislation relies on determining the path to detect anxiety and stress without confusing it with fear or excitement and even more strenuously to define that the actual stressful stimuli or condition comes from the working setting [15]. There have been several approaches to achieving an emotional classification in altered surroundings, such as measuring emotions of a specific spectrum in a controlled environment [16–19], but none of them has received any real application or actual proof. Instead, they intend to detect emotions and emotional states by the use of the emotional range, which can be deceiving and won't work as a reliable method for validating company regulations.

To address the previous situation, it is well known that Industry 4.0 carries some applications that allow the advancement of innovative methods to diagnose emotional conditions in an operating environment. For example, self-learning algorithms have received a significant impact on healthcare data uses in the last 5 years [12], the development of a medical identification or prediction system, particularly one that utilizes deep neural networks and other smart algorithms [20].

Based on that, an approach exercising artificial intelligence to detect an emotional range and posteriorly separate the objective emotion from the ones present in its emotional spectrum enhances a genuine possibility. The goal of this document is to demonstrate a practical application of emotional detection validated by the NOM-035. Because of the legislative nature of the norm in such an approach, the capability of proving stress and anxiety is a must. Therefore, to achieve this aim, a method to identify and classify emotions in a working environment is proposed.

2 Method

Next, the proposed method of how to read a particular emotion instead of the range depends on the technology and application that will be managed. Therefore, it is to construct a rough framework or process that can be employed in a real-world environment; in consequence, such procedure is mentioned in the list below:

1. Selection of sensorial tools
2. Development of data acquisition system
3. System programming:
    - Data repair and preparation
    - External variable analysis
    - Definition of objective values for data
    - Construction of a base classifier
    - Evaluation of base classifier
    - Optimization algorithm coding
    - Learning system for emotional characterization
    - Algorithm integration

2.1 Selection of Sensorial Tools

To begin with, the relevant tools for the development of the project should be distinguished, starting with the data acquisition methods, through a Technique for Order of Preference by Similarity to Ideal Solution (TOPSIS) analysis among the routinely used sensors for stress reading [26, 35]. Following the TOPSIS procedure, the selection process can be shown below:

1. Alternative selection
2. Determination of criteria
3. Weight analysis with AHP (analytic hierarchy process)
4. TOPSIS distance ratio analysis

It should be noted that each section of this procedure is weighted based on previous research and experimentation. Starting with the selection of alternatives, this is based on relevant research on multiple biometric signals and the corresponding sensors [15, 21, 22]. Consequently, the following four sensors were chosen for evaluation:

- Photo plethysmograph. This sensor allows the detection of cardiovascular signals. In that case, it will be employed for HRV (heart rate variability) detection. It was selected mainly for its versatility of readings and its already mature use in the medical industry [23, 24].
- Galvanic response sensor. This sensor allows reading of electrical conductivity in the skin. This sensor was selected for its wide employment in emotional reading [18].
- Electrocardiogram. This system is widely used in medicine and shows the behavior of the cardiac subsystem, so it is considered relevant for emotional changes and for previous research where it was applied [25, 26].
- Electroencephalogram. This system is characterized by obtaining data directly from the cranial vault and is widely utilized for reading emotional changes and medical reactions within the same system [25, 27, 28].

Continuing with the analysis of criteria, it is significant to look for criteria that prove to be relevant to the research. In this case, multiple attributes that are critical at the time of employing a viable stress reading were raised. These are mentioned and expanded in the following list:

- Data reliability. Determining how reliable a body indicator is depends on whether it delivers a clear reading after a nominative time of duration or use under specified conditions.
- Reading speed. A common difficulty in reading biological values is obtaining digital readings at a frequency close to the generation of the signal. This is because biometrically related readings are more functional if they are close to real-time behavior.
- Acquisition flexibility. Within the biometric signals, there are indicators more oriented to specific phenomena, that is, measuring skin sweat is not effective to determine blood acidity; what is measured from the human body must correspond to the occurrence analyzed. But even within the analysis of signals from specific systems, phenomena that demand a more invasive to be read. Consequently, simpler signals to acquire are preferable, that is, signals that handle sensors and methods that are not very invasive and painful.
- Robustness of readout. Any digital system is prone to noise and biased readings, so a critical factor in receiving biological signals must always be the use of robust sensors that are little affected by external phenomena or noise.

2.2 Development of Data Acquisition System

Firstly, to understand how to get biological signals and use them as data, an acquisition system must be constructed; it is important to have the following variables in mind when designing it:

- Where is the signal going to be taken? Meaning, the environment where it gets used, whether heavy duty or just a nonhostile one.
- How much use is the acquisition system going to receive? Frequency, or the amount of use and length of time in which it has to work.
- In which type of climate does the hardware have to perform? If the ambient is hot or cold, the electronic construction would change, demanding different specs from the circuits.
- Which electronic demands have the system and biomarkers that would be read? For all the biological systems that may require a different electrical system to be measured, a previous analysis is recommended.

As a consequence, the development process of the proposed acquisition system is shown below:

1. Sensorial system identification. To read the biological phenomenon and turn it into valuable data, a sensorial system is used.
2. System block diagram. When creating the preliminary production design, you begin by defining the system-level block diagram. This diagram specifies each electronic function and how all functional components are interconnected.
3. Select the production components. Next, you must select the various production components, microchips, sensors, displays, and connectors, based on the desired functions and the selling price to the target audience of your product. This will allow you to create a preliminary bill of materials (BOM).
4. Design the circuit schematic. Now, you continue with the design of the schematic circuit diagram based on the system block diagram created earlier. The schematic diagram shows how each component is connected, from the microchips to the resistors. While a system block diagram focuses primarily on the high-level product functionality, a schematic diagram focuses on the small details. Something as simple as a misnumbered pin on a component in a schematic can cause a complete lack of functionality. In most cases, a separate subcircuit was needed for each block in the system block diagram. These various subcircuits were connected to form the complete schematic circuit diagram.
5. Evaluate, program, debug, and repeat. Now it is time to evaluate the prototype electronics. Keep in mind that the prototype will rarely work perfectly. It will most likely go through several iterations before finalizing the design. In this section, we identified, debugged, and fixed many problems with the prototype.
6. Encapsulation prototyping. For most products, this includes at least the enclosure that holds all the system components together, which functions as the package.

Once the relevant criteria were obtained, the AHP pairwise analysis technique was used. This hierarchical analysis allows determining the importance of the criteria, giving the ability to generate weights in the decision. The following procedure is used for the AHP work procedure:

- Step 1: Determine the final objective of the process. In this case, compare selection criteria.
- Step 2: Make a list of the different options available to achieve your objective.

- Step 3: List the different criteria by which the above alternatives are evaluated. This may include sublevels of criteria and depends entirely on how much control you want, especially when the options are very similar.
- Step 4: Build the hierarchy from step 1 to step 3. The links between each node indicate that they are considered during the evaluation process.

Now, to build the program in a programming environment (Python-based), it is necessary to expand the procedures so that the steps followed for the analysis are explained. These procedures are shown below:

1. Library importation. In this case, the mathematical support libraries pandas and NumPy were used.
   - NumPy consists of a free-use open-source (based on terms of the modified Berkeley Software Distribution (BSD) license) project for numerical computing in the Python language. It is used as a base for mathematical and statistical structures, improving the constructions such as arrays.
   - Pandas library allows the expanded use of NumPy by giving an interface for third-party libraries to define their own personalized data types. In the current study case, Pandas works to improve the use of matrixes for data calculation and allow communication and exchange of data with Excel sheets.
2. Constructing the pairwise comparison matrix. For this purpose, a declaration of the number of options to be compared was used, and an identity matrix was built based on the number of options, filling in the nonidentity elements utilizing loops and user input queries to define levels of importance.
3. Once the matrix is generated, we calculate the priority vector by normalizing the eigenvector of the largest eigenvalue. The elements of this eigenvector are the criteria weights.
4. Both codes are joined in the form of a function, to be called at the time of generating comparison matrices between pairs of each criterion. In addition, the consistency radius is calculated based on the number of options.
5. The last step is to add their weighted calculations to obtain the ranking vector.

2.3 System Programming

Continuing with the approach for system integration, the following paragraph describes the method employed for the programming of the algorithm. Describe every technique required for its construction, and at the end, describe the algorithm flow.

2.3.1 Data Repair and Preparation

This section describes the models and procedures proposed to handle the initial data that could be obtained from the acquisition system. Next, the procedure is explained in the form of a listing of the approaches to the repair and preparation of the signal obtained by means of the Python programming Integrated Development and Learning Environment (IDLE):

1. Pass the initial signal value through the process of normalization of values. This is done by the method of minimum reduction, employing the following equation:

$$y = \frac{(x - \min)}{(\max - \min)} \tag{1}$$

2. Standardize signal values to reduce possible spuriousness and noise:

$$y = \frac{(x - \text{media})}{\text{desviacion estandar}} \tag{2}$$

2.3.2 External Variable Analysis

To perform an analysis of the external variables that affect sensory reading, the use of various statistical tools embedded in machine learning systems is proposed. Expanding on the aforementioned, the first analysis proposed is a Pearson correlation index (included in a confusion matrix in Python). This index allows the evaluation of correlations between variables. This type of review is accurate when analyzing factors like humidity, temperature, hours of sleep, age, etc. Since in the case of a significant correlation between two variables, the less influential one can be disregarded, reducing processing consumption in the algorithm. To be able to analyze the data, a general approach to development through specific systems was taken, in such a way that minor methods can be constructed for a general objective. To analyze the data efficiently, the following sequential method of biometric data processing is proposed:

- Definition of objective values for the data
- Construction of a base classifier
- Evaluation of base classifier
- Construction of final vectors for the deep neural network
- Review of target values

2.3.3 Definition of Objective Values for Data

In any intelligent system based on machine learning, it is necessary to include references or examples of the objective or goal that the system should obtain or seek.

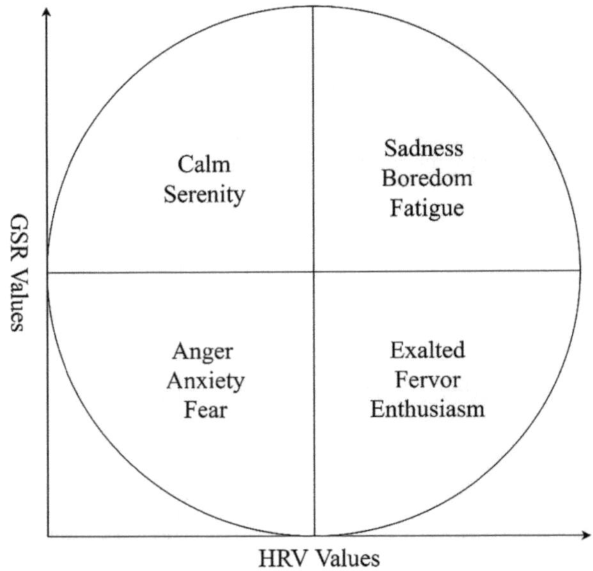

Fig. 1 Emotional spectrum

Therefore, the first thing to be done when processing data employing computational algorithms is to set such objectives. To determine the search for the ideal outputs of the algorithm, the use of common clinical values that define the presence of the emotional range of anxiety was proposed. Because of the utilization of two main signals and a reference signal, the following pattern of target values is proposed:

- Based on Fig. 1, it can be assumed that changes of reduction in the biometric variables present a behavioral pattern within the emotional range of stress anxiety.
- Within the similar emotional range mentioned above, there are several types of emotions with similar biomarkers. Therefore, it is necessary to eliminate possible false positives between emotions, and consequently, it became essential to transfer data verified with the corresponding target emotion, that is, samples of stress, anxiety, excitement, fright, anger, etc.
- After defining the samples, a base classifier was employed to construct the target values mentioned in the previous step.

2.3.4 Construction of a Base Classifier

It started with the processing of the data, utilizing a machine learning classifier that provides an efficient computational processing value without losing efficiency. To achieve this goal, two classification algorithms, K-nn and Naive Bayes, were selected. First, for the construction of the K-nn algorithm, the standard procedure was used in a general way:

1. Calculate the Euclidean distance between the object to be classified and the rest of the objects in the training input data:

$$d\left(x, x'\right) = \sqrt{(x_1 - x\prime_1)^2 + \cdots + (x_n - x\prime_n)^2} \tag{3}$$

2. Select the k closest elements.
3. Perform a majority voting among the k points:

$$P\left(y = j|X = x\right) = \frac{1}{K}\sum_{i \in A} I\left(y^i = j\right) \tag{4}$$

4. Class determination.

Continuing with the classification, the Naive Bayes algorithm also allows the classification of data, using similar methodologies to the K-nn algorithm, but with some critical differences as mentioned in the previous chapter. The procedure for this algorithm is as follows:

1. Calculate the prior probability for class labels:

$$P\left(y|X\right) = \frac{P\left(X|y\right) * P(y)}{P(X)} \tag{5}$$

2. Find the likelihood probability with each attribute for each class:

$$P\left(X|y\right) = P\left(x_1, x_2, \ldots\ldots., x_n|y\right) \tag{6}$$

3. Put these values into Bayes' formula and calculate the posterior probability:

$$P\left(y|x_1, x_2, \ldots.., x_n\right) \alpha P(y) \prod_{i=1}^{n} P\left(x_i|y\right) \tag{7}$$

4. Observe which class has a higher likelihood, given that the input belongs to the highest likelihood class.

2.3.5 Evaluation of Base Classifier

To evaluate the classifier and its performance, the use of a simple data split validation was proposed, utilizing the following procedure:

- A set of previously verified data is given, that is, with a known result.
- The data is divided into two subsets, one containing 80% of the data and the other comprising 20% of the remaining data.
- The subset of data consists of 80% of the sample size and is used to train the classifier.
- The 20% subset of data is used as a test, in such a way that the training done with 80% of the data should predict the remaining 20%.

Based on the above, the efficiency of the algorithm was evaluated by comparing the quantity of accurate predicted data with the amount of inaccurate data. In addition to the common assessment, the use of a confusion matrix accompanied by precision, accuracy, and sensitivity evaluations was considered. This is proposed to avoid unbalanced classes and the biases they have as a consequence, and given the presence of unbalance, methodologies for its treatment and adjustment will be reviewed.

2.3.6 Optimization Algorithm Coding

To code the optimization algorithm, a metaheuristic known as PSO or particle swarm optimization (PSO) is used, which works as follows:

1. Create an initial population swarm of n number of random particles. Each particle within the population generally consists of four fundamental elements:

   (a) Position. The position represents a certain combination of outcome values of the particles (based on the objective function); each particle has a different position value and depends on the particle dimension, that is, a two-dimensional position consists of a two-number outcome (Eq. 8):

$$P_i^t = \left[x_{0,i}^t, x_{1,i}^t, x_{2,i}^t, x_{3,i}^t, \ldots x_{n,i}^t\right] \tag{8}$$

   (b) Objective function. The objective function is the optimization relation to which all particles are subject.

(c) Velocity. The velocity value indicates where each particle moves to (in the form of a vector, as shown in Eq. 9):

$$V_i^t = \left[v_{0,i}^t, v_{1,i}^t, v_{2,i}^t, v_{3,i}^t, \ldots, v_{n,i}^t\right] \tag{9}$$

2. Construct a record of the best position the particle has been in up to that point in the cycle.
3. Evaluate each particle within the objective function.
4. Based on the objective function evaluation result, update the position and velocity of each particle, known as the motion (Eqs. 10 and 11):

$$P_i^{t+1} = P_i^t + V_i^{t+1} \tag{10}$$

$$V_i^{t+1} = mV_i^t + c_1 r_1 \left(P_{\text{best}(i)}^t - P_i^t\right) + c_2 r_2 \left(P_{\text{bestglobal}}^t - P_i^t\right) \tag{11}$$

5. If a high criterion is not met, return to step 2.

2.3.7 Learning System for Emotional Characterization

For the implementation of the deep learning system, the following procedure is necessary (Fig. 2):

1. Define the structure of the learning model. In the case of this analysis, the diagnosis won't rely completely upon the form of the neural network. Therefore, a feedforward artificial neural network is the one that balances performance with computational cost the best.
2. Introduce the essential characteristics and parameters of the model (number of iterations, layers, size of layers, and learning rate).
3. Execute the main loop:

   (a) Error calculation (forward propagation):

      (i) Pre-activation function: The weighted sum of the inputs is calculated in this function.
      (ii) Activation function: Based on the weighted sum, an activation function is applied to create the nonlinear network and make it learn as the computation progresses.

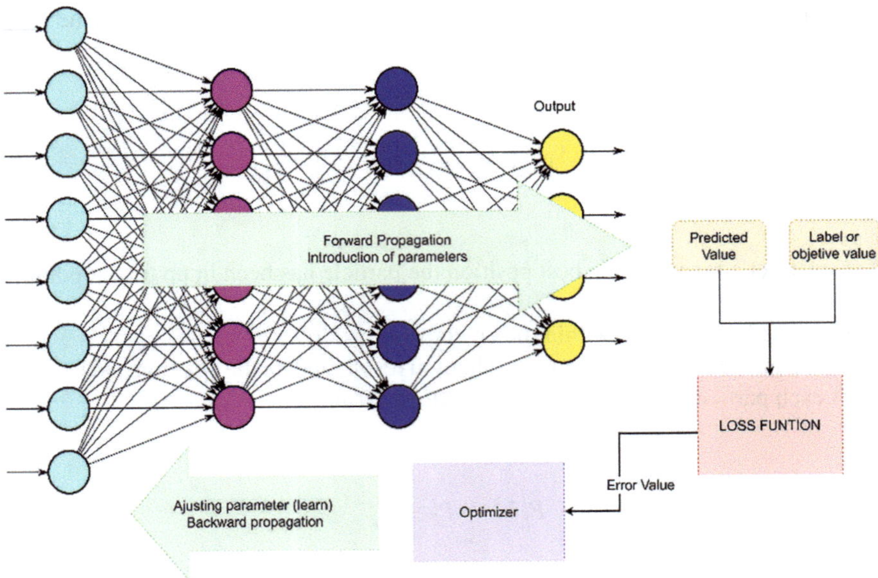

Fig. 2 Deep learning model representation

   (b) Gradient calculation (backward propagation).
   (c) Update parameters.

4. Value prediction based on the model.

It is necessary to specify that the network has supervised learning, so we must give the output parameters or objective values. Consequently, the programming of the learning system is derived from the initial classifier and should be seen as a continuation of the methodology of the first classifier and optimization code.

2.3.8 Algorithm Integration

Lastly, the integration of the algorithm, this section shows the proposed method of joining together to obtain a confirmed diagnosis for a specific emotion, going from the data acquisition to the synergy of the clarification and intelligent algorithms. The main method is displayed in Fig. 3.

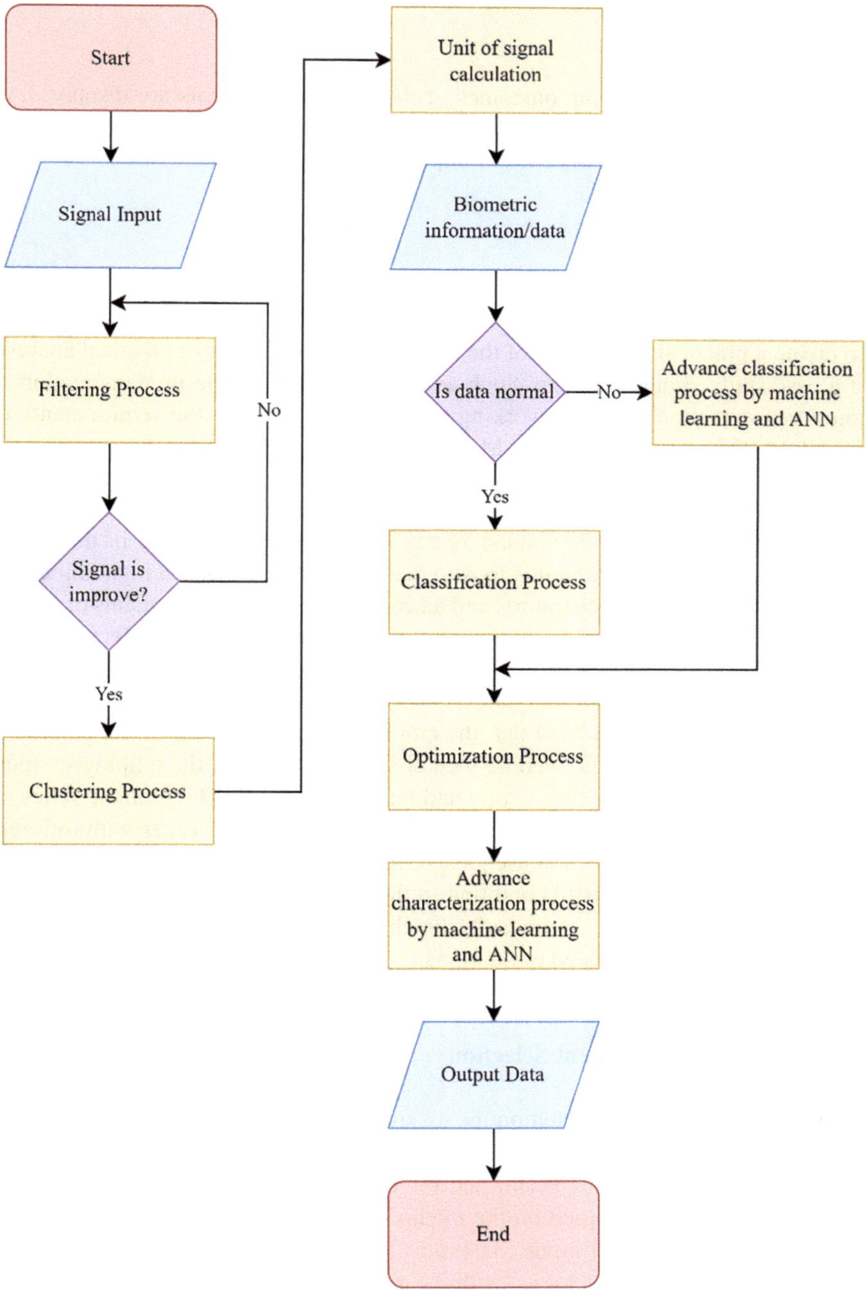

Fig. 3 Proposed methodology for diagnosis

3 Results

In the following section, the outcomes of the previous segments are displayed, as well as some minor notes that don't affect the end result but are important to take into account if replicating the current implementation.

3.1 Case of Implementation

To create a practical framework of the topics mentioned before, a practical analysis of a case study should be done. Such analysis results from the implementation of emotional detection within a working environment to fulfill the requirements of the NOM-035 in Ciudad Juarez, Mexico. First, let's specify the parameters and conditions of the case; this experiment was conducted at a Mexican University, the university located in Ciudad Juarez on the northern frontier of Mexico. The Universidad Tecnologica de Ciudad Juarez plays an important role in the higher education of young people in this locality and therefore demands a high amount of work from students, administrators, and teachers alike. Such requirements of mental and physical load can create various emotional conditions that affect different measures of all they involve.

In Mexico, a new federal law (NOM-035) [14, 29] was created to reduce stress in workers. This law stipulated that the employer must take care of the emotional load of their employees, defined as mental stress. Therefore, the employees must have the less stressful working setting and iteration possible. The problem relies on how you diagnose mental stress in a worker, especially in a worker with different stressful environments. As a consequence, to fulfill the NOM-035, it is necessary to prove or show the stressful condition in the working environment and not in the home or somewhere else. As a result, the characterization of emotions becomes a critical step for the mentioned evaluations.

3.1.1 Result of Equipment Selection

The idea for the implementation is to sort emotional conditions through the diagnosis of stress or anxiety by resorting to technological means instead of a psychological one; this was performed to complete the NOM-035 revision. First, to find the emotion, we based on the methodology (Fig. 3) to recognize a specific emotion; therefore, signal input constitutes the first step to detecting an emotion. Biological signals can be obtained with the use of biometrical sensors. Such devices must be designed and implemented according to the needs of the diagnosis and the biological phenomenon which is to be detected, hence the analysis of sensor selection and hardware design. For this case, the first analysis was done by a multicriteria decision-making approach, applying the Fuzzy TOPSIS (FTOPSIS) method to select the best sensor among the various options possible, because this diagnosis was used to validate the fulfillment of a working law, and the key criteria for sensor selection were data reliability, speed of the data acquisition, the flexibility of the lecture and data acquisition, and robustness, obtaining Table 1:

This table functions as mentioned above as a decision matrix arrangement, in which the source of the information obtained for the construction of its level or criterion values is added. It should be noted that the values are expressed in a nonnumerical way, but to improve the implementation, the values were averaged by literature within the computer program.

Continuing with the analysis of sensor alternatives, Table 2 presents the result of the geometric averages; these consist of the fuzzified result of the ratio of distances to the worst and best solution. The ratio provides the basis of the TOPSIS method. To be able to utilize a scale of values in the FTOPSIS method, the center of the area was sought between the three fuzzy values, the latter being our defuzzied alternative value. Once the fuzzy value was defuzzy, the result was obtained as in a common TOPSIS, as shown in Table 3, using the variables S+ and S- which denote distances to the best and worst solution, respectively, resulting in a performance score, which indicates the final value of preference of one alternative over another.

Table 1 Decision matrix

Criteria	Source	Data reliability	Lecture speed	Flexibility in data acquisition	Lecture robustness
PPG	[22, 27, 30]	Very high	High	High	Low
GSR	[31, 32]	High	Medium	Very high	Low
ECG	[27, 33, 34]	Very high	Medium	Very low	High
EEG	[27, 35, 36]	Very high	Very high	Very low	Medium

Own elaboration

Table 2 Fuzzy result of decision

Fuzzy geometrical mean			Fuzzy weights			Middle of area
2.63214	3.20108	3.72241	0.36602	0.55957	0.80614	0.57724
0.86334	1.00000	1.49534	0.12005	0.17480	0.32383	0.20623
0.28117	0.33031	0.40824	0.03909	0.05774	0.08841	0.06175
0.84089	1.18920	1.56508	0.11693	0.20788	0.33894	0.22125

Own elaboration

Table 3 Result of performance value of sensors

Alternative	Si+	Si-	Performance value
PPG	0.02538570	0.07955816	0.75810209
GSR	0.06374639	0.06480123	0.50410291
ECG	0.06480123	0.06374639	0.49589709
EEG	0.07396403	0.03877186	0.34391761

Own elaboration

3.1.2 Signal Input

Once the data source was reviewed, it proceeded to define the construction of the physical or data reading and processing system. This system is based on three main sections, the first one being the sensory system, followed by the controller, and ending with a computer. Specifically, the controller functions as the main link with the sensors of the sensory system, regulating the operation and inputting information to it, while the computer functions as the storage and advanced processing system, as shown in Fig. 4.

As shown in Fig. 4, the reading or data acquisition system consisted of the following operational base points:

1. Data reception by selected sensors. This section consists of the acquisition of the biometric signals via the most optimal sensory systems obtained through previous analysis.
2. Change of signals to useful data. The second section constitutes the part in which the signals are filtered and converted into useful information results for further analysis, this by means of a suitable controller and computer program.
3. Data transfer from controller to computer. This phase only consists of the communication of valuable data from the controller to the computer, this for analyses that demand more computational power.
4. Obtaining a computational dataset sample.

For the construction of a system that complies with the abovementioned points, the following components are proposed in their respective systems:

- Bill of materials for the PPG system (Table 4).

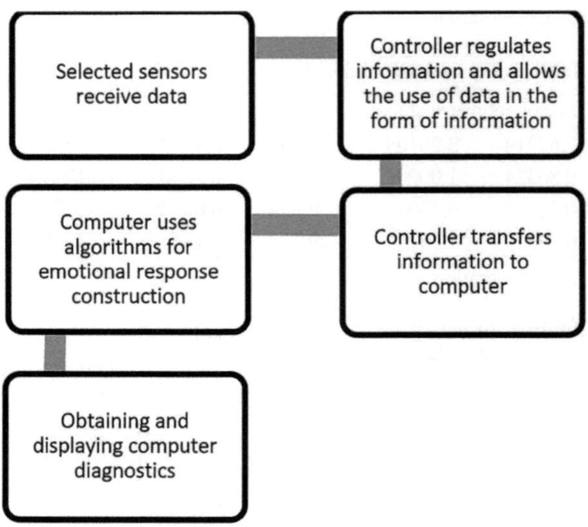

Fig. 4 Data processing system

Table 4 Bill of materials, PPG system

Part number	Equipment name	Value	Quantity	Description	Packages
1	Resistor	1 kΩ	2	Resistor ceramic	N/A
2	Resistor	10 kΩ	3	Resistor ceramic	N/A
3	Resistor	100 kΩ	1	Resistor ceramic	N/A
4	Resistor	100 Ω	1	Resistor ceramic	N/A
5	Resistor	120 Ω	1	Resistor ceramic	N/A
6	Capacitor	1 kF	1	Capacitor ceramic	N/A
7	Capacitor	200 μF	1	Capacitor ceramic	N/A
8	Operational amplifier	20–200	1	Amplifier	LM386
9	Sensor PPG	N/A	1	Analogic sensor	N/A
10	Controller	N/A	1	Controller	Micro PI

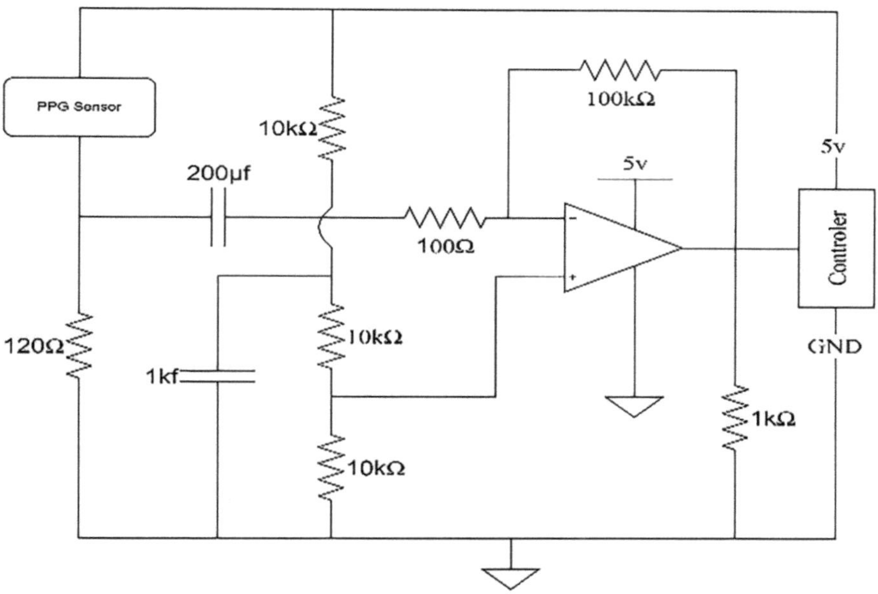

Fig. 5 PPG electronic schematic design

Once the system has been analyzed, it is important to define how to design and build the data acquisition system. Based on the selection of sensors, it is necessary to build complementary electrical systems for the purification of the sensory signal. For this reason, each sensor is accompanied by a filtering circuit, starting with the basic filtering circuit of the PPG sensor (Fig. 5). The circuit generates a basic conditioning, defined by a low-pass filtering, followed by an amplification with a gain factor of about 1000. Similarly, some protection components are placed before entering an analog input to a controller, resulting in a more reliable and smoother signal that will reduce the need for robust digital filters.

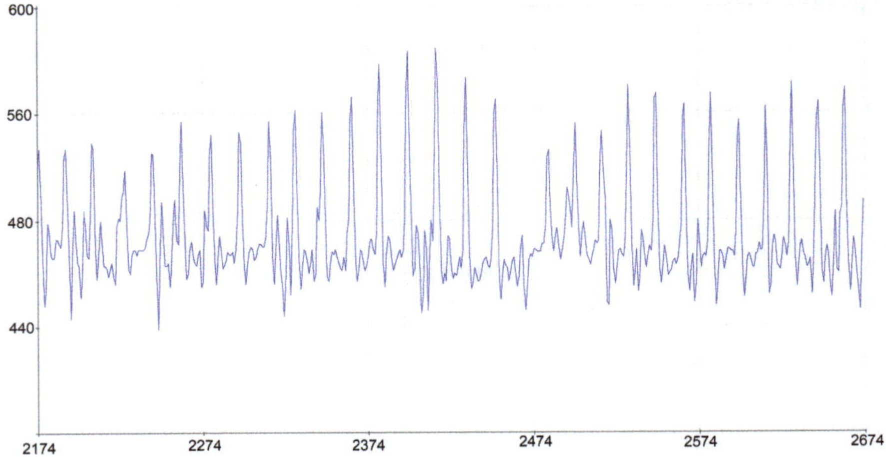

Fig. 6 PPG signal

For the PPG sensor signal, unlike skin conductivity, which could be inferred from a direct measurement of the electrode through voltage, a plethysmograph will only detect the presence of blood flow. Therefore, a more comprehensive analysis of the desired signal is necessary in this case; HRV was selected for the motives mentioned previously in this paper. This behavior in Fig. 6 demonstrates only analogic input behavior, but the signal can be analyzed as rising and falling patterns that represent a basic heart rate cycle, which in turn could be expanded to a beats per minute reading.

Then, to obtain the cardiac variability value of HRV, it is necessary to analyze the cardiac rhythm (Fig. 7). Within the signal, there are some intrinsic variables of the biological process, consisting of specific parts of the beat or flow. In this case, we used the variable P like the use in electrocardiograms to determine the respective peak of the pulse, which comes from the maximum blood flow, while QS is prepared for the reduction of the flow up to T, which should represent the relaxation of the blood flow. Within this reading, several observations can be obtained from the signal that may help its analysis and improvement, being the following:

1. The signal tends to be direct, that is, for a cardiac signal, it appears to be fast and at particularly high rates.
2. The variation in the limits is minimal; an extra filter would be prudent for its analysis.

Once an underlying rhythm has been obtained, the involuntary variability of the cardiac pulse is detected in the form of HRV. To achieve this, the frequency between beats and positive peaks must be estimated by calculating the root mean square difference, or RMSSD, based on the formula shown below, where RR_i represent the number of intervals detected:

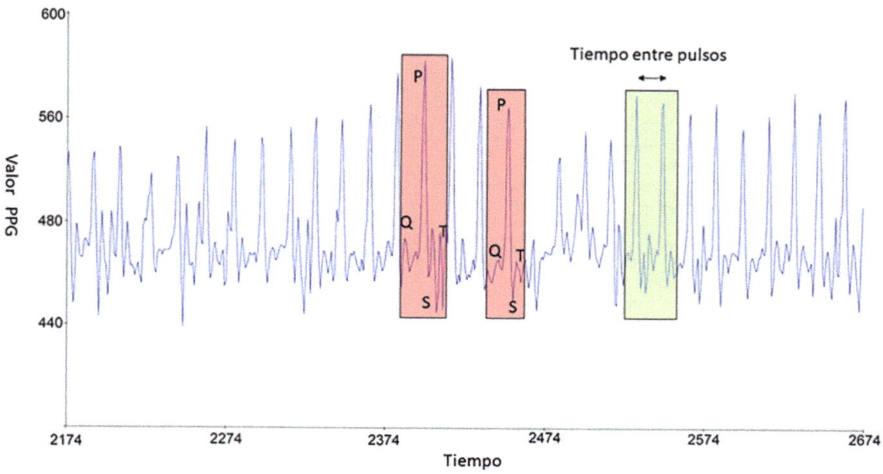

Fig. 7 Relevant sections of the PPG signal

Table 5 Bill of materials, GSR system

Part number	Equipment name	Value	Quantity	Description	Packages
1	Resistor	1 MΩ	1	Resistor ceramic	N/A
2	Resistor	4.7 kΩ	2	Resistor ceramic	N/A
3	Resistor	100 kΩ	2	Resistor ceramic	N/A
4	Resistor	200 kΩ	4	Resistor ceramic	N/A
5	Resistor	120 Ω	1	Resistor ceramic	N/A
6	Capacitor	100 μF	2	Capacitor ceramic	N/A
7	Operational amplifier	10–100	3	Amplifier	LM324
8	GSR sensor	N/A	1	Analogic censor	N/A
9	Controller	N/A	1	Controller encapsuled	Micro pi

$$\mathrm{MSSD} = \sqrt{\frac{\sum_{i=1}^{n-1} (RR_i - RR_{i+1})^2}{n-1}} \tag{12}$$

– Bill of materials for the GSR system (Table 5)

Continuing, the filtering of a GSR signal should be more complete due to the difficulty of acquisition (position and motion). Because of this complexity, for the conditioning circuit of the galvanic response system, the circuit shown in Fig. 8 was proposed.

Once a reliable data source has been constructed, it proceeds to the design and coding of the algorithm for filtering. This algorithm was handled at different levels and with diverse specific functions, that is, the algorithm itself is composed of more than one functional subcode, directly integrated for information reading

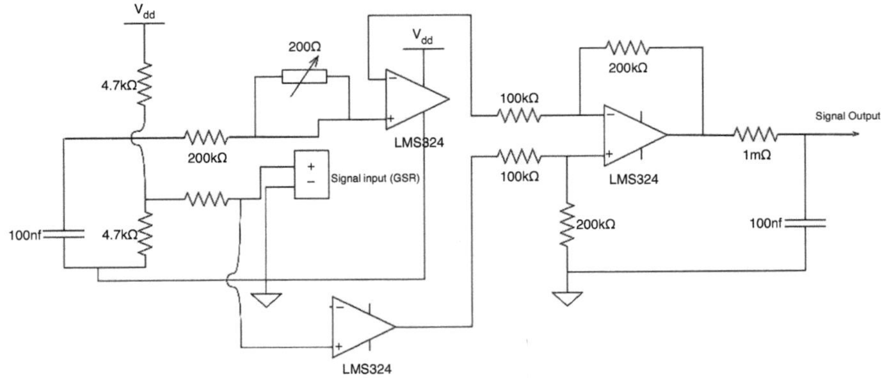

Fig. 8 GSR electronic schematic design

and complementary processing. To begin with, it is necessary to program the data acquisition system, which is built on a sensory network. To achieve this, the code was divided into three primary functions, so that each reading can be evaluated individually within the signal lecture, before being integrated into a communication function for further advanced processing.

In the following section, the sections involved in the reading of biometric signals are shown in detail, starting with the GSR data. Based on the theoretical framework, it is known that the GSR reading corresponds to a signal that evaluates the micro-sweat exposed on the skin surface and is measured according to the electrical conductivity of the skin surface. Based on the above discussion, the first thing to analyze is what the sensor measures; beyond the biological phenomenon, the sensors are digital and therefore electronic, that is, they read electrical values that indicate the presence of a biological phenomenon. In the case of GSR, the sensor seeks to find electrical resistance, but the sensor itself cannot detect electrical resistance on a skin surface. Consequently, the sensor opts for a more practical reading, the voltage measurement. In summary, it can be stated that to identify a GSR parameter, it is necessary to invert the reading value. In other words, the higher the voltage detected by the sensor, the lower the skin resistance and vice versa. As a consequence, a conductivity reading is considered and should be handled and measured in Siemens/meter; this unit is classified as the inverse of resistance, with the following representation:

$$R = \Omega = \text{Resistence}$$

$$\sigma\ \acute{o}\ \Omega^{-1} = \text{conductivity} = \frac{1}{\Omega} \tag{13}$$

Based on the above discussion, Fig. 9 shows how the signal of skin resistance comes into the controller:

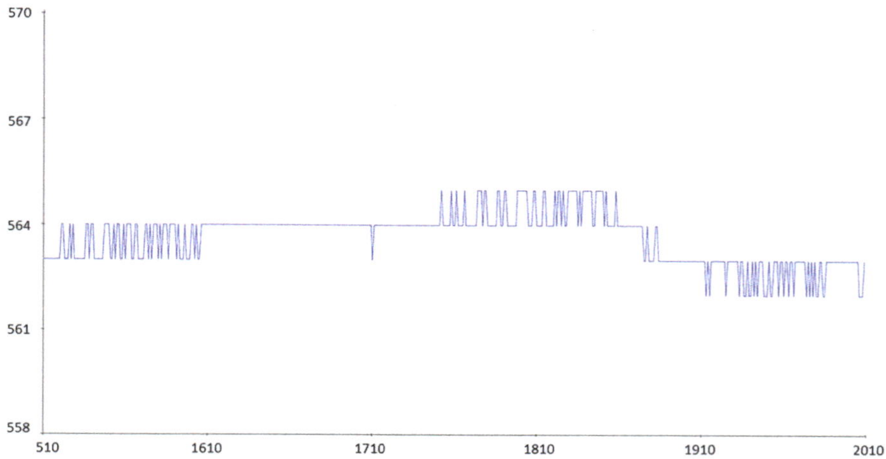

Fig. 9 GSR signal sample

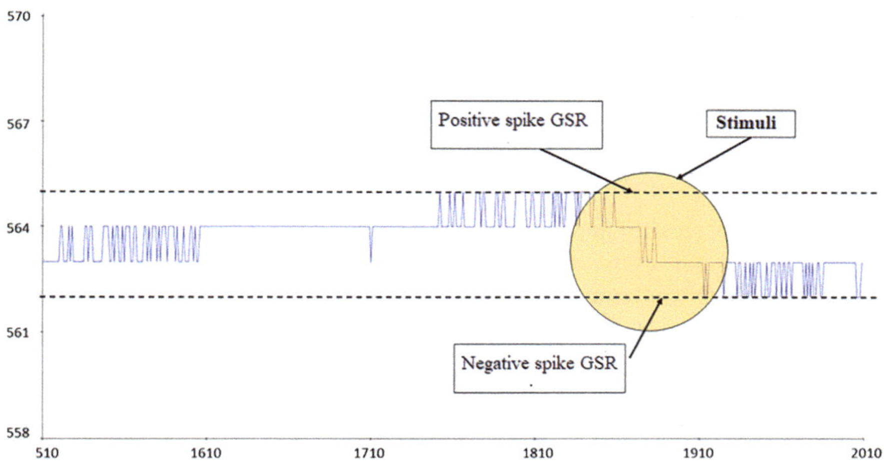

Fig. 10 Relevant sections of the GSR signal

Subsequently, it is necessary to analyze behaviors; in the case of Fig. 9, the graph was made in conjunction with external stimuli to observe a change in the base values. For this, two main aspects are reviewed, the positive peak and the negative peak, which are generally presented a few seconds after the stimulus. Such peaks can identify different emotional responses, with the low amplitude being stress, concentration, or anxiety reaction and the positive peak being fun, calmness, or an emotion within the range of relaxation, as shown in Fig. 10.

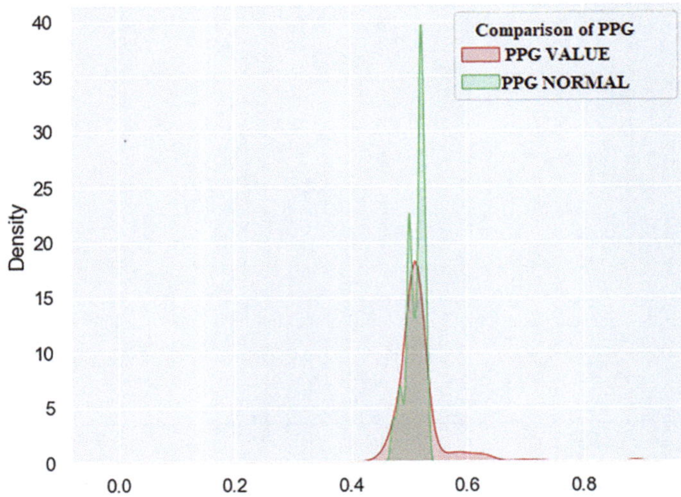

Fig. 11 PPG data normalization

3.1.3 Data Processing

For basic data processing, digital data normalization was performed on both signals, as shown in Fig. 11.

As can be appreciated in both graphs, the noise was considerably reduced, which is especially useful when working with the GSR signal (Fig. 12); although with reference to the PPG density plot (Fig. 11), this procedure generated a reduction of signal peaks, which is quite useful for the reading of heart rate variability.

3.1.4 Data Gathering

The gathering of data comprises an important section of any real application of technological diagnosis, although it's significant to mention that the data gathering depends heavily on the company or industry in which it is applied. In this case of study, the data is collected in a university to define the level of anxiety and stress of workers and administrative personnel. As mentioned before, the system for data input was implemented in the Universidad Tecnologica de Ciudad Juarez to approve a federal normative; in contrast to other diagnosis applications, time remains not an issue because the main focus is to obtain the general status of anxiety and stress within this workplace. Because of this, the university emphasizes the general diagnosis on location and the construction of the lecture environment.

First, in order to obtain a value of emotional health within the workforce, measures must be taken, by using the hardware previously described to obtain the required biological data. To achieve the emotional diagnosis, the biometric lecture must be as controlled as possible, meaning that it requires similar conditions

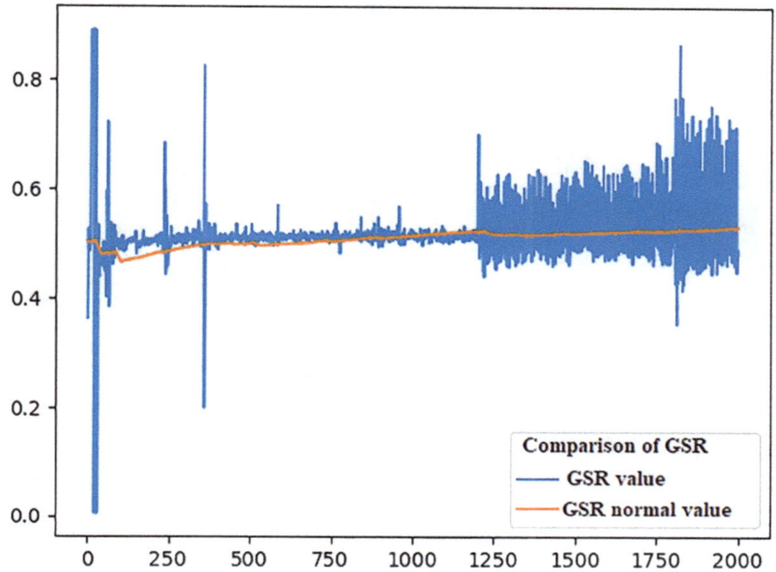

Fig. 12 GSR normalized data

for all measure subjects; as a result of this requirement, all the interviews were conducted in the same room in the same building (Fig. 13). It is important to remark that the increase in control of the data acquisition creates a major restriction on how well an emotion can be detracted. This is because most of the time, the administrative personnel, teachers, and general employees are taken away from their actual workplace.

After the first operational tests of the system and interviews, it was reviewed that the external variables affected the readings quite a lot even with the signal preprocessing procedures; consequently, from this result, it was decided to incorporate them directly in the processing code and emotional classification, but it must be understood that the external variables are those that are not included within the biometric readings and due to the external changes of the environment when taking data lectures, because of their constant shift they are considered as a critical sector of the system and diagnosis algorithm.

To address the variables, a pre-assortment was proposed by using a simple questionary for each interview individual, as a consequence, allowing the measure more flexibility and more adaptability by implementing the answers as a parameter for emotional classification. To acquire such outer values as well as some relevant medical data, a scientifically use only format was created with the assistance of the judicial department of the university and a questionary with mostly numerical values to obtain the data as shown in Table 6.

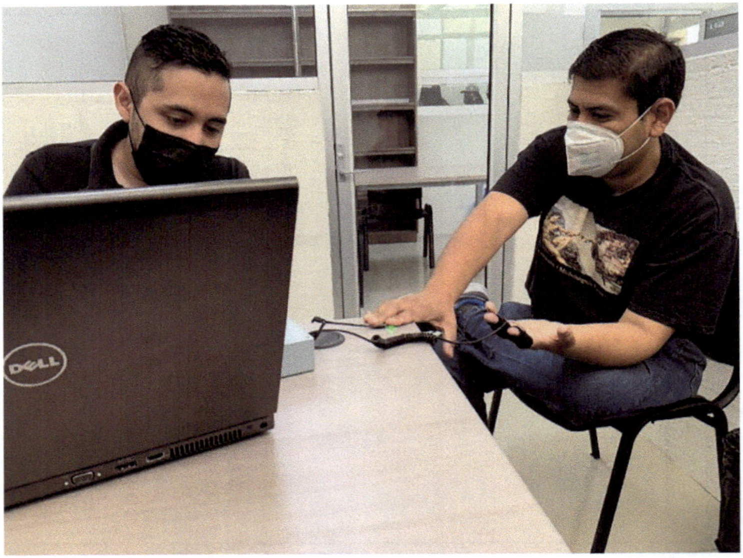

Fig. 13 Data gathering

Table 6 External data acquisition format

Number of samples	V_1	V_2	V_3	V_4	V_5	.....	V_n
1							
2							
3							
...							
n							

where V_1 = pre-existing cardiac conditions; V_2 = previous respiratory conditions/including caused by Covid-19; V_3 = fever in the last 10 days/including caused by Covid-19; V_4 = how good is the subject at handling anxiety, stress?; V_5 = number of hours of exercise per week; V_6 = hours working per day; V_7 = conditions affecting blood flow; V_8 = hours of sleep per day; V_9 = consumption of foods high in caffeine; V_{10} = age; V_{11} = gender; V_{12} = marital status; V_{13} = temperature; and V_{14} = humidity and V_{15} = emotional state (accordant to subject)

3.1.5 Data Classification and Case Results

The classification works in two major ways in the current case: First a minor classification algorithm was used to find goal values; in that case Naïve Bayes, this procedure is important due to the fact that emotions work at ranges; therefore, it's necessary to differentiate emotions like anxiety from fear because both are in the same spectrum. Second, the classification Naïve Bayes is useful for finding similar values for perdition allowing the user to separate and order specific input data into a label, thus separating emotional ranges.

Table 7 Evaluation of anxiety ranges on external variables

Algorithm	Variable relation	Sensitivity	Precision
KNN	Emotional range control predicting emotional anxiety and stress	0.80	0.80
Naïve Bayes		0.85	0.90
KNN	Hours of sleep predicting emotional condition anxiety and stress	0.90	0.80
Naive Bayes		0.80	0.80
KNN	Working hours predicting emotional condition anxiety and stress	0.80	0.70
Naive Bayes		0.80	0.80
KNN	Hours of exercise predicting emotional condition anxiety and stress	0.70	0.70
Naive Bayes		0.70	0.70
KNN	Caffeine or sugar consumption predicting emotional condition anxiety and stress	0.70	0.70
Naive Bayes		0.80	0.80
KNN	Age predicting emotional condition anxiety and stress	0.023	0.023
Naive Bayes		0.5	0.5
KNN	Gender predicting emotional condition anxiety and stress	0.60	0.50
Naive Bayes		0.70	0.70
KNN	Marital status predicting emotional condition anxiety and stress	0.80	0.80
Naive Bayes		0.80	0.80

It's important to remember that in a further clinical diagnosis, the categorization of the emotion would require a more personal analysis through time, but due to the law's requirements in Mexico, in this analysis, the crucial factor is to establish that the working environment is not excessively stressful, as the consequences can divert the focus of the investigation towards understanding the emotional conditions during a typical workday. To identify stress on workers due to working environment or working load, first it is necessary to discover a relation between external variables and the stress/anxiety phenomenon. Furthermore, to acquire more information from the dataset, an extra algorithm (KNN) was proposed to identify opportunities and create a comparison of performance. As a result of the evaluation of anxiety ranges on external variables on all the subjects, the following table was obtained (Table 7):

The result of the previous algorithm shows predictable reactions, between external variables and the emotional range. Now, with this information and by comparing two or more external emotion variable prediction, an assortment of the real specific emotion diagnosis could be obtained, by creating a numerical range for the scale of anxiety/stress; in consequences, by reusing the KNN algorithm, a prudent classification of the emotion can be obtained. As shown in Fig. 14, it can be seen how depending on the range from 0 to 100 the result of classification of a specific emotion is obtained; only the highest ranges are considered stress (100–80)

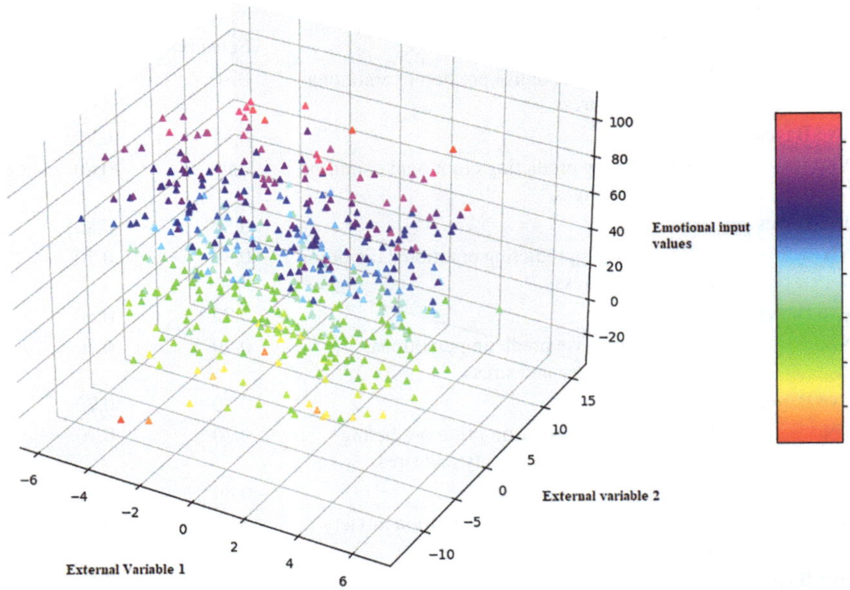

Fig. 14 Results of emotional classification

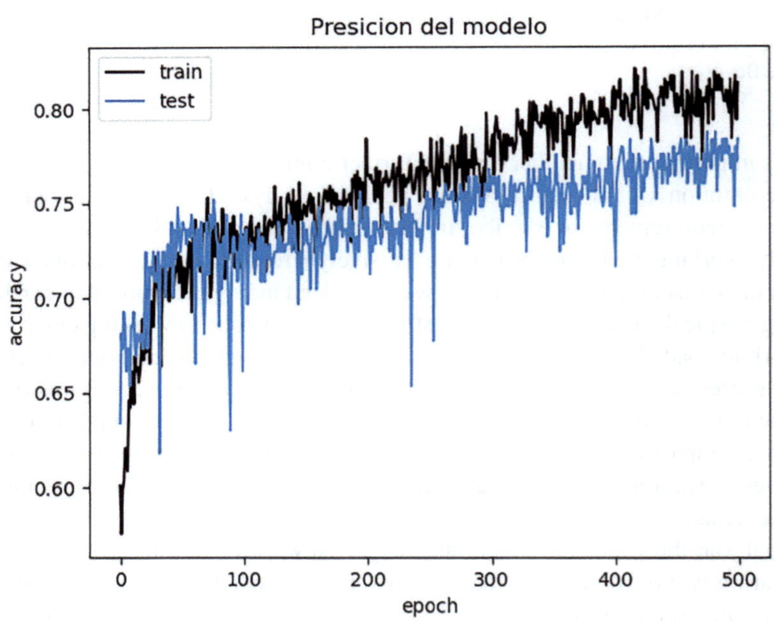

Fig. 15 Accuracy behavior of model

while the higher average (80–60) would be anxiety, average values would be fear (60–40), and finally low values (40–0) would represent emotion; the negative values are considered bias.

3.1.6 Deep Learning Classification

Once the genuine emotion or goal value for emotional diagnosis is obtained, the separation of emotions within the same emotional range is possible. Therefore, using the emotion goal values as training bases for ANN would allow a new classification to detect stress or anxiety. The result of the neural network classifier/diagnosis is shown below (Fig. 15 and Table 8):

3.1.7 Case Final Diagnosis

As a final result from the lecture of 120 individuals, the following information was obtained from the classification and previously declared for the NOM-035 fulfillment (Table 9).

Table 8 Precision and recall of model

	Precision	Recall
Case 1	85.4	0.90
Case 2	80.2	0.70
Accuracy (approx.)	83%	

Table 9 Case results

Workers diagnosed	Quantity	Average percentage of population with anxiety/stress results (control lecture)	Average percentage of population with average anxiety/stress results (stressful condition)
Administrative personnel	30	0.10 ± 0.02, stress; 0.13 ± 0.03, anxiety; and 0.77 ± 0.02, out of emotional range	0.28 ± 0.03, stress; 0.18 ± 0.03, anxiety; and 0.53 ± 0.05, out of emotional range
Teachers	30	0.082 ± 0.03, stress; 0.12 ± 0.3, anxiety; and 0.798 ± 0.02, out of emotional range	0.127 ± 0.02, stress; 0.14 ± 0.3, anxiety; and 0.733 ±, out of emotional range
Laboratory workers	30	0.02 ± 0.001, stress; 0.06 ± 0.003, anxiety; and 0.92 ± 0.01; out of emotional range	0.02 ± 0.002, stress; 0.0625 ± 0.002, anxiety; and 0.917 ± 0.001, out of emotional range
Others	40	0.196 ± 0.005, stress; 0.212 ± 0.002, anxiety; and 0.7 ± 0.02, out of emotional range	0.08 ± 0.003, stress; 0.12 ± 0.002, anxiety; and 0.8 ± 0.02, out of emotional range

4 Discussion

First of all, the system works fine in the application, but the demand of time off from the users has to be focused on a sequential diagnosis, meaning that the system performs better when more data is taken through time. Also, based on the validation of the NOM-035, the company showed good behavior in emotional health from workers, but some variables were hard to pin down and compensate in the algorithm. The main variable that was not taken into account and affected severely the measure and diagnostic was the "second work." This stands for the academic personnel or students that had another job or labor in their free time; some of them had really demanding careers of 8 h a day, creating a bias of stress environment that may be acquired in another working place instead of the university. To address this problem, a new variable should be taken into account when defining the emotional objective in the first classification. Added to earlier mentioned observations, the control of external variables changes deeply the data acquisition capabilities of the system, although not in the expected way, for example, some of the users were increasingly sensitive to heat; as consequence, GSR measures present weird values even with previous signal conditions; in contrast, lack of sweat did not affect the system or algorithm in any significant way. Therefore, the control of the environment has to be stricter to avoid the mentioned problem.

In terms of reliability of the emotional classification, the previous medical conditions (respiratory, fever, and cardiac) and their addition to the main code were efficient when adjusting the ranges of diagnosis. Therefore, health states such as past strokes and Covid-19 weren't as dispersed as expected, but the learning from the subject, as mentioned before, surely be prolonged to truly guarantee a better result in subjects with critical conditions (recent fevers and recently diagnosed Covid-19 patients). Resuming the previous mentioned topic, it's significant to clarify that the biodata from the user comes from a specific set of sensors for a specific biological system. As a consequence, if an attempt is made with a different sensor, a new revision of how to muffle previous health conditions in the user should be revised according to the new biosignal or sensor.

At programming level, the system presents decent accuracy for a real-world application, although the learning demands more data and more time for the diagnostic, but this creates a greater computational load. The proposed method for separating emotions proves to be efficient enough for laws in Mexico; this shows that a multilevel analysis of emotions is functional with relatively simple algorithms and can be accurate. One of the unexpected situations that were placed over the algorithm was the difficulty to join the different phases of it, requiring a modular approach to achieve the best result. In order to improve the system and algorithm, adaptation must be made; this means that all applications have particular hardware requirements, and at the algorithm level, the measurement in different program dates could help immensely in the diagnosis of real industrial applications (e.g., one measure in August, another in October, etc.).

References

1. T. Masood, P. Sonntag, Industry 4.0: Adoption challenges and benefits for SMEs. Comput. Ind. **121**, 103261 (2020). https://doi.org/10.1016/j.compind.2020.103261
2. S. El Hamdi, A. Abouabdellah, M. Oudani, Industry 4.0: Fundamentals and main challenges, in *2019 International Colloquium on Logistics and Supply Chain Management (LOGISTIQUA)*, (2019), pp. 1–5. https://doi.org/10.1109/LOGISTIQUA.2019.8907280
3. E.C. Moraes, H.A. Lepikson, Industry 4.0 and its impacts on society, in *International Conference on Industrial Engineering and Operations Management*, (2017), pp. 729–735
4. A. Badri, B. Boudreau-Trudel, A.S. Souissi, Occupational health and safety in the industry 4.0 era: A cause for major concern? Saf. Sci. **109**, 403–411 (2018). https://doi.org/10.1016/j.ssci.2018.06.012
5. S. Digiesi, V.M. Manghisi, F. Facchini, E.M. Klose, M.M. Foglia, C. Mummolo, Heart rate variability based assessment of cognitive workload in smart operators. Manag. Prod. Eng. Rev. **11**, 56–64 (2020). https://doi.org/10.24425/mper.2020.134932
6. V. Tiwari, R.S. Thakur, *An Extended Views Based Big Data Model Toward Facilitating Electronic Health Record Analytics* (Elsevier Inc, 2019). https://doi.org/10.1016/B978-0-12-816948-3.00013-1
7. H. Khaloufi, K. Abouelmehdi, A. Beni-Hssane, M. Saadi, Security model for big healthcare data lifecycle. Procedia Comput. Sci. **141**, 294–301 (2018). https://doi.org/10.1016/j.procs.2018.10.199
8. A. Selvikvåg Lundervold, A. Lundervold, An overview of deep learning in medical imaging focusing on MRI. Z. Med. Phys. (2018). https://doi.org/10.1016/j.zemedi.2018.11.002
9. M.T.V. Yamuza, J. Bolea, M. Orini, P. Laguna, C. Orrite, M. Vallverdu, R. Bailon, Human emotion characterization by heart rate variability analysis guided by respiration. IEEE J. Biomed. Health Inf. **23**, 2446–2454 (2019). https://doi.org/10.1109/JBHI.2019.2895589
10. K.Y. Ngiam, I.W. Khor, Big data and machine learning algorithms for health-care delivery. Lancet Oncol. **20**, e262–e273 (2019). https://doi.org/10.1016/S1470-2045(19)30149-4
11. P. Hamet, J. Tremblay, Artificial intelligence in medicine. Metabolism **69**, S36–S40 (2017). https://doi.org/10.1016/j.metabol.2017.01.011
12. H. Haick, N. Tang, Artificial intelligence in medical sensors for clinical decisions. ACS Nano **15**, 3557–3567 (2021). https://doi.org/10.1021/acsnano.1c00085
13. Z. Lv, L. Qiao, Analysis of healthcare big data. Futur. Gener. Comput. Syst. **109**, 103–110 (2020). https://doi.org/10.1016/j.future.2020.03.039
14. Norma Oficial Mexicana NOM-035-STPS-2018, Factores de riesgo psicosocial en el trabajo-Identificación, análisis y prevención. | Secretaría del Trabajo y Previsión Social | Gobierno | gob.mx, https://www.gob.mx/stps/articulos/norma-oficial-mexicana-nom-035-stps-2018-factores-de-riesgo-psicosocial-en-el-trabajo-identificacion-analisis-y-prevencion. Last accessed 06 Sept 2020
15. S.S. Panicker, P. Gayathri, A survey of machine learning techniques in physiology based mental stress detection systems. Biocybern. Biomed. Eng. **39**, 444–469 (2019). https://doi.org/10.1016/j.bbe.2019.01.004
16. M.Z. Poh, K. Kim, A. Goessling, N. Swenson, R. Picard, Cardiovascular monitoring using earphones and a mobile device. IEEE Pervasive Comput. **11**, 18–26 (2012). https://doi.org/10.1109/MPRV.2010.91
17. D. Girardi, F. Lanubile, N. Novielli, Emotion detection using noninvasive low cost sensors, in *2017 7th International Conference on Affective Computing and Intelligent Interaction, ACII 2017*, (2018), pp. 125–130. https://doi.org/10.1109/ACII.2017.8273589
18. R. Norman, L. Mendolicchio, C. Mordeniz, Galvanic skin response & its neurological correlates. J. Conscious. Explor. Res. **7**, 553–572 (2016)
19. G. Udovičić, J. Derek, M. Russo, M. Sikora, Wearable emotion recognition system based on GSR and PPG signals, in *MMHealth '17: Proceedings of the 2nd International Workshop on Multimedia for Personal Health and Health Care, Co-located with MM 2017*, (2017), pp. 53–59. https://doi.org/10.1145/3132635.3132641

20. Z. Ghahramani, T.M. Mitchell, A. Notsu, K. Yasuda, S. Ubukata, K. Honda, H. Pei, K. Wang, Q. Lin, P. Zhong, Y. Tazawa, K.C. Liang, M. Yoshimura, M. Kitazawa, Y. Kaise, A. Takamiya, A. Kishi, T. Horigome, Y. Mitsukura, M. Mimura, T. Kishimoto, A. Jeerige, D. Bein, A. Verma, J.G. Carbonell, R.S. Michalski, T.M. Mitchell, K.M. Lee, J. Yoo, S.W. Kim, J.H. Lee, J. Hong, A. Géron, J. Waring, C. Lindvall, R. Umeton, J. Allmer, T. Baltrusaitis, C. Ahuja, L.P. Morency, Multimodal machine learning: A survey and taxonomy. IEEE Trans. Pattern Anal. Mach. Intell. **41**, 423–443 (2019). https://doi.org/10.1109/TPAMI.2018.2798607
21. L. Shu, J. Xie, M. Yang, Z. Li, Z. Li, D. Liao, X. Xu, X. Yang, A review of emotion recognition using physiological signals. Sensors (Switzerland) **18** (2018). https://doi.org/10.3390/s18072074
22. R. Sioni, L. Chittaro, Stress detection using wearable physiological sensors. Artif. Comput. Biol. Med. Lect. Notes Comput. Sci. **360**, 526–532 (2015). https://doi.org/10.1007/978-3-319-18914-7_55
23. J. Zhu, L. Ji, C. Liu, Heart rate variability monitoring for emotion and disorders of emotion. Physiol. Meas. **40** (2019). https://doi.org/10.1088/1361-6579/ab1887
24. Y. Chen, L. Zhang, B. Zhang, C.A. Zhan, Short-term HRV in young adults for momentary assessment of acute mental stress. Biomed. Signal Process. Control **57**, 101746 (2020). https://doi.org/10.1016/j.bspc.2019.101746
25. N.Y. Oktavia, A.D. Wibawa, E.S. Pane, M.H. Purnomo, Human emotion classification based on EEG signals using Naïve Bayes method, in *Proceedings – 2019 International Seminar on Application for Technology of Information and Communication: Industry 4.0: Retrospect, Prospect, and Challenges, iSemantic 2019*, (2019), pp. 319–324. https://doi.org/10.1109/ISEMANTIC.2019.8884224
26. D. Nikolova, P. Petkova, A. Manolova, P. Georgieva, ECG-based emotion recognition: Overview of methods and applications, in *Proceedings of the ANNA 2018 – Advances in Neural Networks and Applications*, (2018), pp. 118–122
27. G. Giannakakis, D. Grigoriadis, K. Giannakaki, O. Simantiraki, A. Roniotis, M. Tsiknakis, Review on psychological stress detection using biosignals. IEEE Trans. Affect. Comput. **1** (2019). https://doi.org/10.1109/TAFFC.2019.2927337
28. Z.N. Zarch, M. Sharifi, M. Heidari, S. Pakdaman, Emotion classification through nonlinear EEG analysis using machine learning methods. Int. Clin. Neurosci. J. **5**, 150–157 (2018). https://doi.org/10.15171/icnj.2018.xx
29. E. Gross, Work, organization and stress. Soc. Stress. **54–110** (2017). https://doi.org/10.4324/9781315129808
30. F. Shaffer, J.P. Ginsberg, An overview of heart rate variability metrics and norms. Front. Public Health **5**, 1–17 (2017). https://doi.org/10.3389/fpubh.2017.00258
31. S. Hassani, I. Bafadel, A. Bekhatro, E. Al Blooshi, S. Ahmed, M. Alahmad, Physiological signal-based emotion recognition system, in *2017 4th IEEE International Conference on Engineering Technologies and Applied Sciences (ICETAS 2017)*, (2018), pp. 1–5. https://doi.org/10.1109/ICETAS.2017.8277912
32. E. Conte, Measurements of electroencephalogram (EEG), galvanic skin resistance (GSR) and heart rate variability (HRV) during the application of a system that gives simultaneously tVNS and brain entrainment on subjects affected by depression and anxiety. Ann. Depress. Anxiety **5**, 2–6 (2019). https://doi.org/10.26420/anndepressanxiety.1095.2019
33. D. Huysmans, E. Smets, W. De Raedt, C. Van Hoof, K. Bogaerts, I. Van Diest, D. Helic, Unsupervised learning for mental stress detection exploration of self-organizing maps, in *BIOSIGNALS 2018 – 11th International Conference on Bio-Inspired Systems and Signal Processing, Proceedings; Part of 11th International Joint Conference on Biomedical Engineering Systems and Technologies, BIOSTEC 2018*, vol. 4, (2018), pp. 26–35. https://doi.org/10.5220/0006541100260035
34. A. Goshvarpour, A. Abbasi, A. Goshvarpour, An accurate emotion recognition system using ECG and GSR signals and matching pursuit method. Biom. J. **40**, 355–368 (2017). https://doi.org/10.1016/j.bj.2017.11.001

35. S. Londhe, R. Borse, Emotion recognition based on various physiological signals – A review. ICTACT J. Commun. Technol. **9**, 1815–1822 (2018). https://doi.org/10.21917/ijct.2018.0265
36. X. Hou, Y. Liu, O. Sourina, Y.R.E. Tan, L. Wang, W. Mueller-Wittig, EEG based stress monitoring, in *The 2015 IEEE International Conference on Systems, Man, and Cybernetics (SMC2015)*, (2016), pp. 3110–3115. https://doi.org/10.1109/SMC.2015.540

Architecture for Initial States Algorithm for Blockchain Scalability in Local OnPrem IIoT Environments

Alfonso José Barroso-Barajas, Jesús Andrés Hernández-Gómez, Roberto Antonio Contreras-Masse, and Salvador A. Noriega-Morales

1 Introduction

Currently, many activities are supported by networked electronic devices and information systems, in the so-called Internet of Things, IoT, or Industrial Internet of Things IIoT. IoT is a network made up of physical electronic devices, integrated software, and connectivity, which allows them to connect to exchange data [4]. The main idea of IoT is to connect millions of devices capable of reacting to stimuli in their environment [16]. The goal of IoT is to enable devices to monitor their environment, communicate, and create a better environment for humans [14]. IoT is present in transportation, smart homes/buildings, public safety, environmental monitoring, and medicine, as well as in industrial processing, agriculture, and livestock, among others [10]. In this way, IoT contributes to working more intelligently, establishing a man-machine and machine-machine relationship.

In this sense, the accelerated growth of IoT devices and the benefits they offer mean that the usual data storage architectures present great technological, performance, and economic challenges. In an IoT network, thousands of transactions from thousands of devices must be verified for authenticity, causing network bottlenecks [1, 12].

On the other hand, blockchain technology is being implemented in various environments more frequently, due to the traceability and immutability of stored

A. J. Barroso-Barajas (✉) · J. A. Hernández-Gómez · S. A. Noriega-Morales
Department of Industrial Engineering and Manufacturing, Universidad Autónoma de Ciudad Juárez, Chihuahua, Mexico
e-mail: al206593@alumnos.uacj.mx; rhernand@uacj.mx; snoriega@uacj.mx

R. A. Contreras-Masse
Instituto Tecnológico de Ciudad Juárez, Chihuahua, Mexico
e-mail: rcontreras@itcj.edu.mx

L. C. Méndez-González et al. (eds.), *Innovation and Competitiveness in Industry 4.0 Based on Intelligent Systems*, EAI/Springer Innovations in Communication and Computing, https://doi.org/10.1007/978-3-031-29775-5_12

data. It can be said that blockchain is a decentralized database, organized in blocks of information, where each connected node stores a copy of the database [6]. The blocks are connected to form a chain, where the transactions are stored, recording the moment they occurred, within a network governed by rules agreed by the participants [8] highlighting its main characteristics such as decentralization, persistence, anonymity, and auditability [18].

Blockchain technology has often been questioned, due to the massive amount of information that is recorded and replicated on the network, thus causing the storage to eventually run out, threatening the stability of the system.

In the words of [7, 13], the storage of data in each blockchain block generates significant difficulties. If blockchain is used within IIoT devices, these difficulties can become a bottleneck for scalability, since a network of IIoT devices can generate a large amount of data in a few minutes or hours; therefore, the requirements of node storage will grow along with reported negative latency and throughput side effects [1, 13].

As a result of these problems, an algorithm is proposed to generate and manage initial states for blockchain. In this way, the blockchain theory and its concepts and variables are also enriched. On the other hand, companies that implement this technique will be able to take better advantage of the storage spaces of their servers, as far as blockchain storage is concerned. Finally, it is important to note that, as more information is stored, more electrical energy is needed to keep the equipment running and, in addition, electrical energy to maintain the cooling of the sites; therefore, by reducing the size of blockchain, these energy costs will decrease considerably, contributing positively against the ecological footprint.

This article is organized into sections. The following section explains the main concepts for the reader, in the form of background. Next, we discuss the methodology with the components required in our architecture. Then, in Sect. 4, the local architecture for IIoT is presented as a materialization of the proposed architecture. Finally, Sect. 5 presents our conclusions and future work.

2 Background

Blockchain, IIoT, and initial states are concepts that are worthy of review to grasp the same context in this paper.

2.1 Blockchain

Blockchain is similar to an accounting ledger, where movements or transactions are recorded in a list of data. Similarly, all transactions are digitally validated and signed, by all blockchain nodes. Data is added to each block, including the date and time the transaction was recorded. In turn, each block is linked to its predecessor,

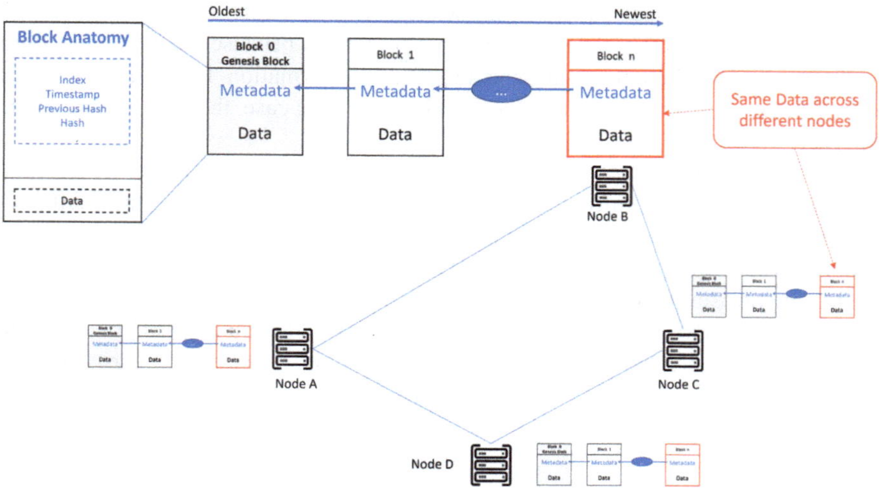

Fig. 1 Blockchain conceptual diagram and block anatomy

thanks to a set of reference data or key. The blockchain uses elliptic curve cryptography (ECC) and a SHA-2 hash scheme to provide strong cryptographic proof for authentication and data integrity [2]. The main idea of blockchain is to add blocks of information regardless of a central authority, which controls the system, with the added value of being a decentralized and encrypted data storage structure [11].

All system nodes share information and record transactions, applying a consensus algorithm to validate that the data to be stored is correct. Any attempt to destabilize the system must include simultaneous attacks on at least 50% plus one of the nodes in the blockchain to affect a single block.

Figure 1 shows the anatomy of a block in the blockchain. At a minimum, each block on the blockchain is made up of the following components: (1) current block index, (2) timestamp, (3) previous block hash in the blockchain, (4) current block hash, and (5) a record of transactions contained in this [9, 15] block.

The first block generated for blockchain is called "Genesis Block" or zero block. Therefore it is not possible to have a block $n + 1$ older than block zero. This in turn contributes to data consistency, since the timestamps of an n block will be greater than the $n - 1$ block and less than the $n + 1$ block. In this way, blockchain technology guarantees the integrity, non-repudiation, and security of the data due to the list of digitally signed transactions.

2.2 *Industrial/Internet of Things (IIoT/IoT)*

IIoT or IoT is one of the nine pillars of Industry 4.0 [5]. IIoT is a network made up of various devices that can be sensors and actuators managed by microcontrollers or household appliances such as a domestic coffee machine or even vehicles and complex machines. It is called the Internet of Things because these devices form an inter-network where they can send and receive information. However, very often these devices send information to the cloud, and, in that case, they do depend on the Internet.

IoT is a general concept. It can be made up of commonly used household devices. The "things" used in different industries and business solutions have different requirements. In this sense, IIoT is a more particular concept, limited to industrial activity, where specific protocols and proprietary networks known as Operational Technology or simply OT are used.

As the use of things grows and becomes more and more popular, the term industrial IoT is created to refer specifically to "things" that are primarily used in industry; however, today, other sectors such as healthcare, logistics, and commerce implement IIoT as well.

2.3 Initial State

As discussed above, each blockchain must have a genesis block. This block must be initialized considering that it has no predecessor and must be replicated in all the nodes in an identical way, that is, the genesis block must be exactly the same in all the nodes of the blockchain networks. As more blocks are added to the chain, the amount of data varies, but it is mandatory that all nodes have the same data stored, in order to carry out the consensus algorithm at the time of a [3] transaction.

On the other hand, when it comes to deploying in the cloud, allocating storage space is easier; in general, all hyperscalers have the ability to easily add blocks. However, what if the blockchain runs locally, on premises, where access to storage space is limited and restricted by physical media or availability time? This is where an algorithm to generate and manage initial states becomes relevant, since it can solve space and availability problems. In this sense, to address this problem in local environments, it is necessary to have an architecture to execute tests and later implement in production. The next section discusses the methodology and approach used to propose a local architecture that fits test and production environments, along with the main concept of a proposed algorithm to handle initial states.

3 Methodology and Materials

For a better understanding of these ideas, two issues must be addressed. The first is the construction of an architecture to implement and test the operation of the blockchain. The second is the algorithm for managing initial states and optimizing storage for the different blocks generated by IIoT devices.

3.1 Network Architecture

The proposed architecture basically comprises three elements: (1) IIoT devices, (2) servers to store the control logic and blockchain, and (3) NoSQL database to store metadata over time. These components must communicate as shown in Fig. 2.

The materials for this architecture are the following:

- IIoT components: Five virtual machines managed by Virtual Box version 6.1 running Raspberry Pi OS version 4.19, each with 2GB memory. These three virtual machines act as clients. Each client has previously been loaded with Python 3.10 programs to transmit the telemetry of CPU temperature and two random numbers in the range [1..100], respectively.
- Three virtual machines with operating system GNU/Linux Ubuntu 20.04, one 1 vCPU each and memory in each server is 4GB (these are referred also as servers). The file system will host the blockchain files. Here is also hosted the initial states algorithm. Each of those three virtual machines is acting as independent blockchain nodes. Each node has the following software:
  - Blockchain Manager. Developed in Python 3.10.2 with the responsibility of constantly reviewing the status of the blockchain in relation to the available capacity on the server.
  - Database MongoDB NoSQL version 5.0.3. MongoDB is a database that stores data in flexible structures called collections. In the database for the present experiment, only one collection is generated to store the data, such as the HASH IDs of the files, date, time, and the IIoT device ID data.

For this experiment, the IIoT device obtains data, either by the processor temperature or by generating a random number ranging from 1 to 100. This data is sent to the server. One of the servers is running a service developed in Python

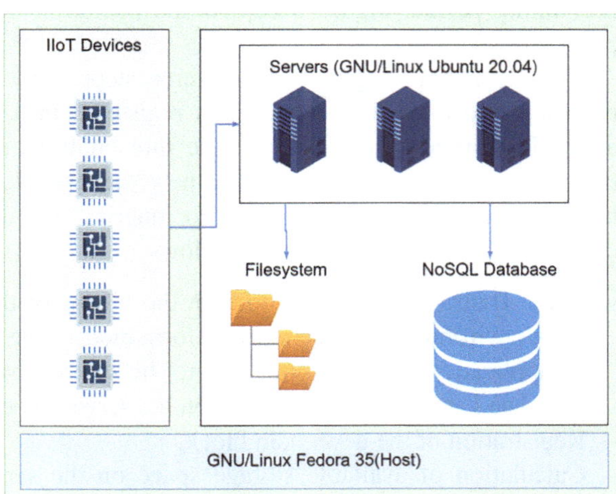

Fig. 2 Materials for this architecture

that, through the POST method, receives the data to process it. The received data is added to the transaction collection. Subsequently, the transactions are stored in the file system, adding an identification HASH and, in turn, as a reference to a previous file, forming the block chain. On the other hand, there is a MongoDB database, in which a backup record of the HASH identifiers of the files is kept, including the date, time, and identification data of the IIoT device. The data registry service has an authentication system in the MongoDB database. The application authenticates its access with the parameters of the connection string: host, port, database, username, and password. On the other hand, the database server has a user account configured with access privileges only to the database and the application's own collections. To carry out the experiment, a temperature sensor was simulated with an application developed in Python. The application runs continuously on the client, sending data every 0.5 s. The devices are too small to store a blockchain. It is important to note that one of the main characteristics of blockchain is decentralization. In this sense, it is decided to have, initially, two servers dedicated to store blockchain information in a decentralized way.

The literature suggests that the amount of data generated will grow considerably over time, and therefore, in a short time there will be numerous blocks in the blockchain and the storage in the file system will quickly become saturated. This is where it is important to develop an algorithm that allows managing initial states over time, in order to optimize storage space.

3.2 Initial States Algorithm

In our algorithm, we adopt the Initial Entries accounting concept. According to [17], an opening entry or initial states is a journal entry to preserve the balances of various assets, liabilities, and equity that appear on the balance sheet from the previous accounting period. Figure 3 depicts the operation of the initial states algorithm proposed.

The algorithm's logic works in three steps. First, transactions arrive and are recorded into a block. Second, it is registered in a new chain. Third, calculate available storage space, and validate thresholds to make a decision to truncate the chain, calculate hashes, and link new "genesis block" pointing to the previous historic blockchain to be available for queries, and do not lose information. The detailed algorithm is explained as follows:

- Start. Transactions are recorded in the blocks of the chain. IIoT devices sense their environment and send the information to the data server. The server runs an application developed in Python. The app receives data from the IIoT device. The data is recorded in a text file on the server to be evaluated.
- Registration of the new chain block.
- Calculation of available storage space on the server(SCL). The server has a secondary storage unit, of limited capacity. Blockchain appends fixed-size blocks

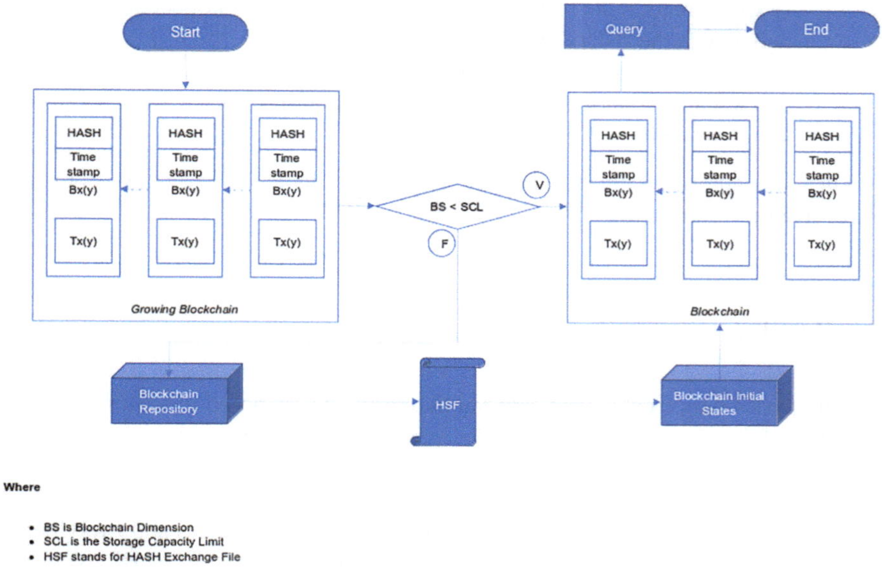

Fig. 3 Blockchain initial states algorithm

with the IIoT data(BS). The algorithm takes both parameters for comparison and decision-making.

- Initial or opening statements are a concept coined in accounting. In the case of the present algorithm, it has the function of generating an "exercise" closure and generating a new one, called the initial or opening state. The closure is executed based on the dimension of the blockchain with respect to the available storage capacity of the server. The algorithm will be constantly reviewing the growth of the blockchain and its relationship with the available space. Once the indicated value is reached, the algorithm will obtain the HASH of the last block. Subsequently, a "swap" file is generated where the HASH of the last block is stored with the HASH of the next block generated, to maintain the continuity of the chain. Finally, the "closing blockchain exercise" is packaged and compressed to be sent to a repository, cleaning the file system of the servers.
- If the remaining storage space is enough to store an additional block, it is added.
- If the storage space is not enough,
  - ∗ Calculate the hash of the last block. A hash key is calculated to allow for the continuity of the blockchain once it is stored in the repository.
  - ∗ Blockchain is compressed, packaged, and stored in the repository. The blockchain compressed package is pushed to the repository. In this way,

space is freed in the storage unit, so that the unit does not become saturated and avoids the risk of system collapse.

* A file is generated with the hash key of the last block (HSF). A new blockchain is started with its first block pointing to the hash previously stored in HSF.

Once this algorithm finishes, the new chain is available for new queries or ready to receive new transactions.

4 Results

For the performance evaluation and validation of the proposed algorithm, a virtualized small blockchain network has been created. This experiment was carried out on a GNU/Linux Fedora 35 machine, with an 8 $\times$ 11th generation Intel® Core™ i7-1165G7 processor at 2.80 GHz and 7.6 GiB of RAM. Three Raspberry Pi virtual terminals are executed on this physical equipment, as IIoT devices and two Ubuntu 20.04 virtual servers, for blockchain storage, connected to a network, using the TCP/IP protocol. IIoT devices run a client application that generates data based on processor temperature and random number generation. One of the servers receives the information and stores it in a MongoDB database.

Figure 4 shows the amount of data generation with a frequency of 0.5 s, with one device, then with two and finally with three IIoT devices. Throughout the experimentation period, it is observed that the data is generated, received, and stored without loss or latency.

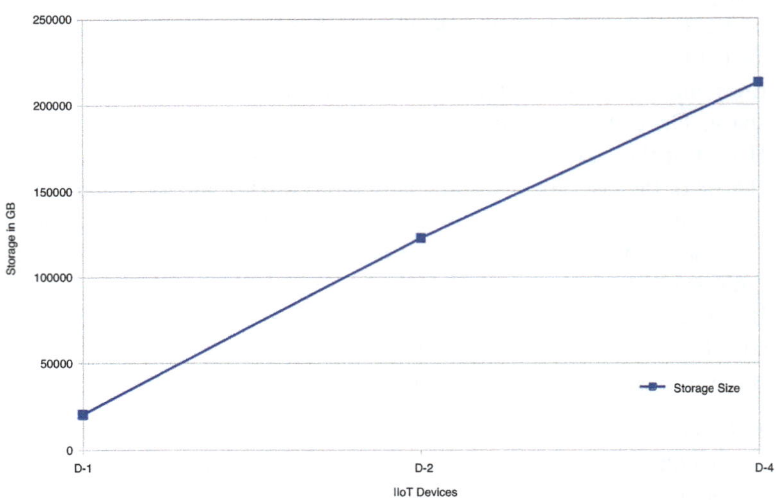

Fig. 4 Storage occupation

At this stage of development, blockchain compression was done manually using the .tar.gz format. Figure 5 shows that the directory occupies 3.1M of disk storage, while the generated compressed file occupies 124K, that is, 4.51% of the total space.

Data generation is done with a frequency of 0.1 s, from five virtualized IIoT terminals. Throughout the experimentation period, it is observed that the data is generated, received, and stored without loss or latency. Figure 6 shows initial state of the database, before system boot.

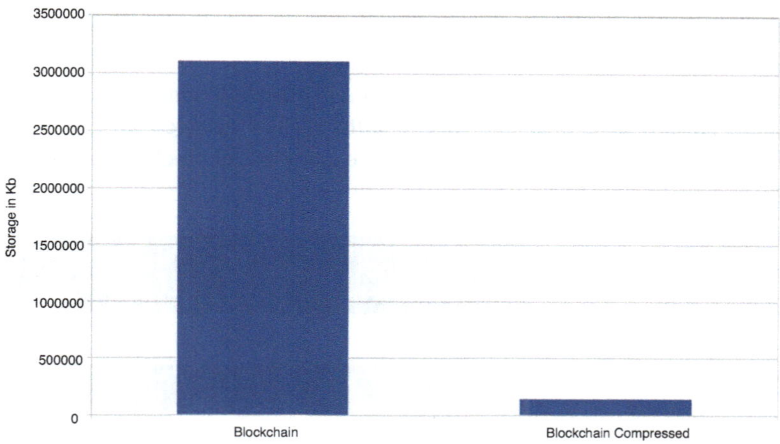

Fig. 5 Compressed blockchain partition

Fig. 6 Database initial state

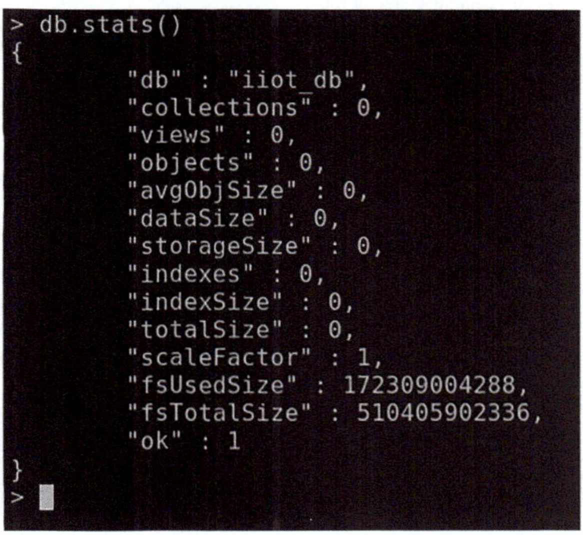

```
> db.stats()
{
        "db" : "iiot_db",
        "collections" : 0,
        "views" : 0,
        "objects" : 0,
        "avgObjSize" : 0,
        "dataSize" : 0,
        "storageSize" : 0,
        "indexes" : 0,
        "indexSize" : 0,
        "totalSize" : 0,
        "scaleFactor" : 1,
        "fsUsedSize" : 172309004288,
        "fsTotalSize" : 510405902336,
        "ok" : 1
}
>
```

Figure 7 shows the initial state of the directory where the blockchain structure is stored.

In a period of 5 min, 770 files were generated for the blockchain on the server, each containing 10 transactions. Figure 8 shows the used command to get the number of generated files.

Figure 9 shows the total occupancy.

The database in MongoDB, reached the following values, as seen in Fig. 10

At this stage of development, compression of the blockchain directory was done manually using the .tar.gz format. The directory occupies 3.1M of storage in disk, while the generated compressed file occupies 124K, that is, 4.51 of the total space, as shown in Fig. 11.

The experiment was repeated in 10 min, giving the following results

Figure 12 shows the difference between the regular storage for blockchain and, on the other hand, the storage required for the compressed blockchain structure.

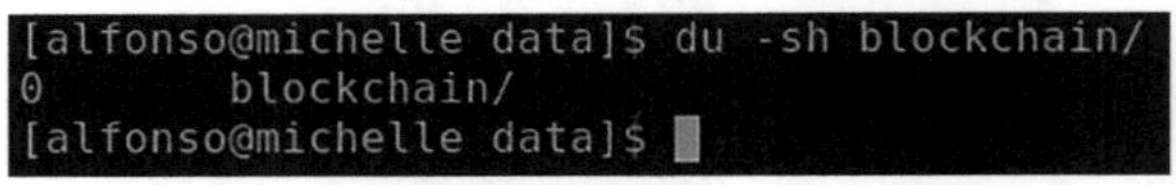

Fig. 7 Directory initial state

Fig. 8 Generated files

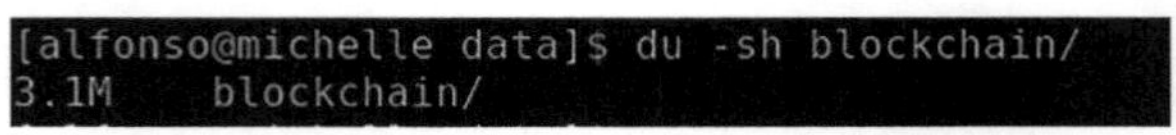

Fig. 9 Used space for blockchain into file system

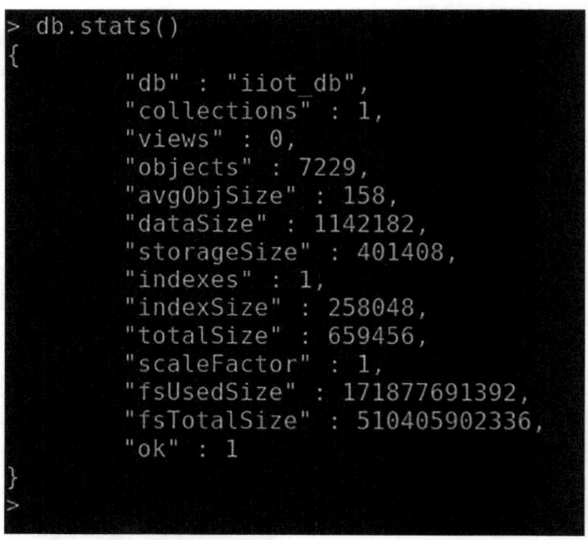

Fig. 10 Database size at the start of the process

Fig. 11 Comparison of space used by blockchain after compression

```
[alfonso@michelle data]$ du -sh *
3.1M    blockchain
140K    blockchain.tar.gz
```

Fig. 12 Database size at the end of the process

```
> db.stats()
{
        "db" : "iiot_db",
        "collections" : 1,
        "views" : 0,
        "objects" : 14233,
        "avgObjSize" : 158,
        "dataSize" : 2248814,
        "storageSize" : 790528,
        "indexes" : 1,
        "indexSize" : 495616,
        "totalSize" : 1286144,
        "scaleFactor" : 1,
        "fsUsedSize" : 170770812928,
        "fsTotalSize" : 510405902336,
        "ok" : 1
}
>
```

5 Conclusions and Future Work

The experiment, with the indicated materials, characteristics, and configurations, showed that, once compressed, the blockchain package occupies only 4.51% of the original space; this means that the application of the algorithm could effectively solve the storage problems. On the other hand, data generation and storage did not present any problem. It will proceed to increase both the amount of execution time and the frequency of the data, to look for a point where the service collapses. Future work is to determine what the optimal storage value is and what the optimal size of the blockchain is, so that the algorithm can decide when to generate the initial inputs, pack the original chain, and generate a new chain.

References

1. A. Abdelmaboud, A.I.A. Ahmed, M. Abaker, T.A.E. Eisa, H. Albasheer, S.A. Ghorashi, F.K. Karim, Blockchain for IoT applications: taxonomy, platforms, recent advances, challenges and future research directions. Electronics **11**(4), 630 (2022). https://doi.org/10.3390/electronics11040630
2. R. Almadhoun, M. Kadadha, M. Alhemeiri, M. Alshehhi, K. Salah, A user authentication scheme of IoT devices using blockchain-enabled fog nodes, in *Proceedings of IEEE/ACS International Conference on Computer Systems and Applications, AICCSA*, vol. 2018-Novem (2019), pp. 1–8. https://doi.org/10.1109/AICCSA.2018.8612856

3. F.M. Benčić, I.P. Žarko, Distributed ledger technology: blockchain compared to directed acyclic graph, in *Proceedings—International Conference on Distributed Computing Systems*, vol. 2018-July (2018), pp. 1569–1570. https://doi.org/10.1109/ICDCS.2018.00171
4. R. Contreras, A. Ochoa, E. Cossío, V. García, D. Oliva, R. Torres, Design and implementation of an IoT-based háptical interface implemented by Memetic Algorithms to improve competitiveness in an industry 4.0 model for the manufacturing sector. Adv. Intell. Syst. Comput. **939**, 103–117 (2019). https://doi.org/10.1007/978-3-030-16681-6_11
5. R. Contreras-Masse, A. Ochoa-Zezzatti, V. García, L. Pérez-Dominguez, M. Elizondo-Cortés, Implementing a novel use of multicriteria decision analysis to select IIoT platforms for smart manufacturing. Symmetry **12**(3) (2020). https://doi.org/10.3390/sym12030368
6. P. Danzi, M. Angjelichinoski, C. Stefanovic, P. Popovski, Distributed proportional-fairness control in microgrids via blockchain smart contracts, in *2017 IEEE International Conference on Smart Grid Communications, SmartGridComm 2017*, vol. 2018-Janua (2018), pp. 45–51. https://doi.org/10.1109/SmartGridComm.2017.8340713
7. A. Gorkhali, L. Li, A. Shrestha, Blockchain: a literature review. J. Manag. Anal. **7**(3), 321–343 (2020). https://doi.org/10.1080/23270012.2020.1801529
8. F. Gürrmann, B. Banusch, Fundamentals of Blockchain (October) (2018)
9. D. Khan, L.T. Jung, M.A. Hashmani, Systematic literature review of challenges in blockchain scalability. Appl. Sci. **11**, 9372 (2021); **11**, 9372 (2021). https://doi.org/10.3390/APP11209372
10. A. Khanna, S. Kaur, *Internet of Things (IoT), Applications and Challenges: A Comprehensive Review*, vol. 114. (Springer US, 2020). https://doi.org/10.1007/s11277-020-07446-4
11. B. Lashkari, P. Musilek, A comprehensive review of blockchain consensus mechanisms. IEEE Access **9**, 43620–43652 (2021). https://doi.org/10.1109/ACCESS.2021.3065880
12. K. Peng, M. Li, H. Huang, C. Wang, S. Wan, K.K.R. Choo, Security challenges and opportunities for smart contracts in Internet of Things: a survey. IEEE Internet of Things J. **8**(15), 12004–12020 (2021). https://doi.org/10.1109/JIOT.2021.3074544
13. A. Reyna, C. Martín, J. Chen, E. Soler, M. Díaz, On blockchain and its integration with IoT. Challenges and opportunities. Fut. Gener. Comput. Syst. **88**(2018), 173–190 (2018). https://doi.org/10.1016/j.future.2018.05.046
14. A.K. Singh, N. Firoz, A. Tripathi, K. Singh, P. Choudhary, P.C. Vashist, Internet of Things: from hype to reality (2020). https://doi.org/10.1016/b978-0-12-821326-1.00007-3
15. S. Singh, A.S. Sanwar Hosen, B. Yoon, Blockchain security attacks, challenges, and solutions for the future distributed IoT network. IEEE Access **9**, 13938–13959 (2021). https://doi.org/10.1109/ACCESS.2021.3051602
16. E. Sisinni, A. Saifullah, S. Han, U. Jennehag, M. Gidlund, Industrial internet of things: challenges, opportunities, and directions. IEEE Trans. Ind. Inform. **14**(11), 4724–4734 (2018). https://doi.org/10.1109/TII.2018.2852491
17. P. Tulsian, Financial Accounting, 2nd edn. (Tata McGrawHill Publishing, New Delhi, 2007)
18. Z. Zheng, S. Xie, H.N. Dai, X. Chen, H. Wang, Blockchain challenges and opportunities: a survey. EC Tax Review **28**(2), 83–89 (2019)

Distribution Route Optimization Using Floyd-Warshall Weighted Graph Analysis Algorithm with Google Maps Integration in Industry 4.0 Context

Uriel Ángel Gómez Rivera, Iván Juan Carlos Pérez Olguín, Luis Asunción Pérez Domínguez, Luis Alberto Rodríguez-Picón, and Luis Carlos Méndez-González

1 Introduction

Industry 4.0 is an organizational model that maintains the value chain throughout the product and manufacturing life cycle employing existing data and physical technologies [1]. Furthermore, in the context of Industry 4.0, digitization has gained popularity, because digitization is defined as the massive adoption of purely digital technology through connected services and devices [2].

Particularly concerning Industry 4.0, the transition from the most recent industrial age to the technology age has resulted in an increased demand for vertical and horizontal and from beginning to end of digital integration. Previous research indicates that the adoption of Industry 4.0 has an important and significant impact on the sustainability aspects of a supply chain network [3]. Furthermore, supply chain organizations inside today's global environment work in the middle of an increasingly difficult, complex, strong, and dynamic marketplace. Therefore, a sustainable supply chain becomes unavoidable to fulfill the rapid change in client expectations.

It should be noted that, according to some reviews, manufacturing companies must accelerate the shift in focus toward sustainability and use technology such as the Internet of Things (IoT) to achieve the organization's objectives and goals [3].

In the environment of Industry 4.0, the supply chain is now principally focused on industries that use modern technology to process their data, standardize and/or start solving current problems, and provide limitless alternatives to optimize certain

U. Á. Gómez Rivera · I. J. C. Pérez Olguín (✉) · L. A. Pérez Domínguez · L. A. Rodríguez-Picón · L. C. Méndez-González
UACJ, Doctorado en Tecnología, Juárez, Chihuahua, México
e-mail: al199110@alumnos.uacj.mx; ivan.perez@uacj.mx; luis.dominguez@uacj.mx; luis.picon@uacj.mx; luis.mendez@uacj.mx

L. C. Méndez-González et al. (eds.), *Innovation and Competitiveness in Industry 4.0 Based on Intelligent Systems*, EAI/Springer Innovations in Communication and Computing, https://doi.org/10.1007/978-3-031-29775-5_13

activities. These previous implementations are mainly aimed at transportation and distribution, storage, and inventory, making their use on the Internet of Things more diverse [4].

The benefits of the supply chain in the transport and distribution area range from improving customer communication through data analysis to applications to optimize transport routes and actions on useful information and going to accelerate information exchange in the supply chain. Subsequently, monitoring processes that require monitoring the status of metric devices like sensors are automated [4].

As a result, the concepts discussed above apply to sustainability, which is why it is essential when Industry 4.0 and its associated technologies are integrated into the supply chain, in specific, in transportation and distribution.

Considering the aforementioned discussion, intelligent waste management systems are highly relevant in this context because it is possible to integrate real-time information analysis tools that facilitate autonomous decision-making in terms of container location and distribution, analysis, and route optimization, among other critical functions [5].

The following is a list of the fundamental tools that comprise an intelligent waste management system:

(a) Data analysis: It is mainly based on the analysis of nodes (locations) that require collection. The route optimization algorithm enables data-driven decision-making.
(b) The route planner determines the route distribution based on the optimization generated by such an algorithm centered on map management systems.
(c) The foregoing section focuses on the computational tool developed, whose communication interface communicates the decision obtained based on the data collected to the user or carrier that is part of a clean supply chain in Industry 4.0.

It is important to note that the need for a management platform that optimizes operating costs is a source of motivation that drives the incorporation of technology in the monitoring, control, and planning of industrial waste collection routes, as it will allow for the reduction of transportation costs, the best allocation of vehicles used in collection processes, adequate planning of maintenance programs (based on their occupancy rate), and a reduction in environmental impact (due to the reduction in the emission of gases into the atmosphere and the overflow of solid waste from used industrial containers).

2 Literature Review

2.1 Supply Chain in Transportation and Applications

Transportation and mobility are two of the most common supply chain issues today. Furthermore, it is one of the most significant challenges that those large cities face, and there is a need to ensure that these cities do not saturate traffic and have clean air and that mobility management costs are optimized because it is a necessity. Regarding the improvement of large companies' supply chains, there are currently technology implementation techniques that include the incorporation of intelligent systems such as sensors in product transport whose task is to provide information or exchange data remotely to the fleet manager and/or its clients on fuel consumption, vehicle location, and other types of information that will minimize the overall cost of the supply chain.

Radio frequency identification (RFID) labels and sensors are employed in inter-organizational supply chain applications which are examples of this type of innovation. The purpose of this contribution aims to investigate the effects of sensor-based emission policies on product quality in the perishable goods supply chain. This enables the examination of the quality of perishable products under various distributor emission policies, the results of which show that the policies are based on automatically collected expiration dates and product quality, with the goal of reducing future costs of said supply chain [6].

In other applications analogous to the prior example, wireless sensors have been applied to pharmaceutical supply chain procedures because continuous, real-time monitoring of high-value commodities can considerably improve the dependability, reliability, and efficiency of this type of supply chain. Technology capabilities for continuous detection and response in the cold chain are provided by wireless sensor networks (WSNs). The results also suggest that WSNs can significantly improve process quality and decrease waste in this cold chain [7].

A real-time monitoring system for the supply chain of perishable items using smartphone-based sensors and NoSQL databases is another novel and innovative use. This application uses smart phone-based sensor devices to gather data on temperature, humidity, GPS, and images. It also features a powerful big data design to manage sensor data produced by the Internet of Things. The system and gateway performance of this product was tested in the Korean kimchi supply chain. The suggested system can process a significant input/output of sensor data efficiently as the number of sensors and customers grows [8].

However, in terms of supply chain applications through the use of algorithms, there is the use of a genetic algorithm for the multi-objective optimization of supply chain networks (SCN). This algorithm's goal is to find the Pareto set and optimal solutions for multi-objective SCN design problems. For this, a larger number of alternative solutions are evaluated, and the proposed solution procedure incorporates two separate weight approaches. Losses are thus avoided throughout the chain process [9].

Using a relief transport planning, heuristic algorithm for emergency supply chain management is one of the other ways that algorithms are used in transportation planning. Given that emergency supply chain operations must satisfy all needs quickly and with a finite amount of resources, a well-known technique for addressing challenges with emergency supply chain planning is mixed-integer programming (MIP). However, as the complexity of the problem increases, the MIP model becomes impossible to solve due to the time and computing resources required (Np-hard problem) [10].

As a result, the algorithm groups and orders the demands based on the required products, the mandated expiration dates, the potential shared capacities, and the distances between demand nodes and warehouses. Following that, the algorithm plans each demand individually, using a shortest travel time tree and a lowest cost production tree [10]. Both supply chain production and transportation planning have similar mathematical programming models. The primary goal is to recognize, identify, and propose a taxonomy framework based on the following components: the supply chain framework, decision level, process model, intent, information exchange, boundaries or constraints, and implementation [11].

As a result of evaluating each one of the examples of applications seen toward the supply chain, it is possible to conclude that the advantaged factors would be transportation fuel costs, time, and efficient service compliance [12].

2.2 Internet of Things and Its Impact on Supply Chain Management

A new network of connected software-enabled devices called the Internet of Things allows for data sharing. Internet usage is still evolving, or in development, from the IoC (Internet of Computers) to the IoT. IoT deployments have become crucial to protect data and resources and reduce associated expenses, particularly in the business and industrial environments that directly impact supply chain management [13].

Since IoT can share important data through a wireless network, the data is safe and secure under the control of the user or a corporation that is entitled to a smart device. The data collection, exchange, and development processes that support them are all business processes.

IoT may manage the data in this case and document the business process' pattern. The Internet of Things network can also control user engagement with gadgets. IoT technology can be used to efficiently start the business process.

It is important to note that with this sophisticated technology system, which calls for the usage of tools with this type of technology, better information may be paired with high quality and productivity to grab the attention of consumers or potential clients. IoT devices offer a number of advantages that can increase business value [13].

Due of logistics' contribution to the supply chain during the past 20 years, researchers and practitioners have now recognized a new trend of supply chain integration and collaboration. This approach has gained a lot of interest in the development of logistics, known as supply chain administration.

Deep learning and neural networks have therefore been used to collaborate on logistics supply chain information. It should be noted that when using traditional methods, it has been found that there is a high level of structural complexity and version control, which makes it difficult to maintain. As a result, it is necessary to integrate the Internet of Things for further process simplification and to use optimization algorithms to get quick and effective results.

On the other hand, supply chain management–based logistics collaboration that employs Internet of Things–based smart solutions provides support to reduce overall costs while improving services for end users in the supply chain.

Due to rapid innovation, increased regional influence, and global growth, the supply chain is now a crucial element. From the firm level, the level of competition is now rising to the industry level. Data influences the market as a whole in addition to undermining the firm. More than half of the population is also supported by the industrial sector, and the automobile industry is the one with the quickest growth. Cooperation in the supply chain increases performance across industries and geographical areas [14].

It is important to keep in mind that the supply chain is a vast area that is constantly changing and that the organization's main logistical job is the movement of products. As academics and researchers advance, they gradually contribute value in areas like marketing, inventory control, finance, advertising, and new product creation.

Supply chain management also regulates resources, money, people, and information both within and outside the supply chain to increase customer satisfaction and gain an advantage over the competitors. The fact that it is not just a simple chain but is more entangled with suppliers, logistics providers, customers, and other participants, depending on the size of the organization and its operations, is another key concern [14].

Studies have been conducted where automotive companies incorporate various criteria such as manufacturers, assemblers, distributors, retailers, and suppliers where information is gathered by questionnaires and distributed as a self-administered survey [14]. Cooperative decision-making, electronic data exchange, and information exchange in supply chains are operational functions with all types of variables (independent and dependent).

Similar to this, effective collaboration in information technology has been recognized as a crucial element in fostering the flow of information. It illustrates how electronic cooperation is encouraged within the context of its supply chain, expanding beyond the information sharing between two supply chain partners to incorporate clients, business partners, the Internet, and other entities [15].

Given the landscape of the Internet of Things applied to the industry in supply channels, an intelligent tool design that enables connection with the elements for excellent management in organizations and enterprises is now necessary.

Virtual appliances are significant because supply chain management is applied and firmly supported by both supply chain management and the Internet of Things to the organization. This is because supply chain management and the Internet of Things operate hand in hand.

As a result, the foundation of the supply chain communication system is the information sharing between all organizations in the platform's supply chain node and the strengthening of their connections. Encourage the process to be transferred throughout the supply chain. Furthermore, the supply chain's overall operating efficiency is improved [15].

Therefore, any organization or business that integrates an intelligent system into its supply chain will increase overall operational effectiveness and improve the supply chain management strategy.

Consequently, for an intelligent supply chain to function properly, information from technical resources, personnel, and members of the same chain must be included to generate synergies in order to benefit and take advantage of competition [16].

In this respect, it is critical to remember that collaboration is about more than just problem-solving; it is also about maintaining relationships to improve creativity and increase and expand opportunities for long-term improvement.

The main objective is to collaborate creatively, solve problems through improved execution, and deliver methods of value that meet customer demands and expectations while taking accurate technical factors into account and optimizing resources that will benefit the company.

Of course, optimization algorithms can have a significant role in the supply chain. They are used in Industry 4.0 to perform important tasks that significantly influence all aspects of the processes involved, with the main task being in the field of routing or supply routes [17].

2.3 *Optimization Algorithms*

Since communication systems are critical in supply chains, wireless technologies can have a significant role in their incorporation. Similarly, the usage of wireless technologies will be determined by the application [17].

However, when approaching the optimization of supply chain routes, factors such as distance and geographical points must be considered when deciding how the data generated by the computer program will be transmitted.

The distribution and transport supply chain are critical because transport is involved, especially when there is an opportunity to implement intelligent systems with the aim of optimizing their routes. Thus, the problem to be addressed from a beginning point to a finish point is essential in this type of problem.

As a result, from the perspective of Industry 4.0, the supply chain problem can be solved by using optimization based on graphs. By definition, a graph is an abstract illustration of a mathematical structure containing edges and vertices connected by pairs of other vertices.

Similarly, the lengths from the edges to the vertices are frequently referred to as weights, which are used to determine the shortest route between two points (location). When this is done in practice, graph theory can be applied to several situations. For instance, a map can be represented by a graph, with the vertices being cities and the edges denoting the routes between the cities or within a city; locations or sites can be connected based on their coordinates [18].

Currently, there are several types of optimization algorithms that solve the shortest path problem; unfortunately, not all of them are compatible with Industry 4.0 supply chain processes.

One of them is the ACO (ant colony optimization), a bio-inspired algorithm used for solving the shortest path problem. In this algorithm, experimental analyses are carried out to assess the efficiency of the algorithm and the quality of the obtained solutions. In this algorithm, two sets of randomly generated small and large problems are typically solved. The solutions of the problems are compared to the label correction solutions; this is one of the most efficient algorithms for solving MOSP (multi-objective shortest path problem) [19].

These results produce some performance distance measures, under uniform distribution. These results can be extended using Pareto frontier analysis and compared to the Pareto optimal frontiers produced by the ACO algorithm and the label correction algorithm. The results of the example problems demonstrate that the suggested approach generates high-quality, non-dominated solutions while saving time in the computation of large-scale bio-inspired shortest path problems [19].

It is worth noting that because of outstanding advancements in computer science and software, the use of graph theory has grown significantly, widely, and rapidly, even expanding to be one of the most significant sciences that is crucial for solving many challenges and problems. Problems with a wide range of applications, computer protocols, Internet platforms seeking route optimization, games, and other applications are examples of these applications. Two very popular algorithms widely utilized are Dijkstra's algorithm and the Bellman-Ford algorithm.

Dijkstra's algorithm was developed in 1956 by Edsger Dijkstra and published in 1959. It is based on a graph and is used to solve single-source shortest path network difficulties and to generate a shortest path network tree. Dijkstra's algorithm has been widely used to calculate and find the shortest path in a wide range of practical applications [20].

This algorithm is considered the best to provide a general solution for optimization problems, but it can sometimes find less than ideal solutions for other problems. When the intended destination node is reached, one of the key advantages of Dijkstra's algorithm is that it does not visit the remaining undesired nodes. On the other hand, the disadvantage of Dijkstra's algorithm is that it consumes a lot of computational resources when running on a large number of nodes. In addition, it has the limitation of being used only on positive weight graphs; it cannot handle negative weights or flanks [20].

In this way, the central principle in this algorithm is to find all the shortest paths that begin at the starting vertex and lead to all the other vertices. When the shortest route between the starting vertex and the remaining vertex in the graph is discovered, the method is finished.

Its pseudocode is as follows:

```
Dijkstra (N, s)
  Begin
       For each ν belonging to V[N]
       Do Dm[ν] = infinite
```

```
        Pm[ν] = null
        Dm[s] = 0
    S = empty
    Q = V[N]
    While Q is not empty
        Do υ = node ν with min Dm[ν]
        S = S union υ \is added to the set of finished nodes
        for each ν belonging adjacent to υ
        Relaxation
        If Dm[ν] > Dm[υ] + ω (υ, ν)
        Then Dm[ν] = Dm[υ] + ω (υ, ν)
        Pm(ν) = υ
```

This pseudocode generates a sequence of subsets of the total set of nodes (N) that satisfy two conditions. First, because the distance between the nodes (number of arcs) in each computed subset is reduced, the distances in the first set will always be greater than those in the following subsets. Second, the operation of this algorithm results in the systematic calculation of all possible paths, until the shortest path necessary to travel from one node to another is obtained [20].

Bellman-Ford algorithm, on the other hand, obtains the shortest paths among a weighted graph's origin node to the rest of its nodes. Because it allows negative values in the arcs, it provides solution to the problem of the shortest or minimum path from an origin node in a more general way than Dijkstra's algorithm. If it finds a negative loop or weight loop, the algorithm returns a Boolean (dichotomous) value. Otherwise, it computes and returns the shortest path with the lowest cost [21].

The attribute Dm[ν] is kept as the upper bound on each vertex ν in V or the cost of the shortest path from the origin s to the vertex ν.

The pseudocode of the algorithm is shown below:

```
Bellman-Ford (N, s)
  Begin
       For each ν belonging to V[N]
       Do Dm[ν] = infinite
   Pm[ν] = null
   Pm[s] = 0
   For i = 1 to V[N]-1
       Do for each arc (υ, ν) belonging to A[N]
  Relaxation
       If dm[ν] > Dm[υ] + ω (υ, ν) then
       Dm[ν] = Dm[υ] + ω (υ, ν)
       Pm(ν) = υ
  For each arc (υ, ν) check negative weight loop
       Do if Dm[ν] > Dm[υ] + ω (υ, ν)
       Then return FALSE the algorithm does not converge
  Return TRUE
```

Finally, Dijkstra's algorithm accomplishes the same task in less time but requires that the weight of the edges not be negative. Consequently, the Bellman-Ford procedure is only applied in the case of negative edges in the graph.

Furthermore, the Bellman-Ford procedure is not qualified to analyze and optimize multiple paths with graph characteristics to achieve the shortest path or distance reduction; however, this algorithm is superior to Dijkstra's algorithm because it accepts negative lengths. It could only be used for two paths to a target or end node because it was built on arcs. However, it suffers from the same limitation as the previous algorithm: it only computes the shortest paths from one node to the rest, ignoring the set [21].

2.4 Fundamental Principles of Graph Theory

To begin, a graph is defined as a mathematical structure composed of two sets whose elements are linked by a function. The graph is written as follows:

$$G = \{V, A\} \tag{1}$$

V stands for vertex or node, and A stands for arc or edges. Geometrically, the graph is made up of circles for nodes and lines for edges or arcs. When the relationship defined by the arcs is given from one node to another in a nonreciprocal way, the arc is called directed, and its representation is made by a line ending in an arrow pointing toward the node that is satisfactory [22].

It is important to note that the arcs have a specific function that relates them with a value commonly known as weight, the value of which is used as distances from one geographic point to another.

In this way, all the information stored in a graph is usually carried out through a matrix representation, which is carried out fundamentally by the relationship of the arcs with the nodes, in such a way that the matrix input corresponds to the intersection of the column that represents the arc with the row that represents the node that contains the value of the weight or the distance to travel [23].

2.5 Floyd-Warshall Algorithm and Its Application

This is an algorithm that analyzes a graph to search the shortest path in a minimal and weighted manner. In a single run, the algorithm finds the path of all pairs of vertices and compares all feasible routes along the network among every pair of vertices.

An estimate of the shortest path between the vertices is iteratively improved until it is determined that the estimate is accurate.

As a direct result, this algorithm provides the shortest path between nodes and locations within a graph and creates an optimal path across the entire graph. In this context, the Floyd-Warshall algorithm is a dynamic programming variation that generates a solution as an interconnected decision. It is crucial to note that in some instances, there may be more than one option, so the input of this algorithm is a weighted and directed graph [24].

The Floyd-Warshall procedure exceeds other algorithms in determining the shortest route, such as Dijkstra's algorithm, which has a disadvantage in route search because it performs a blind search, that is, it does not consider the entire graph with its corresponding nodes and weights [25]:

$$\delta^0 = \begin{cases} \delta_{ij} & \text{if } Ad > 0 \\ \infty & \text{if } Ad = 0 \end{cases} \tag{2}$$

Given the previous formula, we proceed to calculate the number of iterations using the following equation (resulting in several iterations equal to the number of nodes):

$$\delta^m = \text{Min}\left(\delta_{ij}^{m-1}, \delta_{im}^{m-1} + \delta_{mj}^{m-1}\right) \tag{3}$$

This means that the algorithm outputs an array of route values:

$$\delta^{m+1} \tag{4}$$

Consequently, the value of the minimal path connecting them is saved in the intersection between source and destination nodes.

Now, to obtain the specification of the optimal routes between the points of interest of some users, the algorithm operation is based on a matrix built from the following equation, which records the changes in each iteration of the algorithm until the construction of a second matrix, which is required to obtain the route matrix:

$$H_0 = \begin{cases} I & \text{if } ij > 0 \\ \text{""} & \text{if } ij = 0 \end{cases} \tag{5}$$

Therefore, the Floyd-Warshall procedure is an accurate tool for calculating the shortest paths to be traced for the purpose of route optimization, and even more so if it is focused on practical and intelligent supply chain applications, as demonstrated by its ability to analyze a large network of nodes.

3 Methodology

Given previous equations, the extended Floyd-Warshall algorithm is presented as pseudocode, including its input and the conditions for its fulfillment visible.

```
Extended Floyd-Warshall algorithm
 Entry or input: Matrix δ⁰= [M, M]
 Where M = |V| cardinality of the set of vertexes of the graph
 that presents the system
```

```
  Return: array of minimum path values δ^{M+1}
  Beginning
//matrix processing
   For θ = 1 until M do
        For i = 1 until M do
            For j = 1 until M do
          If δ$_ [i, j]^ θ − 1$ < δ$_ [i, θ]^ θ − 1$ < δ$_ [θ, j]^ θ − 1$
          Then δ$_ [i, j]^ θ$ = δ$_ [i, θ]^ θ − 1$ + δ$_ [θ, j]^ θ − 1$
                     Else δ$_ [i, j]^ θ$ = δ$_ [i, j]^ θ − 1$
//Post treatment of the path values matrix
    For i = 1 until M
     Do δ$_ [i, i]^ M + 1$ = 0
  End
```

In the same way, it is of vital importance to expose the algorithm for obtaining the matrix of routes from the generation of the second matrix or output matrix:

```
  Entry or input: Matrix H^M= [M, M]
  Where M = |V| cardinality of the set of vertexes of the graph
  that presents the system
  Return: Minimal path matrix H^{M+1}
  Beginning
//process all entries in the array
    For i = 1 until M do
     For j = 1 until M do
      If H$_ [i, j]^ M$ ≠ ""
      Then:
          Auxiliary variable = i-th node + ? + H$_ [i, j]^ M$ + ?
 + j-th node
        While auxiliary variable contains a question mark (?) do
           P = position first question (?)
          If H$_ {Node{T − 1}, Node[T + 1]}^ M$ = Node[T − 1] or
H$_ {Node{T − 1}, Node[T + 1]}^ M$ = Node[T + 1]
          Then:
                     Remove the question mark (?) in position n
                     by T
                    Else:
                           Replace the question mark (?) by:? +
H$_ {Node{T − 1}, Node[T + 1]}^ M$ + ?
    Else:
H$_ [i, j]^ M + 1$ = " "
H$_ [i, j]^ M + 1$ = auxiliary variable
  End
```

Given previous algorithms, the Floyd-Warshall algorithm is an iterative optimization algorithm that can be composed of multiple nodes and will always generate all the shortest routes to start from a point (*i*) to any point (*j*), providing the path required for completing the optimized path, by which the path to be generated will be a data matrix resulting from an initial matrix given by the user, which will be equal to the number of vertices or loci. Furthermore, the number of iterations (*k*) is equivalent to the matrix size [26].

The process is represented below in computational pseudocode, also in mathematical computational process:

1. Produce a square matrix using the previously established graph.
2. Customize the matrix based on the routes' compliance with the following protocols (d_ik):
   (a) If there is a direct path connecting the nodes in a straight line, record each distance.
   (b) In case of no direct path connecting available between node i and node j, mark with infinity (∞) consecutively.
   (c) Write down a zero when indicating the distance of node i with node i (same node).
3. Begin iterations based on the algorithm's own formula to identify the shortest route between each node and, in general, the optimized path network:

$$\text{minimumpath } (i, j, k) = \min (\text{minimumpath } (i, j, k-1);$$
$$\text{minimumpath } (i, k, k-1) + \text{minimumpath } (k, j, k-1));$$
$$\text{minimumpath } (i, j, 0) = \text{edgeweight } (i, j);$$

In order to determine the minimal path $(i, j, 2)$ for every pair (i, j), this algorithm first executes the minimum path $(i, j, 1)$ for every pair (i, j). This process is repeated until $k = n$, and the minimum path for every pair of vertices (i, j) is found applying the procedure in some intermediate vertex. It should be noted that the algorithm must use quadratic memory when translated into a computational language, so when the k – th iteration is calculated, it is translated to $k-1$.

Moreover, because this is a two-dimensional matrix, path ij represent the minimum path from node i to node j using intermediate values ranging from 1 to $k-1$, where each path ij is initialized to edge weight (where it records the distance from node to node). Given the preceding procedure, we have the mathematical structure:

```
Floyd-Warshall procedure ()
For k: = 0 to n - 1
For all (i, j) in (0, 1, ..., n - 1)
Path (i, j) = Min(Path(i, j), Path(i, k)+Path(k, i))
```

In this manner, the purpose of the algorithm is to establish that there is a path from vertex i to vertex j that does not pass through a number of vertexes greater than k if there is already a path from i to k that avoids passing through several vertexes greater than $k-1$ and also if there is a path from i to k that does not pass through a number of vertexes greater.

3.1 Floyd-Warshall Algorithm Applied to Information Technology and the Supply Chain in Transportation and Distribution

Next, the use of the Floyd-Warshall algorithm and the generation of data in applied computing are presented with the objective of its practice in the supply chain within an Industry 4.0 in the sustainable context. The supply chain is of vital

0	4	5.9	7.6	10.9	11	13.4	13.4	15.2	11.9		
Inf	0	2.5	4.8	8.4	8.5	10.4	10.4	12.2	9.7		
Inf	Inf	0	5.7	6.7	6.8	8.7	8.7	10.5	7.2		
Inf	Inf	Inf	0	7.6	7.859999999999999			10.2	10.2	12	8.8
Inf	Inf	Inf	Inf	0	0.26	2.6	2.6	4.4	2.6		
Inf	Inf	Inf	Inf	Inf	0	2.7	2.7	4.5	2.9		
Inf	Inf	Inf	Inf	Inf	Inf	0	1.5	2.5	4.7		
Inf	Inf	Inf	Inf	Inf	Inf	Inf	0	3.6	4.8		
Inf	Inf	Inf	Inf	Inf	Inf	Inf	Inf	0	3.1		
Inf	Inf	Inf	Inf	Inf	Inf	Inf	Inf	Inf	0		

Fig. 1 Output matrix

importance in transportation and distribution, since critical information is released through a computerized system to track the situation of the supply chain, increasing transparency throughout the process and the chain and allowing its users to make decisions with more accurate information, in real time and subject to any changes.

Therefore, the way is to develop a computer program that is complementary to the main software. The purpose of this first computer program is to generate the node matrix based on the Floyd-Warshall algorithm with the goal to search and find the shortest path connecting nodes or locations within a graph creating an optimal path across the overall graph.

First, a node graph of a specific case of a transportation and distribution supply chain must be addressed. The computer program must optimize the path of these nodes by generating the order matrix based on the algorithm. Subsequently, the node-to-node distances are related, and all distances are recorded. Once all distances have been recorded, the route sequencing process continues to optimize the overall graph. In which the distance data is entered in the computational subprogram elaborated in the C# language, in this way, the consecutive path of the nodes that make up the graph in general will be known according to the Floyd-Warshall process.

In this way, the data generation is completed based on the Floyd-Warshall optimization algorithm. Then the following computing tool is presented, elaborated in the same way in C#, which aims to focus on a supply chain oriented toward sustainability within Industry 4.0 (Fig. 1).

The characteristic of this computer tool is that it is positioned in the field of transport and distribution at an industrial level, so it has adequate monitoring functions for various transports and their respective routes optimized under the Google Maps interface.

As shown in Fig. 2, it is necessary to have a satellite overview on the main screen where you can also search for and add nodes to have a detailed image of their location through their coordinates and generate a database with information such as node, identification, latitude, longitude, name, address, driver, and day. It should be noted that the coordinates of the starting point will automatically be displayed in the main window.

Consequently, a route plan for the nodes or locations must be established for all drivers in a period of 7 days that make up a week, so the software function called "Route plan" must be entered.

Fig. 2 Main window

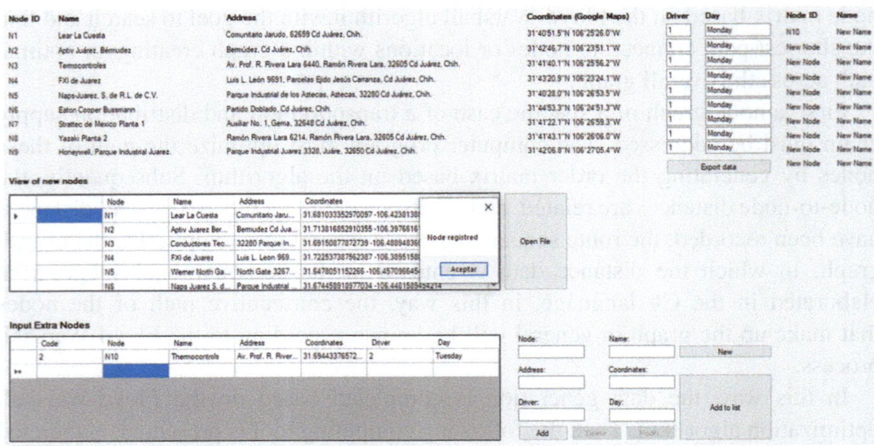

Fig. 3 Route plan

This function, called "Route plan," shows the pre-programmed nodes in the system together with general information such as node name, address, coordinates, driver, and day blanks, in which the user will assign the driver's number and the day of the week that will carry out the collection within the supply chain. Next, the window with the features mentioned above will be displayed (Fig. 3).

Subsequently, the desired information is exported to where the exported data is displayed. This way, there are two important functions, which are validating nodes and generating optimization.

Now, when carrying out the function of generating optimization, the series of conditions of the computer program that enforce the route optimization of the Floyd-Warshall algorithm is executed.

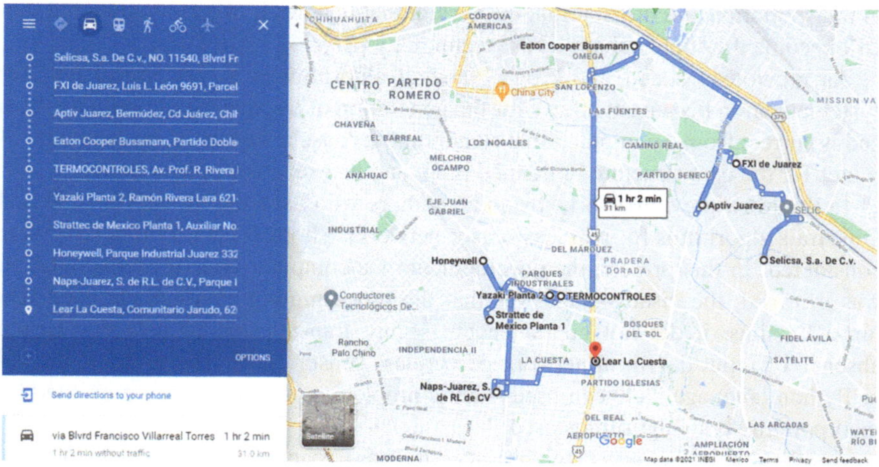

Fig. 4 Optimized route

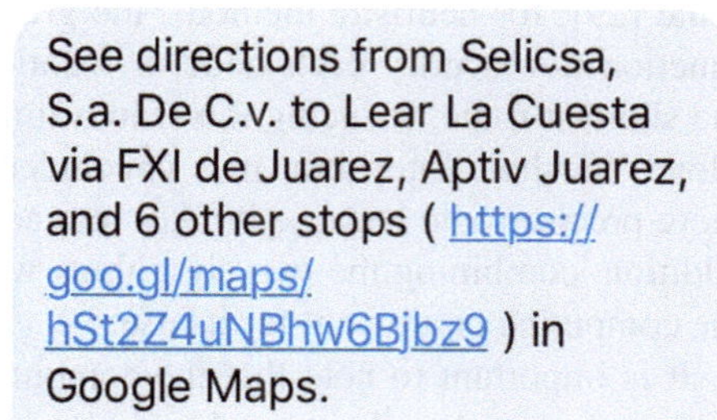

Fig. 5 Message generated from the platform

When executing this algorithm, the system automatically generates a pop-up window which transmits the node tracking data or the ordered sequence of nodes to follow that makes up the general graph of the route (Fig. 4).

In this way, the system will automatically link the Google Maps platform with the routes programmed by the system and will return to the optimal route according to the sequence. It should be noted that the optimized route is sent directly to the driver assigned in the function that is within the same platform; for this, you must previously have the telephone number of each driver or an email that the company or driver handles such as shown in Fig. 5.

4 Conclusion

Supply chain management is a complex procedure that yields better results when performed using mathematical models such as the Floyd-Warshall algorithm. It should be noted that the more data that is handled in a supply chain, particularly in transportation and distribution, the system that handles it as a computational tool will become the supply chain's base technology [27].

As previously stated, the Floyd-Warshall algorithm is a very useful tool for systematization because it allows for the generation of a sequence of optimal routes and is correct from a computational standpoint because it solves the management of routes for the data structure of defined input and its execution time is finite.

In this way, previous works by some authors who used the Floyd-Warshall and Dijkstra's algorithms for wireless sensor networks' shortest path routing [28] were considered. In this project, the development of a computer code for its simulation was involved; the disadvantage was that the programming environment used was turbo C, since it did not guarantee the secure transmission of data due to the absence of input data in a network of wireless sensors [28]. Therefore, if the C# or Python language had been used, such a problem would have been prevented by incorporating both algorithms used.

Another factor that was taken into account was to capture said algorithm in travel routes as found in some works related to practical application, so it was found that some authors use a series of combinations of algorithms to find a precise minimum route [29]; for heuristic methods, the greedy algorithm is found, whose particular function is to verify each node or location through which it passes to calculate the shortest route, reducing the search time. In this way, when compared with the Floyd-Warshall algorithm, it is concluded that the latter generates a shorter and more precise route because it takes into account all the points of the graph [29]. In addition, combining the two algorithms would require more hardware power from the computing equipment for its use.

It is important to note that the computational performance that this computing tool is designed to have enables it to be used in low-capacity systems, as the technical computer specifications are surpassed by a limited computer, but a valuable tool in terms of computational intelligence must be considered.

Subsequently, in Industry 4.0, the supply chains applied in the transport and distribution areas will gain outstanding flexibility, as digital transformation will introduce the formation of more agile chains that are radial, multidirectional, and intelligent, while also allowing a clean and sustainable environment [30].

Similarly, the excellent management of the mechanisms that intervene in said tool is illustrated due to the ease of use and transparency toward the company or organization and, most of all, the value of the algorithm implemented in said intelligent tool, whose code grants the best optimization toward the transport routes that must be configured.

As a result, the Internet of Things guarantees substantial improvements in supply chain management via route optimization tools. The maturity of several key technologies, such as the one just presented, contributes to the viability of the Internet of Things implementation. In this sense, the use of a computing tool with an embedded algorithm necessitates the fulfillment of various tasks and responsibilities of some of the supply chain team members, most notably the maintenance of a broadband network of users in numerous places, and carriers must certify the entire operational process.

Furthermore, the collected measurement data should be stored in an auditable and confidential manner. Aside from active support, licenses must be granted: owners and operators of transport vehicles, warehouses, and storage facilities must grant permissions for the use or installation of the computing tool.

It is important to note that transportation supply chains can use a variety of scenarios to assign responsibilities to their actors. For example, certain configurations provide specific access privileges; for example, the computer administrator owners' role will always have access to the driver's data, routes, and location (unless there are specific mechanisms that regulate access to the data). Control over operational performance data increases power by allowing you to point out quality errors in others' processes (assignment of responsibilities), report them to customers (carriers), and compare them. As a result, good rights management is required to prevent long-term problems.

A good control system or smart computing tool can save additional transportation costs, shipping handling, and possibly claims processing depending on the application and will demonstrate a proactive quality approach for carriers or drivers.

Finally, even if real-time information sharing represents a technological advance, the organizational and political challenges are more difficult to overcome in this context.

Using a multi-stakeholder's perspective and a process adoption model, practitioners know the implications and interaction of multiple factors, but these insights can also help practitioners effectively manage the initiation of intelligent systems through optimization algorithms in transportation and distribution supply chains.

References

1. J.L. Del Val Roman, Industria 4.0: la transformación digital de la industria, in *codiinforme. Revista Deusto ingeniería*, vol. 2016, (Conferencia de directores y decanos de ingeniería informática, Deusto Bilbao, 2016), pp. 1–10
2. C. Ramirez, La digitalización y la industria 4.0: *Impacto industrial y laboral*, in *Industria CCOO*, 1st edn., (Madrid, 2017), p. 93
3. E. Manavalan, K. Jayakrishna, A review of Internet of Things (IoT) embedded sustainable supply chain for industry 4.0 requirements. Comput. Ind. Eng. **127**, 925–953 (2019). https://doi.org/10.1016/j.cie.2018.11.030
4. D. Makris, Z. Nadja, L. Hansen, O. Khan, Adapting to supply chain 4.0: An explorative study of multinational companies, in *Supply Chain Forum: An International Journal*, (Taylor & Francis, 2019), p. 17
5. Q. Zhang, H. Li, X. Wan, M. Skitmore, H. Sun, An intelligent waste removal system for smarter communities, in *MDPI: Sustainability*, (2020), p. 26
6. A. Dada, F. Thiesse, Sensor applications in the supply chain: The example of quality-based issuing of perishables, in *The Internet of Things*, Lecture Notes in Computer Science, ed. by C. Floerkemeier, M. Langheinrich, E. Fleisch, F. Mattern, S.E. Sarma, vol. 4952, (Springer, Berlin/Heidelberg, 2008). https://doi.org/10.1007/978-3-540-78731-0_9
7. G. Hendrik Haan, J. van Hillegersberg, E. de Jong, K. Sikkel, Adoption of wireless sensors in supply chains: A process view analysis of a pharmaceutical cold chain. J. Theor. Appl. Electron. Commer. Res. **8**(2), 21–22 (2013) MDPI AG. Retrieved from https://doi.org/10.4067/S0718-18762013000200011

8. G. Alfian, M. Syafrudin, J. Rhee, Real-time monitoring system using smartphone-based sensors and NoSQL database for perishable supply chain. Sustainability **9**(11), 2073 (2017) MDPI AG. Retrieved from https://doi.org/10.3390/su9112073
9. F. Altiparmak, M. Gen, L. Lin, T. Paksoy, A genetic algorithm approach for multi-objective optimization of supply chain networks. Comput. Ind. Eng. **51**(1), 196–215 (2006). https://doi.org/10.1016/j.cie.2006.07.011
10. C.C. Chern, Y.L. Chen, L.C. Kung, A heuristic relief transportation planning algorithm for emergency supply chain management. Int. J. Comput. Math. **87**(7), 1638–1664 (2010). https://doi.org/10.1080/00207160802441256
11. J. Mula, D. Peidro, M. Díaz-Madroñero, E. Vicens, Mathematical programming models for supply chain production and transport planning. Eur. J. Oper. Res. **204**(3), 377–390 (2010). https://doi.org/10.1016/j.ejor.2009.09.008
12. J. Macaulay, L. Buckalew, G. Chung, Internet of things in logistics: A collaborative report by DHL and Cisco on implications and use cases for the logistics industry, in *DHL Trend Research*, (2020), pp. 1–10
13. V. Vemuri, Priya, V. Naik, V. Chaudhary, K. RameshBabu, M. Mengstie, Analyzing the use of internet of things (IoT) in artificial intelligence and its impact on business environment. Mater. Today Proc. **51**, 2194–2197 (2022). https://doi.org/10.1016/j.matpr.2021.11.264
14. Z. Zhou, Y. Liu, H. Yu, Q. Chen, Logistics supply chain information collaboration based on FPGA and internet of things system. Microprocess. Microsyst., 80 (2021). https://doi.org/10.1016/j.micpro.2020.103589
15. Z. Wang, Y. Liu, Information sharing system for logistics service supply chains based on XML, in *2014 11Th International Conference on Service Systems and Service Management (ICSSSM)*, (2014). https://doi.org/10.1109/icsssm.2014.6874049
16. V. Vijaya Kumar, M. Devi, P. Vishnu Raja, P. Kanmani, V. Priya, S. Sudhakar, K. Sujatha, Design of peer-to-peer protocol with sensible and secure IoT communication for future internet architecture. Microprocess. Microsyst. **78**, 103216 (2020). https://doi.org/10.1016/j.micpro.2020.103216
17. J. Lloret, J. Tomas, A. Canovas, L. Parra, An integrated IoT architecture for smart metering. IEEE Commun. Mag. **54**(12), 50–57 (2016)
18. M. Kairanbay, M.J. Hajar, A review and evaluations of shortest path algorithms. Int. J. Sci. Technol. Res. **2**(6), 99–104 (2013)
19. K. Ghoseiri, B. Nadjari, An ant colony optimization algorithm for the bi-objective shortest path problem. Appl. Soft Comput. **10**(4), 1237–1246 (2010). https://doi.org/10.1016/j.asoc.2009.09.014
20. S. AbuSalim, R. Ibrahim, M. Zainuri Saringat, S. Jamel, J. Abdul Wahab, Comparative analysis between Dijkstra and bellman-ford algorithms in shortest path optimization. IOP Conf. Ser.: Mater. Sci. Eng. **917**(1), 012077 (2020). https://doi.org/10.1088/1757-899x/917/1/012077
21. K. Hutson, T. Schlosser, D. Shier, On the distributed Bellman-Ford algorithm and the looping problem. INFORMS J. Comput. **19**(4), 542–551 (2007). https://doi.org/10.1287/ijoc.1060.0195
22. P.L. Restrepo, L.F. Sepulveda, A computational method to obtain optimal paths in road networks. DYNA **78**, 1–11 (2011)
23. K. Ross, C. Wright, *Discrete Mathematics*, 5th edn. (Prentice-Hall, Englewood Cliffs, 2003), p. 612
24. Risald, A.E. Mirino, Suyoto, Best routes selection using Dijkstra and Floyd-Warshall algorithm, in *2017 11th International Conference on Information & Communication Technology and System (ICTS)*, (2017), pp. 155–158
25. A. Singh, P. Kumar Mishra, Performance analysis of Floyd Warshall algorithm vs rectangular algorithm. Int. J. Comput. Appl. **107**(16), 23–27 (2014). https://doi.org/10.5120/18837-0372
26. A. Aini, A. Salehipour, Speeding up the Floyd–Warshall algorithm for the cycled shortest path problem. Appl. Math. Lett. **25**(1), 1–5 (2012). https://doi.org/10.1016/j.aml.2011.06.008
27. M. Merlino, I. Sproģe, The augmented supply chain. Procedia Eng. **178**, 308–318 (2017). https://doi.org/10.1016/j.proeng.2017.01.053

28. P. Khan, G. Konar, N. Chakraborty, Modification of Floyd-Warshall's algorithm for shortest path routing in wireless sensor networks, in *2014 Annual IEEE India Conference (INDICON)*, (2014), pp. 1–6. https://doi.org/10.1109/indicon.2014.7030504
29. H. Azis, R.D. Mallongi, D. Lantara, Y. Salim, Comparison of Floyd-Warshall algorithm and greedy algorithm in determining the shortest route, in *2018 2nd East Indonesia Conference on Computer and Information Technology (EIConCIT)*, (2018), pp. 294–298. https://doi.org/10.1109/EIConCIT.2018.8878582
30. A. Calatayud, R.L. Katz, *Cadena de suministro 4.0: Mejores prácticas internacionales y hoja de ruta para América Latina* (2019), p. 178

Feature Selection in Electroencephalographic Signals Using a Multicriteria Decision Analysis Method

Alexis Edmundo Gallegos Acosta, María Dolores Torres Soto, Aurora Torres Soto, Eunice Esther Ponce de León Sentí, and Carlos Alberto Ochoa Ortiz Zezzatti

1 Introduction

Since the introduction of the industry 4.0 concept in 2011, changes in technology and society have been taking place. This new paradigm favors the integration of technologies and their digitalization using intelligent systems [1–3]. This new universe of applications and technologies brings with it support in the daily life of users with new infrastructures for access to culture, leisure, education, commerce, and healthcare services, to mention just a few approaches [1, 4].

Technology has evolved enough to have the ability to process large amounts of information, which is a crucial trend in the industry 4.0 approach [5]. However, it also resulted in the generation and availability of an uncountable amount of data. The complexity of data processing of the dataset increases, and, therefore, the greater the challenge to detect and exploit the relationships between the features of the dataset [6]. Consequently, the processing and analysis of information require increasingly complex models with higher computational costs [6, 7]. Under this context, there are two fundamental pillars in the development of this study: feature subset selection and multicriteria decision-making.

This chapter focuses on the acquisition and analysis of a database of electroencephalographic (EEG) signals with motor imagery (MI), a conscious mental representation of movement without any body movement [8]. In this case, the motor

A. E. Gallegos Acosta (✉) · M. D. Torres Soto · A. Torres Soto · E. E. Ponce de León Sentí
Benemérita Universidad Autónoma de Aguascalientes, Aguascalientes, Mexico
e-mail: alexis.gallegos@edu.uaa.mx; mdtorres@correo.uaa.mx; atorres@correo.uaa.mx; eponce@correo.uaa.mx

C. A. Ochoa Ortiz Zezzatti
Universidad Autónoma de Ciudad Juárez, Ciudad Juárez, Mexico
e-mail: alberto.ochoa@uacj.mx

L. C. Méndez-González et al. (eds.), *Innovation and Competitiveness in Industry 4.0 Based on Intelligent Systems*, EAI/Springer Innovations in Communication and Computing, https://doi.org/10.1007/978-3-031-29775-5_14

imagery of opening and closing the right hand was used. The database was collected with the support of 6 test subjects and consisted of 23,846 records described by 14 features, which are the available nodes in the Emotiv EPOC+ EEG device. This device can describe the EEG signal with 14 electrodes distributed according to the 10–20 system. Therefore, the first objective was to reduce the number of electrodes using feature subset selection.

Feature subset selection is part of pattern recognition and is responsible for reducing the number of features describing objects or phenomena, identifying those that provide relevant information for classification purposes [9]. For this case, the combinatorial logic approach to pattern recognition is used by obtaining a set of typical testors. As described below, a typical testor represents a subset with the minimum number of features with the ability to distinguish objects belonging to different classes, just as the full set would [10]. At the end of this phase, a set of five typical testors was obtained, reducing the description of the EEG-MI signals from 14 features to 9 or 10 depending on the typical testor.

Each typical testor was used to train and validate an artificial neural network (ANN) model to classify the EEG-MI signals. The ANN model presented a variable quality of classification for each typical testor. For this reason, the second objective of this chapter was the use of a multicriteria decision-making method to decide which typical testor is the best for classification purposes according to the classification quality metrics: accuracy, positive precision, negative precision, false-negative rate, and false-positive rate.

Meanwhile, the ability to generate information in the era of industry 4.0 has given rise to the creation of methods and algorithms to support decision-making in favor of finding the optimal solution for process design, scheduling, planning, and control [11, 12]. Furthermore, industry 4.0 offers the opportunity to create global systems that coordinate machines, human collaborators, and smaller systems for decentralized real-time operation and decision-making, promoting flexibility, temporality, and autonomy [13]. In this way, the creation of "smart factories" is one of the key features within the industry 4.0 paradigm [14], which are challenged to face dynamic environments, "full of uncertainties, complexities, and ambiguities that demand fast and trustworthy decisions" [15].

Multicriteria decision-making methods are an approach that allows the evaluation of alternatives for decision-making, especially in complex problems where quantitative and qualitative problems are involved [16]. The selected method was TOPSIS (Technique for Order Preference by Similarity to Ideal Solution). This method represents a strategy that concludes with the best solution from a set of alternatives; the set of typical testors, described by certain attributes; and the classification quality metrics. TOPSIS proves to be a good alternative due to its advantages such as the possibility of working with many alternatives and decision criteria that may conflict with each other. In addition to choosing the best alternative, it is possible to rank the rest in descending order toward the least favorable alternative.

At the end of the chapter, the selection of the most appropriate typical testor to train and validate a neural network model is shown. Additionally, it is a subset of features that allows such a model to improve its performance as a faster and better cost-benefit classifier by using only the most relevant information.

2 Related Concepts

2.1 Industry 4.0

The Fourth Industrial Revolution or industry 4.0 brings a fusion of many important approaches such as artificial intelligence, robotics, the Internet, and genetic engineering [17]. Specifically, industry 4.0 uses electronics, software, and connectivity for the computerization and digital transformation of production, where technology applications are capable of self-managing themselves by obtaining information from their environment and their state for autonomous and decentralized decision-making [18].

According to Marc Sachon in [19], "industry 4.0 promises to deliver reconfigurable production tools, machines, facilities, and products capable of connecting with each other and with workers through augmented reality or other means to transmit data quickly, intuitively, and efficiently." The impact of industry 4.0 on both society and industry and their different approaches is measured against the following pillars [1]:

1. *Big data:* Ability to store and process large, diverse, and complex volumes of data for monitoring, measuring, and managing [1, 20]. The crucial point is the extraction of valuable information for decision-making in real time [18].
2. *Autonomous robots:* Automation and process control through autonomous, flexible, and cooperative robots with the ability to solve complex problems which cannot be solved easily by a human [18, 20, 21].
3. *Simulation:* Simulation in industry 4.0 allows the modeling, design, and optimization of processes involving machines, humans, and products in virtual environments (2D or 3D) as a mirror of the real world to prevent errors, reduce costs, and improve quality [1, 21].
4. *Horizontal and vertical system integration:* It seeks cohesion between departments, functions, and capabilities within the company in terms of automation communication and cooperation [20–22]. Horizontal integration refers to close and high-level collaboration between companies (i.e., inter-company integrality) to enrich the product's life cycle. Meanwhile, vertical integration refers to the manufacturing system (intra-business integrality), that is, the information exchange and collaboration between the different levels of the company [2].
5. *The industrial internet of things:* Connecting devices, factories, offices, or companies using sensors and actuators that allow the capture of real-time data during the manufacturing process with the interest of analyzing the behavior of the industrial environment through the big data pillar [1, 2], resulting in

agile, competitive, and integrated operations, as well as effective and automatic decision-making [20].

6. *The cloud:* It refers to the ability to store data in the cloud allowing faster access to information, eliminating IT infrastructure, and optimizing resources dynamically. Thus, it is possible to have access to the cloud, from different devices across the factory sharing information between themselves [1, 2, 21].
7. *Additive manufacturing (3D print):* With additive manufacturing, companies are capable of customizing products with faster and cheaper production [1]. According to the book *Industry 4.0: Current Status and Future Trends* [2], additive manufacturing allows "greater customization without the need of additional tools or manufacturing costs, maximizing the use of the material, fostering a 'zero waste' motto."
8. *Augmented reality:* It comprises the integration of physical elements with virtual elements [1], that is, the use of the virtual world as part of real environments [18]. Landa exposes in [23] three basic applications for augmented reality: (1) to bring a static element to life, (2) to enrich information, and (3) to simulate the reality of a new design in different situations.
9. *Cybersecurity*: It protects the information (either on a device or on the network) to generate secure and reliable communications in industrial systems and manufacturing lines [2, 22].

The current technological innovation advances brought about by industry 4.0 provided access to a universe of applications that support the daily lives of users [4, 24]. This represents a radical change as the availability of smarter technology and a larger infrastructure propose a new paradigm of access to leisure, cultural, educational, commercial, tourism, and healthcare services [1, 4].

As discussed so far, industry 4.0 is a proper industrial concept; however, new application scenarios have been proposed in which cost can also be reduced and the quality of products and services can be improved [2, 25]. Such is the case of healthcare, even though called health 4.0 [25], it makes use of different technologies "including digitization, artificial intelligence, ergonomics, human psychology, the internet of things, big data, and augmented reality to name a few" [26]. The growth of the population has as consequence a greater demand in the healthcare field, requiring effective treatments and better quality. This is why healthcare remains one of the most important challenges for modern society [27].

With health 4.0, industry 4.0, and health industry 4.0 [28], crucial advances in technology have been generated and, in turn, have changed healthcare systems with improved diagnoses, new drug development, new treatments, and advances in transplant surgeries [26]. According to Ahsan and Siddique [29], healthcare has evolved dramatically, and the advances are divided into four main stages. Healthcare 1.0 was limited to physical health services (e.g., vital sign check). Healthcare 2.0 introduced surgery services and medical machines. In turn, healthcare 3.0 distinguished itself by providing patient-centric services. Finally, healthcare 4.0 introduces digital health technologies such as online health, mobile health, telemedicine, medical information technology, digital medicine, health information system, etc. to offer personalized and flexible services.

Considering some pillars of industry 4.0, augmented reality makes it possible for medical students to practice surgery by interacting with virtual human models [2]. For its part, big data analytics improves the diagnosis and treatment process by reducing medication errors and even has great consideration in the prediction of epidemics and resource requirements [30]. With the Internet of Things, it is possible for patients to monitor their health status and transmit the data in real time to the specialist, facilitating remote care [31]. In this chapter, the focus is on the use of electroencephalographic sensors to record motor imagery signals and the analysis of the information they provide. Section 2.2 describes EEG signals in more depth, as well as their potential applications for industry 4.0.

2.2 Electroencephalographic Signals

As mentioned in Sect. 2.1, sensors are part of the Internet of Things pillar in the industry 4.0 era. Sensors are, in fact, the ones who provide information that feeds the industry 4.0 paradigm [32]. In this way, healthcare is being revolutionized by focusing its ecosystem on industry 4.0 technologies (big data, the Internet of Things, blockchain, cloud computing, artificial intelligence, etc.) [31].

In the healthcare and medicine domain, sensors can measure and store information such as steps, calories, hydration, blood pressure, fatigue levels, electrocardiogram (ECG), and electroencephalogram (EEG) results, to name a few [26]. The study in this chapter focuses on the analysis of electroencephalographic signals.

Electroencephalography (EEG) is a noninvasive measurement of the electrical activity of the brain [33] recorded by electrodes (sensors) placed on the scalp [34]. This electrical activity is reported as electroencephalographic (EEG) signals and is created by the interconnected neurons in the brain which function as information carriers [34, 35].

EEG is applied in healthcare, neuroscience, and biomedical engineering "because of its high temporal resolution, non-invasiveness, and relatively low cost" [36]. Along with the progress in deep learning techniques, EEG has become an important diagnostic tool for brain- and mental-related diseases and abnormalities such as epilepsy, brain tumors, sleep disorders, dementia, attention disorders, learning problems, and behavioral disturbances [37, 38].

Consequently, the study of EEG signals has been closely related to classification tasks [36]. Among the most outstanding research projects includes emotion recognition, used by industry 4.0 in neuromarketing to get a better understanding of the customers [36, 39]; mental workload tasks, the measurement of mental activity while the user is solving complex tasks [36, 40]; seizure detection in epileptic patients and early detection of Alzheimer's disease [36, 41]; sleep stage, to find sleep disorders [36, 37]; detection and assessment of pain (type, intensity, and origin), to discover medical conditions and improve their treatment [42]; and motor imagery (MI) tasks, to detect the movement of certain muscles and help to learn or improve complex movements [36, 43]. In this chapter, EEG signals are focused on motor imagery recording of opening and closing the right hand. To learn more about MI, the reader can refer to the topic in Sect. 2.3.

2.3 Motor Imagery

Motor imagery (MI) or mental practice [44] is an important concept related to sports and rehabilitation research [43]. In papers such as those of Ruffino et al. [45], Dickstein and Deutsch [46], and Pilgramm et al. [8], motor imagery is defined as the mental representation of movement from a personal perspective without any body movement. It is a "process during which a participant is asked to recall the sensorimotor representations that are normally generated during actual execution" [45].

MI practice represents a complex cognitive operation (internal, conscious, and self-intended [8]) that triggers sensory and perceptual processes that activate specific motor actions within the working memory [46]. This means that motor imagery involves activity in the same brain regions typically associated with physical movement [47].

In the literature, it is possible to find different studies that demonstrate the close relationship between the mental representation of movement by MI and the improvement of complex physical movements [43]. Moreover, they show positive effects on motor performance and learning in professional sports (e.g., swimming, golf, gymnastics, windsurfing [43]), people with neurological conditions (e.g., stroke, spinal injury, Parkinson's disease) [46], and preventing declines in mobility for older adults [47].

Aside from sports and rehabilitation, MI research is also related to biological engineering. MI signals can be recorded by EEG (electroencephalogram) techniques, which are normally noninvasive using sensors placed on the scalp. Each sensor detects changes in the electrical activity of the brain [48].

The use of MI signals is common in the design of brain-computer interfaces (BCI) [48, 49]. BCI systems allow brain-machine interaction by translating thoughts into external device commands [48] "to carry out specific tasks, such as controlling wheelchairs, home appliances, robotic arms, speech synthesizers, computers, and gaming applications" [49]. In this way, a new alternative to nonmuscular communication is provided [49].

2.4 Multicriteria Decision-Making Methods

Today, the world is in constant change; dynamic environments full of uncertainties, complexities, ambiguities, and the growing expectations of users and customers present the challenge for industry and business to become more efficient and agile [13, 15]. In this sense, the technology offered by industry 4.0 allows the opportunity to improve operational and decision-making processes [13].

As mentioned above, industry 4.0 introduces some approaches such as the Internet of Things, wireless sensor networks, big data, cloud computing, and mobile Internet, among others [14]. This implies a great facility to generate information, which makes the control and operation process increasingly complex [11, 12]. Decision-making requires the comparison of different alternatives and the consideration of multiple criteria. This makes it a complicated task, especially dealing with multi-expert and multicriteria environments [50] and making necessary real-time decisions [51].

Multicriteria decision-making (MCDM) methods have grown as part of operations research and represent an approach that allows evaluation and decision-making in complex problems involving both quantitative and qualitative factors [16]. These kinds of problems are characterized by high uncertainty, conflicting criteria and objectives to be considered simultaneously, different types of data, multiple perspectives and interests, and complex systems [52, 53]. Formally, "MCDM is a procedure that combines the performance of decision alternatives across several, contradicting, qualitative, and/or quantitative criteria and results in a compromise solution" [54].

In the literature, there are different formal techniques and methods for decision-making that define the process to design the best decision alternative [52]. In this way, MCDM "is a generic term for all methods that exist for helping people make decisions according to their preferences" [16]. Such techniques have a wide range of applications such as engineering, risk management, manufacturing design, high technology market sectors, problem design, material selection problems, computational modeling, etc. [53, 55]. Thus, MCDM methods are applicable in real-life problems and industrial activities where it is necessary to consider several alternatives whose evaluation criteria may conflict with each other [54].

According to Ceballos et al. [56], a common MCDM problem is characterized by ordering a given set of alternatives under a set of specific characteristics, usually based on a decision matrix with the scores of each of the alternatives. In general, a multicriteria decision-making method follows the following three steps:

1. "Determine the relevant criteria and alternatives."
2. "Attach numerical measures to the relative importance of criteria and the impacts of the alternatives on these criteria."
3. "Process the numerical values to determine a ranking of each alternative" [57].

Table 1 shows some of the most used MCDM techniques. Each of them represents a different perspective to address a given problem, and therefore, they have their requirements [54, 58].

In the literature, it is possible to find different applications of multicriteria decision-making methods in environments oriented to the development of industry 4.0. For example, Damidavicius et al. [59] conducted a study on sustainable transport system planning in which they applied MCDM methods for the evaluation

Table 1 Multicriteria decision-making methods [52, 54, 57]

Approach	Method	Author(s)
Value-based theory methods	AHP (analytic hierarchy process)	Saaty (1980)
	WSM (weighted sum model)	Fishburn (1967)
	WPM (weighted product model)	Bridgman (1922) and Miller and Starr (1969)
	ELECTRE (elimination and choice translating reality; English translation from the French original: Elimination et choix traduisant la realité)	Benayoun et al. (1966)
	TOPSIS (technique for order preference by similarity to ideal solution)	Yoon and Hwang (1980)

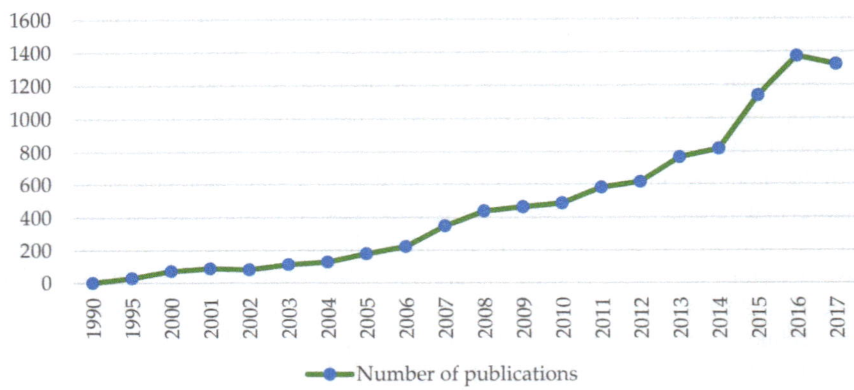

Fig. 1 The publications on MCDM methods, 1990–2017. (Extracted from the paper: "An Overview of Multi-Criteria Decision-Making Methods in Dealing with Sustainable Energy Development Issues" [60])

of different urban transport systems and their potential in terms of sustainable mobility in Lithuanian cities. This is to promote investments in no-motorized and public transport infrastructure, aligned with the importance of environmental care.

In the same line, Siksnelyte et al. [60] gave an overview of the use of MCDM methods for the solution of sustainability issues in the energy sector. The authors made evident the increase in studies based on applications of MCDM methods from 1990 to 2017 (Fig. 1). This, thanks to their universality and their wide selection of specific problems, has become increasingly popular.

On the other hand, Esra Aytaç Adalı and Ayşegül Tuş [61] conducted a study on hospital site selection employing distance-based MCDM methods: TOPSIS (Technique for Order Preference by Similarity to Ideal Solution), EDAS (evaluation based on distance from average solution), and CODAS (combinative distance-based assessment). Due to population growth in cities, it is of vital importance to have easy access to different hospital services. Thus, the location of a hospital is a strategically important decision in terms of being accessible to a large part of the population. The criteria used were cost, transportation, geological factors, land strategy, financial support by the government, environmental consideration, and demographic consideration. At the end of the study, the authors obtained the same ranking in the three MCDM methods; therefore, they concluded that these methods can be used one in substitution of the other.

In this chapter, TOPSIS (Technique for Order Preference by Similarity to Ideal Solution) model is used to select the feature subset (typical testor) that allows the best performance in an artificial neural network classifier.

2.5 TOPSIS

TOPSIS (Technique for Order Preference by Similarity to Ideal Solution) method is a solid multiattribute decision-making strategy that determines the best solution from a collection of alternatives described by a certain number of attributes [62, 63]. TOPSIS is a commonly MCDM method due to its advantages over other methods [58, 64] such as [63] the following:

1. Many alternatives and attributes support
2. Limited input requirements
3. Logical and programmable nature
4. Comparative consistency in the alternative ranking

   The classical TOPSIS approach is based on the concept of closeness to ideal and anti-ideal solutions in decision-making problems [52]. That is, the best solution minimizes the geometric distance from the positive ideal solution (PIS, best theoretical solution) and maximizes the distance from the negative ideal solution (NIS, worst theoretical solution) [52, 58]

$$A = \begin{matrix} a_1 \\ a_2 \\ \vdots \\ a_m \end{matrix} \begin{bmatrix} x_{1,1} & x_{1,2} & \dots & x_{1,n} \\ x_{2,1} & x_{2,2} & \dots & x_{2,n} \\ \vdots & \vdots & \ddots & \vdots \\ x_{m,1} & x_{m,2} & \dots & x_{m,n} \end{bmatrix} \tag{1}$$

$$C = c_1 \; c_2 \; \dots \; c_n \tag{2}$$

$$W = w_1 \; w_2 \; \dots \; w_n \tag{3}$$

TOPSIS method is composed of the following steps:

1. The first step is the construction of the decision matrix (see Eq. 1). It is made of m alternatives $A = \{a_1, a_2, \ldots, a_m\}$ evaluated in terms of n criteria C (see Eq. 2). Equation 1 shows the decision matrix, where the x_{ij} value represents the performance measure for the alternative a_i under the criteria c_j and W is the vector of weights (see Eq. 3) associated with the set of criteria C (Eq. 2) [52, 56, 57].
2. Considering that the criteria may not be in the same domain, the second step is to normalize them. Then, the normalized decision matrix is obtained by the following expression [56, 57, 63]:

$$r_{ij} = \frac{x_{ij}}{\sqrt{\sum_{i=1}^{m} x_{ij}^2}} \quad j = 1, 2, \ldots, n; i = 1, 2, \ldots, m \tag{4}$$

being r_{ij} an element of the normalized decision matrix (Eq. 5) and x_{ij} is an element of the original decision matrix (Eq. 1) [62, 63].

$$R = \begin{bmatrix} r_{11} & r_{12} & \cdots & r_{1n} \\ r_{21} & r_{22} & \cdots & r_{2n} \\ \vdots & \vdots & \vdots & \vdots \\ r_{m1} & r_{m2} & \cdots & r_{nm} \end{bmatrix} \tag{5}$$

3. The third step consists of the construction of the weighted normalized decision matrix V (Eq. 6) using the normalized decision matrix R from Eq. 5 and a set of weights W from Eq. 3, where $\Sigma w_i = 1$ [56, 57, 62]. Each w_i is assigned by the decision-maker, and it is associated with criteria c_i:

$$V = \begin{bmatrix} v_{11} & v_{12} & \cdots & v_{1n} \\ v_{21} & v_{22} & \cdots & v_{2n} \\ \vdots & \vdots & \vdots & \vdots \\ v_{m1} & v_{m2} & \cdots & v_{nm} \end{bmatrix} \tag{6}$$

Each value v_i is given by Eq. 7 [56, 58, 62, 63, 65]:

$$v_{ij} = r_{ij} * w_j \tag{7}$$

4. The fourth step calculates the ideal solution A^+ and the anti-ideal solution A^- [62, 63]. According to Triantaphyllou [57], these alternatives are both fictitious. A^+ is the ideal solution, that is, it has the maximum value concerning the other alternatives according to its benefit criteria. On the other hand, A^- is the minimum value for cost criteria, that is, the least preferable alternative (anti-ideal solution). Equations 8 and 9 determine the A^+ and A^- values, respectively [56, 57]:

$$A^{+} = \{v_1^{+}, \cdots, v_n^{+}\} = \left\{\left(\max_i v_{ij}, j \in J\right)\left(\min_i v_{ij}, j \in J'\right)\right\}; i = 1, \ldots, m \tag{8}$$

$$A^{-} = \{v_1^{-}, \cdots, v_n^{-}\} = \left\{\left(\min_i v_{ij}, j \in J\right)\left(\max_i v_{ij}, j \in J'\right)\right\}; i = 1, \ldots, m \tag{9}$$

Value $J = \{j = 1, 2, \ldots, n\}$ is associated with benefit criteria, and $J' = \{j = 1, 2, \ldots, n\}$ value is associated with cost criteria [56, 57, 65].

5. The fifth step corresponds to the distance calculation [52], that is, measure the Euclidian distance from the *i-th* alternative to the ideal solution (Eq. 10) [52, 54, 57]:

$$s_i^{+} = \sqrt{\sum_{j=1}^{m}\left(v_{ij} - v_j^{+}\right)^2} \tag{10}$$

And the Euclidean distance from the *i-th* alternative to the anti-ideal solution (Eq. 11) [52, 54, 57]:

$$s_i^{-} = \sqrt{\sum_{j=1}^{m}\left(v_{ij} - v_j^{-}\right)^2} \tag{11}$$

6. The sixth step calculates the relative closeness of each alternative a_i to the ideal solution by using the next expression [56, 57]:

$$R_i = \frac{s_i^{-}}{s_i^{+} + s_i^{-}}, i = 1, \ldots, m \tag{12}$$

Thus, the more the R_i*-value* tends to 1, the closer the alternative a_i is to the ideal solution [56, 62, 65].

7. As a final step, the hierarchy of preferences is established by ordering the alternative a_i starting with those with the $R_i -$ value closest to the ideal solution [52, 65], that is, "the best alternative has the shortest distance to the ideal solution" [57].

2.6 Testors Theory

Sections 1 and 2.1 discussed the generation of high voluminous data that reaches big data systems every second, making big data one of the pillars of industry 4.0 [5, 6]. This amount of data is generated in real time and is a consequence of the growth of the number of interconnected smart devices (Internet of Things data, cybersecurity data, mobile data, business data, social media data, health data, etc.) [6, 66, 67]. Thus, the volume and velocity of data will keep growing exponentially. In this context, it can be observed that how decisions are made is influenced by the amount of data available. This is from the point of view of mathematics, advanced statistics, and computer science since data with multidimensional structures make visualization, inference, and learning tasks more complex and computationally demanding (processing time and storage space) [6, 68, 69].

Working with high-dimensional raw datasets implies the existence of redundant information, and in many cases, most of the features that compose the dataset are irrelevant [6]. To address this situation, there are tools available that have the capability of extracting insights or useful information from raw datasets in a timely and intelligent way [67]. Artificial intelligence, mainly machine learning, has emerged as the most important way method for data analysis and computation providing intelligent applications. Within the machine learning field, there are dimensionality reduction techniques in charge of pattern recognition [67, 68].

Pattern recognition is a scientific discipline that addresses two main goals: classification and feature selection [70], where the latter consists of extracting those features or attributes that provide relevant information, usually for classification processes [71]. That is, feature selection makes it possible to discard "noisy, redundant and irrelevant data" [72]. According to Rodríguez-Diez et al. [71], a feature selection process provides important benefits for the classification process such as reducing its computational cost and, in some cases, could even allow better classification accuracy. However, feature selection has an important role in expert and intelligent systems [71], getting the following three goals:

1. "Improving prediction performance."
2. "Providing faster and more cost-effective predictors."
3. "Providing a better understanding of the underlying process that generated the data" [73, 74].

Among the tools used to perform feature selection is the testors theory, which is part of the logical-combinatorial approach to pattern recognition [71, 75]. The concept of testor related to pattern recognition is formulated by Zhuravlev et al. in 1966 for classification in the geology field [76].

Zhuravlev's testor computation [76] requires a dataset (LM, learning matrix) with k objects described by n features $R = \{x_1, \ldots, x_n\}$ and grouped into r classes [77]. Those classes must be disjoint sets, every attribute has Boolean comparison criteria, and the similarity criterion between objects assumes that two objects are different if at least one of their features is also different [77, 78]. In this context, a *CM* is a binary comparison matrix, and it is obtained by comparing every feature of a pair of objects in *LM* belonging to different classes. That is, the comparison of every feature adds a *0* if both objects are equal in that feature and a *1* if they are different [77].

Now, let $T \subseteq R$ be a subset of features from *LM*. *T* is a testor of *LM* if in the sub-matrix of *CM* created only by the columns in *T*, there are not any zero rows [10,

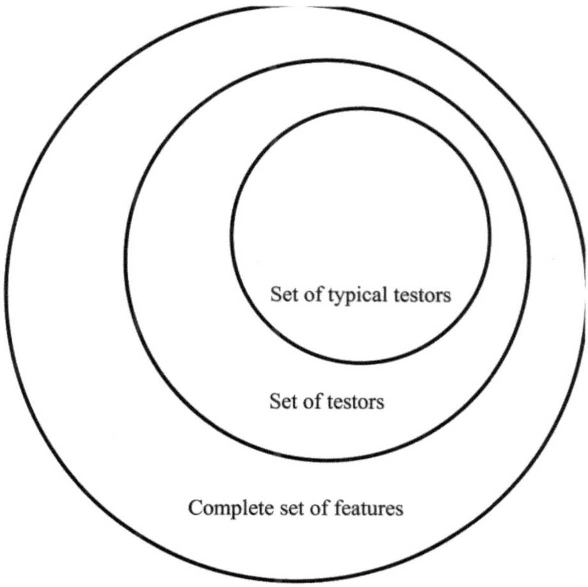

Fig. 2 Sets of testors and typical testors [80]

77]. That is, a testor is a subset of features of *LM* in which all objects from different classes can be discerned according to the comparison criteria established for each feature [10, 79].

As can be seen in Fig. 2, within the set of testors are the typical testors. The set of typical testor is extracted from the basic matrix (BM) where BM is a reduction of *CM* without loss of relevant information [77]. Then, a feature subset $T \subseteq R$ is a typical testor in BM if T is a testor and does not exist any other testor T'such that $T' \subset T$ [10, 77, 81]. Therefore, a typical testor is an irreducible testor, that is, each feature in a typical testor is essential to qualify as a testor. In other words, if a feature were removed from a typical testor, the resulting feature subset will not be a testor [71, 81].

Finally, it is important to mention that each typical testor preserves the ability to discern the class from the original set of features [77]. Besides, finding the complete set of typical testors has a high computational cost. However, this set is an important "tool for solving feature selection problems, especially those problems in which objects are described by both quantitative and qualitative features" [79].

2.7 Artificial Neural Network

Artificial neural networks (ANNs) provide an important framework for the implementation of machine learning with a wide range of applications [82] in different fields including computing, science, engineering, medicine, environment, agricul-

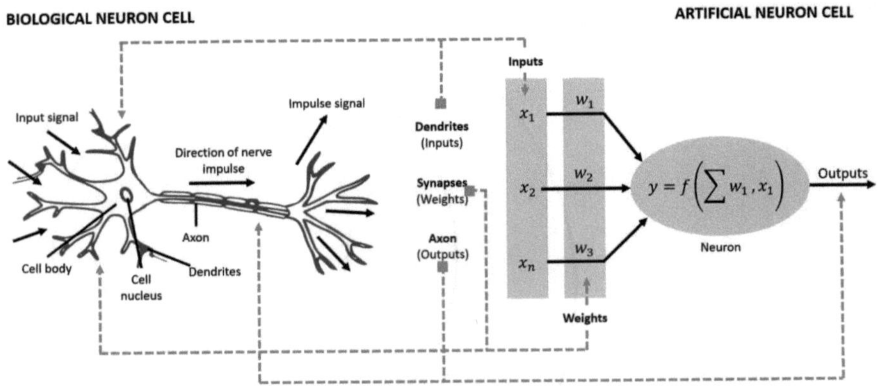

Fig. 3 Comparison between a biological neuron (left) and an artificial neuron (right). (English version from the original published in [89])

ture, mining, technology, climate, business, arts, nanotechnology, etc. Thus, the ANNs represent a popular model "for classification, clustering, pattern recognition, and prediction in many disciplines" [83].

ANNs have had a major impact within industry 4.0 in terms of automating organizational and business processes and improving the execution of automatic information extraction, recognition, classification, forecasting, and optimization tasks [84]. The development of ANN models is closely related to machine learning and, especially, deep learning [85]. Among the applications of ANN in industry 4.0 are data security, stock market predictions, medical diagnostics, fraud detection, search engines, natural language processing, virtual assistants, intelligent vehicles, quality control, artificial visions, and scrap reduction [85–88], just to name a few.

Asanza et al. [90] define ANN as a family of information processing techniques inspired by the biological information processing performed by the nervous system of living beings. That is, "artificial neural networks are referred to as non-linear statistical data models that replicate the role of biological neural networks" (BNN) [91].

ANNs also possess nonlinearity, robustness, high parallelism, fault and failure tolerance, learning, and imprecise and fuzzy information handling and generalize capability, which all are inherited characteristics of biological systems [92].

Figure 3 shows both a biological neuron and an artificial neuron, which are processing units in both models [83]. As shown, an ANN model is a simplified model of a BNN's unit [93]. It is worth mentioning that the main goal of an ANN is not to exactly replicate the way a BNN works but to abstract how biological neurons receive, process, and communicate information with each other [92]. Thus, ANN became "a mathematical representation of human neural architecture, reflecting its "learning" and "generalization" abilities" [94].

An ANN is made up of a set of artificial neurons organized in layers [94], where each artificial neuron has a certain number of input-output connections to other

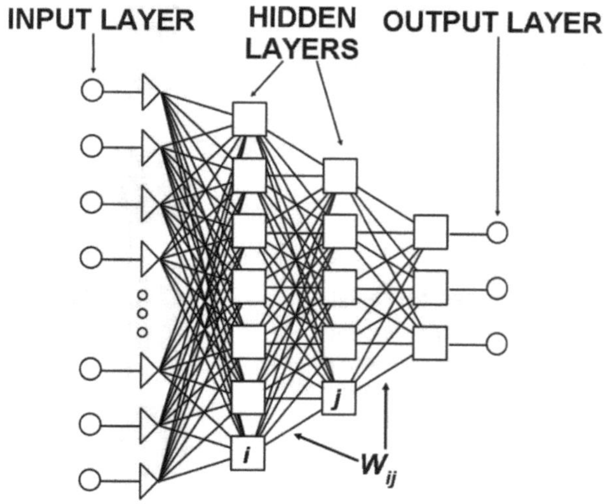

Fig. 4 Artificial neural network [94]

neurons. According to Fig. 4, the neural network has an "input" layer where the data is received. Next, the data are transferred to one or more "hidden" layers. The hidden layers are responsible for the mathematical processing of the information by multiplying the input value by a weight *W* (see Fig. 4), and the result is propagated to the neurons of the following layers [94, 95]. Finally, the "output" layer provides the result of the network.

Each layer in a neural network is independent of the rest of the layers, that is, a specific layer can have a particular number of artificial neurons (nodes) [83, 94]. However, the number of layers, as well as the number of neurons in each layer, depends entirely on the problem to be solved [94]. For example, if a network is creating a classifier, the input and output neurons correspond to input features and output classes [83].

In the end, artificial neural networks have found wide-ranging acceptance due to their ability to provide easy-to-use models with accurate solutions for complex real-world problems in a variety of applications by modeling "highly non-linear systems in which the relationship among variables is unknown or very complex" [83, 92, 94].

3 Methodology

As mentioned above, the main objective is to choose from a set of typical testors, the one that allows the best classification performance in an artificial neural network model. This is done using the TOPSIS method described in the previous section.

The methodology used to perform the respective decision-making by the TOPSIS method is described in Fig. 5.

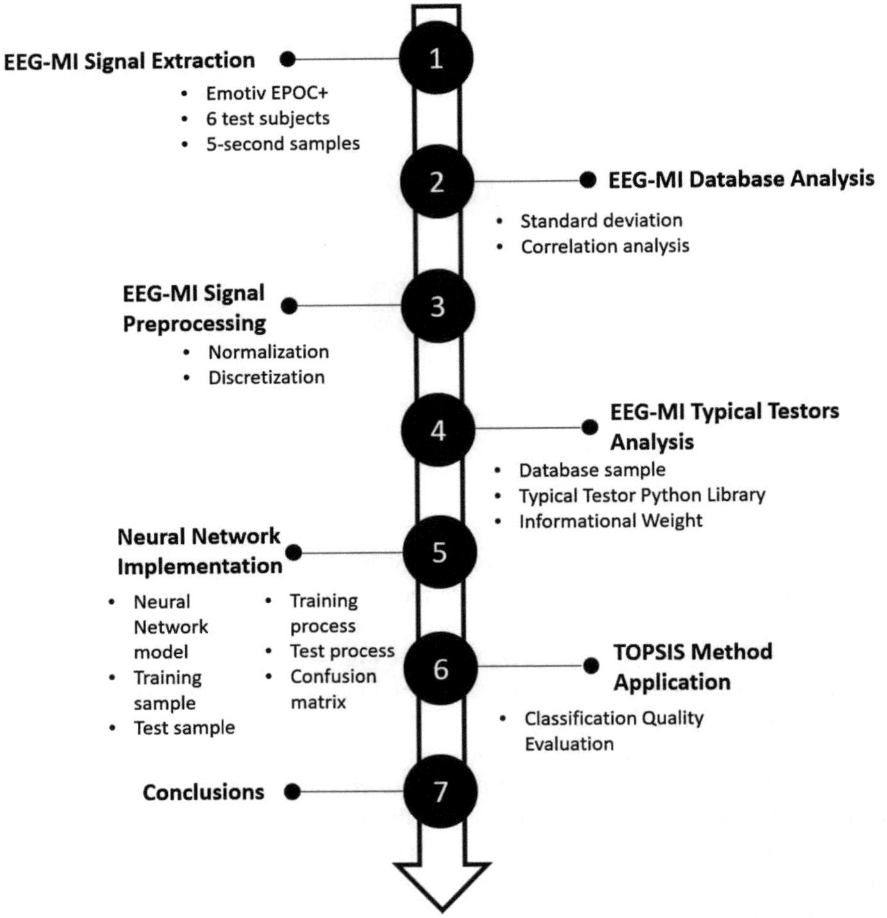

Fig. 5 Methodology

As can be seen in Fig. 5, phase 1 of the methodology consisted of the acquisition of EEG signals with motor imagery (see Sect. 2.5) of opening and closing the right hand. For this purpose, the Emotiv EPOC+ EEG device was used. This device is a wireless EEG headset composed of 14 channels (AF3, F7, F3, FC5, T7, P7, O1, O2, P8, T8, FC6, F4, F8, AF4) located according to the 10/20 international standard, and it can produce 128 samples per second (see Fig. 6). For this part of the methodology, six five-second samples were taken from six test subjects. These samples consisted of three sets with the imagery of opening the right hand and three more with the imagery of closing the right hand. Moreover, the sampling was made possible thanks to the open-source application Cykit, which allowed storage of these samples in CSV files used in the following phases of the methodology described below.

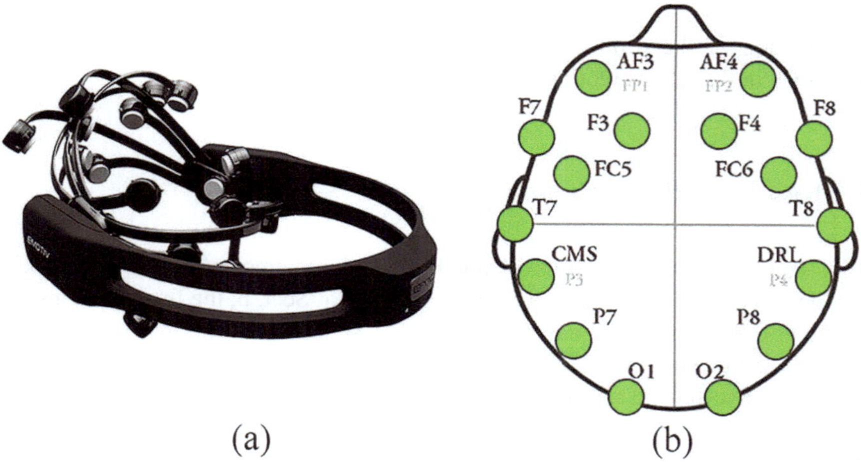

Fig. 6 (**a**) Emotive EPOC+ device and (**b**) device's electrode arrangement on 10–20 standards. (Extracted from the paper "Artificial Neural Networks to Assess Emotional States from Brain-Computer Interface" [96])

Phase 2 consisted of the analysis of the data obtained in phase 1 of the methodology. The first step was to calculate the standard deviations of each node/feature of the database, to know its behavior concerning its respective mean. For the second step, a Pearson correlation analysis was carried out, which made it possible to observe the behavior between features.

In phase 3, the data was preprocessed, starting with a normalization using the Z-score method, and included a data discretization process. This was done to facilitate data processing in the following phases of the methodology.

In phase 4, testor analysis was performed on a representative sample of the data set. This analysis made it possible to obtain a set of typical testers (see Sect. 2.3). In addition, the informational weight of each characteristic was calculated to determine its impact on the classification process.

Phase 5 included the training of a dense neural network model (see Sect. 2.4). For this purpose, the model was trained once for each typical testor with a random sample of 50% of the dataset. Once the neural network was trained, a set of 30 validations was carried out using random samples of 25% of the dataset not used in the training phase. The quality of each validation was measured using the metric obtained from their corresponding confusion matrices (accuracy, positive precision, negative precision, false-negative rate, and false-positive rate). At the end of phase 5, an integrated matrix was available consisting of the set of typical testors and the means of each metric calculated from each set of replicates.

In phase 6, the typical testor that achieved the best classification quality was chosen using the TOPSIS multicriteria decision-making method (see Sects. 2.1 and 2.2). For this, the resulting matrix from phase 5 was used as the decision matrix, which functions as the input to the TOPSIS method.

Finally, phase 7 of the methodology, as its name implies, corresponds to the presentation of the results and conclusions achieved. The complete phase 7 is described in the results section below.

4 Results and Conclusions

This section presents the results obtained from the implementation of the methodology described in Sect. 3. As can be seen in Fig. 4 of Sect. 3, the first step consisted of the construction of the database with EEG signals with the motor imagery of opening and closing the right hand. For this task, six test subjects were used, who performed five-second samples: three corresponded to opening and three to closing the right hand.

As mentioned above, the Emotiv EPOC+ device was used, which allowed the recording of signals and their storage in CSV files. Each record is described by 14 features/channels/electrodes (AF3, F7, F3, FC5, T7, P7, O1, O2, P8, T8, FC6, F4, F8, AF4) measured in μV (see Fig. 6). As a result of phase 1, there was a database composed of 23,846 records, of which 12,190 are related to the motor imagery of opening and 11,656 corresponded to the motor imagery of closing the hand.

As a result of phase 2, the standard deviations of each feature describing the EEG signals were calculated. As can be seen in Table 2, although the 14 nodes of the Emotiv EPOC+ device measure the same type of variable (μV), the values of their standard deviations show a large mean dispersion concerning each other. Therefore, in phase 3, it was decided to normalize the data for the classification process performed by the neural network model described later.

As mentioned above, the second part of phase 2 consisted of a correlation analysis, the results of which are shown in Fig. 7.

For the correlation analysis, Pearson's correlation coefficient was calculated, which allowed the construction of the graph shown in Fig. 7. The graph shows that the coefficients obtained are greater than 0 (tendency to color white), which means that there is a positive correlation between the features. However, there are values extremely close to 0 (black color tendency), which shows that there is no linear relationship between the features involved.

Table 2 Standard deviation results

Node	STD	Node	STD
AF3	42.6789182478843	*O2*	57.3289657
F7	53.7523558	*P8*	71.0371007909142
F3	80.187847	*T8*	45.6327024207423
FC5	82.04517812232	*FC6*	93.0843388893872
T7	27.2264746337066	*F4*	122.134084225115
P7	70.5757321082372	*F8*	75.6765945803192
O1	40.6529441779588	*AF4*	53.3759259517592

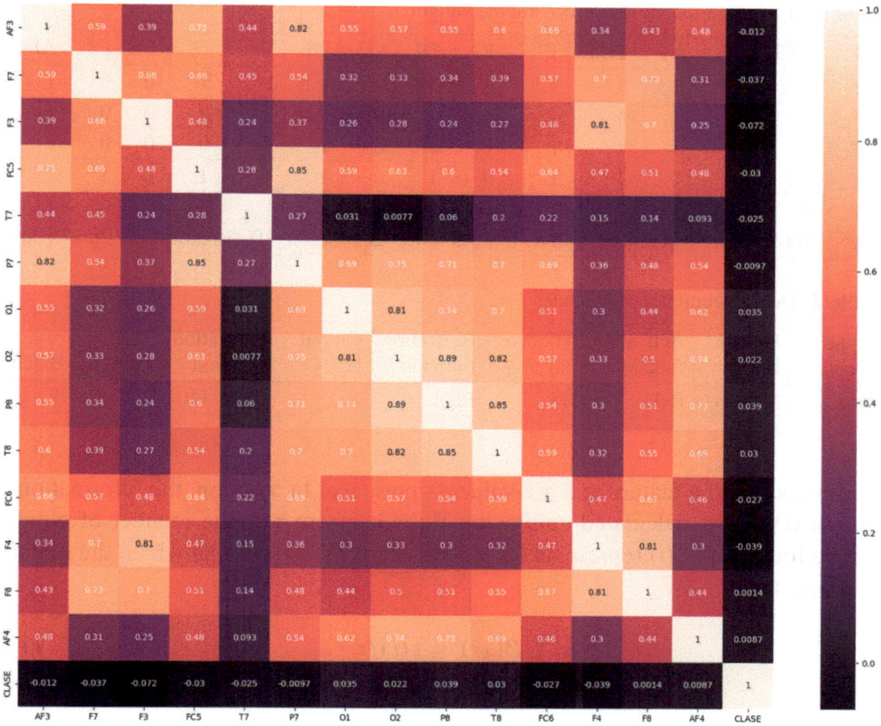

Fig. 7 Correlation analysis

According to the proposed methodology, phase 3 consisted of a normalization and discretization process as part of a preprocessing mechanism. Normalization was performed using the Z-score normalization approach. The goal of normalizing the data was to prepare the data to facilitate the analysis to be performed by the neural network model in the subsequent phases of the methodology. In addition, phase 3 also includes a process of discretization of the variables, which was performed using the preprocessing package of the Sklearn Python library, using specifically KbinsDiscretizer with the k-means strategy. The main goal of discretizing the data was precisely to preprocess the data for the computation of a set of typical testors.

In phase 4, a typical testor analysis (see Sect. 2.3) was performed. For this purpose, a random sample of 200 hand-opening and 200 more hand-closing records was drawn. That is, since the calculation of typical testors is an exponential algorithm, therefore the computational cost of extracting a set of typical testors is reduced. The calculation of typical testors was carried out by implementing the TypicalTestors library developed by Daniel Barajas, a PhD student at the Autonomous University of Aguascalientes, México.

Table 3 shows a set of five typical testors found by the Python library. As can be seen, each typical testor is represented by binary values, where a 0-value indicates

Table 3 Typical testor set and informational weight

ID	AF3	F7	F3	FC5	T7	P7	O1	O2	P8	T8	FC6	F4	F8	AF4
TT_1	1	1	1	1	1	0	0	0	1	1	1	0	0	1
TT_2	1	1	1	1	1	1	1	0	0	1	0	0	0	1
TT_3	1	1	1	1	1	1	0	0	1	1	0	0	0	1
TT_4	1	1	0	1	1	0	1	0	1	1	1	0	1	1
TT_5	1	1	0	1	1	0	1	1	1	1	1	0	0	1
IW	100%	100%	60%	100%	100%	40%	60%	20%	80%	100%	60%	0%	20%	100%

Table 4 Dense neural network model

Input layer	Hidden layer 1	Activation function	Hidden layer 2	Activation function	Output layer	Activation function	Epochs	Batch
Var	75	Relu	15	Relu	1	Sigmoid	100	5

the absence of the feature and a 1-value its presence. In addition, the informational weight (IW) was calculated employing Eq. 13, where τ is the number of typical testors found and $\tau(i)$ represents the number of typical testors in which the feature xi appears:

$$P(xi) = \tau(i)/\tau \tag{13}$$

The IW value is interpreted as a measure of significance for each of the features. That is, the higher its information weight value, the greater is the relevance in a classification process. Such is the case of the features in Table 4, where six of them (AF3, F7, FC5, T7, T8, AF4) with an IW value of 100% can be observed, which means that knowing their values is indispensable to deciding whether it is a signal to open or to close the hand. On the other hand, feature F4 with an IW value of 0 means that it can be discarded because its value does not generate relevance in the classification process.

Phase 5 is the training and validation of a dense neural network. This model is composed of two hidden layers, and its configuration is based on the model tested in [97] for the classification of EEG signals. The proposed model was trained five times, once for each typical testor found, with a 50% sampling of the available data, and the rest was used for a validation process. As mentioned in Sect. 3, each training of the model was replicated 30 times using random samples of 25% of the remaining sample not used in the training phase.

Each replica was evaluated using the metrics: accuracy, positive precision (open hand), negative precision (close hand), false-negative rate, and false-positive rate that can be extracted from its respective confusion matrix. At the end of each set of replicates, the values of the metrics were averaged to construct Table 5, which is shown below. Finally, the resulting data from phase 5 constitute the decision matrix

Table 5 Training results and decision matrix

ID	Accuracy	TP precision	TN precision	FP rate	FN rate
TT_1	83.5379627	84.0354695	83.0625839	16.9374161	15.9645305
TT_2	82.9173655	87.2011396	78.8755443	21.1244557	12.7988604
TT_3	83.4317343	83.6756158	83.2012539	16.7987461	16.3243842
TT_4	83.4015431	85.0647696	81.8122399	18.1877601	14.9352304
TT_5	82.5785531	83.5208287	81.6811839	18.3188161	16.4791713
Weights	0.2	0.35	0.35	0.05	0.05

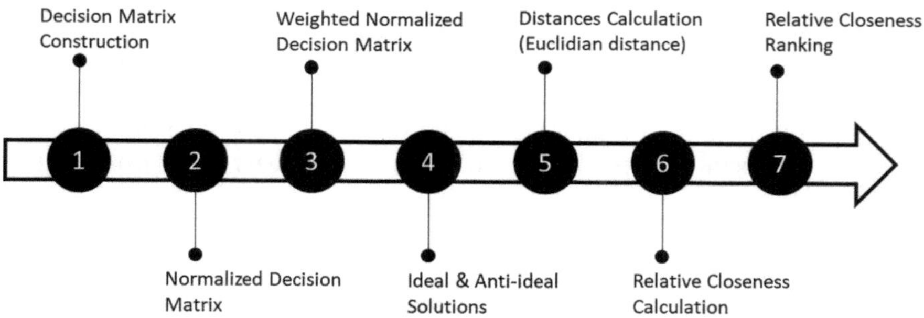

Fig. 8 TOPSIS method overview

which, in turn, is the input for the application of the TOPSIS method to choose the typical testor with the best classification performance.

In phase 6, the classification performance of each typical testor is evaluated by TOPSIS to choose the one that allowed the neural network model to better classify motor imageries. Figure 8 shows an overview of the TOPSIS multicriteria decision-making method described in Sect. 2.2.

As mentioned above, the decision matrix was obtained because of phase 5. In addition, the vector of weights was added as shown in Table 5. The values of the vector of weights were chosen based on the importance of each decision criterion (metrics from the confusion matrix). That is, the positive and negative precisions describe if the neural network model can distinguish each class of motor imagery, and therefore, a higher weight is assigned to them.

Second is the performance metric, which describes the ratio of the correctly classified objects to the total number of classified objects. However, it may not be a completely reliable metric as it can sometimes have a high value, but there is a possibility that the classifier only identifies one class and ignores the rest. Therefore, it is given a lower weight. Finally, for the false-negative and false-positive ratios, it was decided to give them a low weight because they have an opposite meaning to the positive and negative precisions. That is, if positive and negative precision values increase, false-positive and false-negative values decrease.

According to Fig. 7, the results of the next steps in the implementation of the TOPSIS method are presented below:

Table 6 Normalized decision matrix

ID	Accuracy	TP precision	TN precision	FP rate	FN rate
TT_1	0.44916928	0.44364985	0.45444211	0.41302293	0.46478745
TT_2	0.44583243	0.46036243	0.43153448	0.51512489	0.3726229
TT_3	0.44859811	0.44175007	0.45520078	0.40964143	0.47526414
TT_4	0.44843577	0.44908385	0.44760137	0.44351286	0.43482066
TT_5	0.4440107	0.4409329	0.44688436	0.4467087	0.47977057

Table 7 Weighted normalized decision matrix

ID	Accuracy	TP precision	TN precision	FP rate	FN rate
TT_1	0.08983386	0.15527745	0.15905474	0.02065115	0.02323937
TT_2	0.08916649	0.16112685	0.15103707	0.02575624	0.01863115
TT_3	0.08971962	0.15461252	0.15932027	0.02048207	0.02376321
TT_4	0.08968716	0.15717935	0.15666048	0.02217564	0.02174103
TT_5	0.08880214	0.15432651	0.15640952	0.02233544	0.02398853

Table 8 Ideal and anti-ideal solutions

Solution	Accuracy	TP precision	TN precision	FP rate	FN rate
A+	0.08983386	0.16112685	0.15932027	0.02048207	0.01863115
A−	0.08880214	0.15432651	0.15103707	0.02575624	0.02398853

Table 9 Distances

ID	D+	D−
TT_1	0.00745321	0.00963717
TT_2	0.00984245	0.00866481
TT_3	0.00829383	0.00986928
TT_4	0.00593451	0.00764308
TT_5	0.00937646	0.00636909

- *Step 2*: Normalization of the decision matrix (Table 6).
- *Step 3*: Computing the weighted normalized decision matrix (Table 7).
- *Step 4*: Creation of the ideal and anti-ideal solutions (Table 8).

It is important to mention that to establish the values of the ideal solution A^+, the accuracy, positive precision, and negative precision metrics had a positive sense, that is, the highest values were chosen. While the metrics false-positive rate and false-negative rate had a negative sense, that is, the minimum values were chosen. For the anti-ideal solution A^-, the opposite logic was followed.

- *Step 5*: Distances to the ideal and anti-ideal solution calculation (Table 9).
- *Step 6*: Relative closeness calculation (Table 10).

Table 10 Relative closeness

ID	D_j
TT_1	0.56389437
TT_2	0.46818427
TT_3	0.54336975
TT_4	0.56291884
TT_5	0.40450083

Table 11 Typical testor ranking

Ranking	ID	Features	Accuracy	TP precision	TN precision	FP rate	FN rate
1	TT_1	9	83.537962	84.035469	83.062583	16.937416	15.964530
2	TT_4	10	83.401543	85.064769	81.812239	18.187760	14.935230
3	TT_3	9	83.431734	83.675615	83.201253	16.798746	16.324384
4	TT_2	9	82.917365	87.201139	78.875544	21.124455	12.798860
5	TT_5	10	82.578553	83.520828	81.681183	18.318816	16.479171

By sorting the set of typical testors in descending order according to the relative closeness values obtained in Table 10, the best and worst typical testor can be known as shown in Table 11.

In conclusion, it is important to emphasize that the most critical points of the presented study were achieved. The first of them was the calculation of a set of typical testors. As can be recalled, the full set of features is 14 nodes, and after the application of phase 4, a set of five typical testors was found (see Table 3). In this way, the objectives of feature subset selection and the application of the typical testor theory were achieved.

In addition, the informational weight (IW) was calculated for the 14 features (see Table 3). This value is a measure of significance that is interpreted as the relevance of each feature in a classification process. In this case, there were six features (AF3, F7, FC5, T7, T8, AF4) with an IW value of 100%, making their values indispensable for a correct distinction of classes. Also, feature F4 got an IW value of 0% found as it was not present in any typical testor and, therefore, does not provide relevant information for signal classification.

After obtaining a set of typical testors and the informational weight of the features describing EEG signals, the question would be which typical testor is the best for classifying EEG signals? On one hand, at first sight, all typical testors satisfy the feature subset selection. However, it is questionable whether one can be better than another when training a classification model.

To solve the above question, an artificial neural network model was trained according to each typical testor found. Next, the ANN performance was measured by the metrics: accuracy, positive precision (open hand), negative precision (close hand), false-negative rate, and false-positive rate.

With the above information, it was possible to assemble a decision matrix for the application of the TOPSIS method for multicriteria decision-making, to know the most suitable typical testor to distinguish EEG signals. The decision matrix is composed of the performance metrics, while the typical testors are the alternative to be selected.

The use of a decision-making method such as TOPSIS is an important mathematical tool for working with criteria and objects that may be in conflict, in addition to the fact that it makes it possible to work with both qualitative and quantitative criteria. In the case of the present study, there were three criteria (accuracy, positive precision (open hand), negative precision (close hand)) whose values had to be maximized and two more (false-negative rate and false-positive rate) had to be minimized to select the best typical testor.

As shown in Table 11, the TOPSIS method not only allowed to know the most suitable typical testor but also allowed to order the alternatives from the best to the worst option. At the end of the study, it was observed that the typical testor 1 presents high values in the criteria to be maximized and low values in those to be minimized, resulting in being the best alternative. However, it can be seen in Table 11 that the typical testor 2 finished in fourth place with better positive precision and a lower false-negative ratio. Therefore, and finally, it is noteworthy that TOPSIS selects the alternative with the shortest distance to the ideal solution and the longest distance to the anti-ideal solution, both constructed from the values of each criterion in the weighted normalized decision matrix.

Finally, the industry 4.0 era not only comes with new technologies to improve processes in organizations but also with opportunities to integrate users by adapting to their different needs and capabilities through adaptability and flexibility. In this sense, industry 4.0 breaks down boundaries between different areas of knowledge and integrates them to facilitate the development of solutions in less time, with good quality and high flexibility.

In the case of this chapter, the focus is on the use of EEG signals, whose use is predominantly clinical. However, the influence of industry 4.0 creates an infinite number of possibilities for its use in different contexts. From the point of view of the pillar mentioned in Sect. 2.1, the sensors allow the recording of brain activity, and big data allows the analysis of the information obtained and its categorization.

In the current literature, it is already possible to find reports of applications related to the study of electrical brain activity (EEG signals). In the field of healthcare, EEG signals are useful in the analysis of sleep quality, detection of brain damage, analysis of concentration and fatigue in different contexts, and the detection of diseases and disorders related to the brain. Also, there is the introduction of brain-computer interfaces that provide a connection between the brain and a machine, taking human-robot interaction to a new level. This type of interface has great potential in rehabilitation, robotics, video games, communications, and the industry itself. Thus, the flexibility that industry 4.0 implies would allow the introduction of new ways of interacting with devices, meaning an interesting advance for the inclusion of people with disabilities by assisting them in their daily lives.

References

1. L. Habib Mireles, Presencia de los pilares de la industria 4.0 en la formación de ingenieros en el noreste de México. Rev. Cuba. Educ. Super. **41**(2), 380–396 (2022)
2. J. Hamilton Ortiz (ed.), *Industry 4.0 – Current Status and Future Trends* (IntechOpen, London, 2020)
3. M. Lom, O. Pribyl, M. Svitek, Industry 4.0 as a part of smart cities, in *2016 Smart Cities Symposium Prague (SCSP)*, May 2016, pp. 1–6. https://doi.org/10.1109/SCSP.2016.7501015
4. F. Bonete, M. Sánchez Martínez, 'Smart City' y patrimonio cultural: las aplicaciones móviles de ciudades patrimonio de la humanidad españolas, in *Ciudad y Comunicación*, 1st edn., (Grupo de Investigación Arte, Arquitectura y Comunicación en la Ciudad Contemporánea. Universidad Complutense de Madrid, Madrid, 2016), pp. 475–482
5. J. Yan, Y. Meng, L. Lu, C. Guo, Big-data-driven based intelligent prognostics scheme in industry 4.0 environment, in *2017 Prognostics and System Health Management Conference (PHM-Harbin)*, July 2017, pp. 1–5. https://doi.org/10.1109/PHM.2017.8079310
6. P. Chhikara, N. Jain, R. Tekchandani, N. Kumar, Data dimensionality reduction techniques for industry 4.0: Research results, challenges, and future research directions. Softw. Pract. Exp. **52**(3), 658–688 (2022). https://doi.org/10.1002/spe.2876
7. J. Cai, J. Luo, S. Wang, S. Yang, Feature selection in machine learning: A new perspective. Neurocomputing **300** (2018). https://doi.org/10.1016/j.neucom.2017.11.077
8. S. Pilgramm et al., Motor imagery of hand actions: Decoding the content of motor imagery from brain activity in frontal and parietal motor areas. Hum. Brain Mapp.**37**(1) (2016). https://doi.org/10.1002/hbm.23015
9. V.I. González-Guevara, S. Godoy-Calderon, E. Alba-Cabrera, J. Ibarra-Fiallo, A mixed learning strategy for finding typical testors in large datasets, in *Progress Pattern Recognition, Image Analysis Computer Vision, Application CIARP 2015*, Lecture Notes Computer Science, vol. 9423, (Springer, Cham, 2015), pp. 716–723. https://doi.org/10.1007/978-3-319-25751-8_86
10. J. Ruíz Shucloper, E. Alba Cabrera, M. Lazo Cortés, *Introducción a la Teoría de Testores* (Departamento de Ingeniería Electrica, CINVESTAV-IPN, 1995), p. 197
11. A. Bousdekis, K. Lepenioti, D. Apostolou, G. Mentzas, A review of data-driven decision-making methods for industry 4.0 maintenance applications. Electronics **10**(7), 828 (2021). https://doi.org/10.3390/electronics10070828
12. F. Rosin, P. Forget, S. Lamouri, R. Pellerin, Impact of industry 4.0 on decision-making in an operational context. Adv. Prod. Eng. Manag. **16**(4), 500–514 (2021). https://doi.org/10.14743/apem2021.4.416
13. F. Rosin, P. Forget, S. Lamouri, R. Pellerin, Enhancing the decision-making process through industry 4.0 technologies. Sustainability **14**(1), 461 (2022). https://doi.org/10.3390/su14010461
14. N. Medic, Z. Anisic, B. Lalic, U. Marjanovic, M. Brezocnik, Hybrid fuzzy multi-attribute decision making model for evaluation of advanced digital technologies in manufacturing: Industry 4.0 perspective. Adv. Prod. Eng. Manag. **14**(4), 483–493 (2019). https://doi.org/10.14743/apem2019.4.343
15. M.L. Hoffmann Souza, C.A. da Costa, G. de Oliveira Ramos, R. da Rosa Righi, A survey on decision-making based on system reliability in the context of industry 4.0. J. Manuf. Syst. **56**, 133–156 (2020). https://doi.org/10.1016/j.jmsy.2020.05.016
16. A. Mardani, A. Jusoh, K. MD Nor, Z. Khalifah, N. Zakwan, A. Valipour, Multiple criteria decision-making techniques and their applications – A review of the literature from 2000 to 2014. Econ. Res. Istraživanja **28**(1), 516–571 (2015). https://doi.org/10.1080/1331677X.2015.1075139
17. M. Cinteza, Maedica-a journal of clinical medicine editorial point of ViEw editorial point of ViEw industry 4.0 and medicine. Maedica A J. Clin. Med. **16**(2) (2021). https://doi.org/10.26574/maedica.2021.16.2.161

18. J.L. Sampietro-Saquicela, Transformación Digital de la Industria 4.0 Digital Transformation of Industry 4.0 Transformação digital da indústria 4.0. Polo del Conocimiento, **5**, 1344–1356 (2020). https://doi.org/10.23857/pc.v5i8.1666
19. M. Sachon, Cuando personas y máquinas trabajan juntos: Los pilares de la industria 4.0. IEEM Rev. Negocios **21**(2), 46–54 (2018)
20. G. Erboz, How to define industry 4.0: Main pillars of industry 4.0, in *Managerial Trends in the Development of Enterprises in Globalization era*, ed. by I. Košičiarová, Kádeková, Zdenka, (Nitra, Slovak: Slovak University of Agriculture in Nitra, 2017), pp. 761–777
21. S. Vaidya, P. Ambad, S. Bhosle, Industry 4.0 – A glimpse. Procedia Manuf. **20**, 233–238 (2018). https://doi.org/10.1016/j.promfg.2018.02.034
22. L. Joyanes Aguilar, *Industria 4.0 La cuarta revolución industrial*, 1st edn. (Alfaomega, Mexico City, 2017)
23. A. Landa, Los 9 pilares de la industria 4.0, *Industria 4.0*. Linkedin, Jun. 02, 2021. Accessed 05 July 2022. [Online]. Available: https://www.linkedin.com/pulse/los-9-pilares-de-la-industria-40-aitor-landa/?originalSubdomain=es
24. D.-N. Le, C. Le, J. Tromp, N. Nhu, *Emerging Technologies for Health and Medicine*, (United States: Scrivener Publishing, 2018). https://doi.org/10.1002/9781119509875
25. C. Thuemmler, C. Bai, Health 4.0: Application of industry 4.0 design principles in future asthma management, in *Health 4.0: How Virtualization and Big Data Are Revolutionizing Healthcare*, (Springer, Cham, 2017), pp. 23–37
26. V.V. Popov, E.V. Kudryavtseva, N. Kumar Katiyar, A. Shishkin, S.I. Stepanov, S. Goel, Industry 4.0 and digitalisation in healthcare. Materials (Basel). **15**(6), 2140 (2022). https://doi.org/10.3390/ma15062140
27. G. Aceto, V. Persico, A. Pescapé, Industry 4.0 and health: Internet of Things, big data, and cloud computing for healthcare 4.0. J. Ind. Inf. Integr. **18**, 100129 (2020). https://doi.org/10.1016/j.jii.2020.100129
28. N. Mohamed, J. Al-Jaroodi, The impact of industry 4.0 on healthcare system engineering, in *2019 IEEE International Systems Conference (SysCon)*, April 2019, pp. 1–7. https://doi.org/10.1109/SYSCON.2019.8836715
29. M.M. Ahsan, Z. Siddique, Industry 4.0 in healthcare: A systematic review. Int. J. Inf. Manag. Data Insights **2**(1) (2022). https://doi.org/10.1016/j.jjimei.2022.100079
30. L. Lhotska, Application of industry 4.0 concept to health care. Stud. Health Technol. Inform. **273**, 23–37 (2020). https://doi.org/10.3233/SHTI200613
31. S. Paul et al., Industry 4.0 applications for medical/healthcare services. J. Sens. Actuator Netw. **10**(3). 2021. MDPI AG. https://doi.org/10.3390/jsan10030043
32. A. Tedesco, D. Dallet, P. Arpaia, Augmented Reality (AR) and Brain-Computer Interface (BCI): Two enabling technologies for empowering the fruition of sensor data in the 4.0 era, in *AISEM Annual Conference on Sensors and Microsystems* (Springer Link, 2021), pp. 85–91
33. A. Biasiucci, B. Franceschiello, M.M. Murray, Electroencephalography. Curr. Biol. **29**(3) (2019). https://doi.org/10.1016/j.cub.2018.11.052
34. G. Coro, G. Masetti, P. Bonhoeffer, M. Betcher, Distinguishing violinists and pianists based on their brain signals, in *Lecture Notes in Computer Science (Including Subseries Lecture Notes in Artificial Intelligence and Lecture Notes in Bioinformatics)*, LNCS, vol. 11727, (Springer Cham, Edinburgh 2019), pp. 123–137. https://doi.org/10.1007/978-3-030-30487-4_11
35. J.S. Kumar, P. Bhuvaneswari, Analysis of electroencephalography (EEG) signals and its categorization–a study. Procedia Eng. **38** (2012). https://doi.org/10.1016/j.proeng.2012.06.298
36. A. Craik, Y. He, J.L. Contreras-Vidal, Deep learning for electroencephalogram (EEG) classification tasks: A review. J. Neural Eng. **16**(3), 031001 (2019). https://doi.org/10.1088/1741-2552/ab0ab5
37. M. de Bardeci, C.T. Ip, S. Olbrich, Deep learning applied to electroencephalogram data in mental disorders: A systematic review. Biol. Psychol. **162**, 108117 (2021). https://doi.org/10.1016/j.biopsycho.2021.108117
38. S. Siuly, Y. Li, Y. Zhang, Electroencephalogram (EEG) and its background, in *EEG Signal Analysis and Classification*, (Springer, Cham, 2016), pp. 3–21

39. A. Ioanid, C. Scarlat, Neuromarketing tools in industry 4.0 context: A study on the Romanian market, in *The 15th International Conference Interdisciplinarity in Engineering*, 2022, pp. 370–381
40. J. Villalba-Diez, X. Zheng, D. Schmidt, M. Molina, Characterization of industry 4.0 lean management problem-solving behavioral patterns using EEG sensors and deep learning. Sensors **19**(13), 2841 (2019). https://doi.org/10.3390/s19132841
41. K.D. Tzimourta et al., Analysis of electroencephalographic signals complexity regarding Alzheimer's disease. Comput. Electr. Eng. **76**, 198–212 (2019). https://doi.org/10.1016/j.compeleceng.2019.03.018
42. L.M. Peñuela Calderón, N.E. Caicedo Gutierrez, Detección de dolor apartir de señales de EEG. Rev. EIA **19**(38) (2022). https://doi.org/10.24050/reia.v19i38.1577
43. T. Schack, K. Essig, C. Frank, D. Koester, Mental representation and motor imagery training. Front. Hum. Neurosci. **8** (2014). https://doi.org/10.3389/fnhum.2014.00328
44. V. Nicholson, N. Watts, Y. Chani, J.W.L. Keogh, Motor imagery training improves balance and mobility outcomes in older adults: A systematic review. J. Physiother. **65**(4), 200–207 (2019). https://doi.org/10.1016/j.jphys.2019.08.007
45. C. Ruffino, J. Gaveau, C. Papaxanthis, F. Lebon, An acute session of motor imagery training induces use-dependent plasticity. Sci. Rep. **9**(1), 20002 (2019). https://doi.org/10.1038/s41598-019-56628-z
46. R. Dickstein, J.E. Deutsch, Motor imagery in physical therapist practice. Phys. Ther. **87**(7), 942–953 (2007). https://doi.org/10.2522/ptj.20060331
47. V.P. Nicholson, J.W.L. Keogh, N.L.L. Choy, Can a single session of motor imagery promote motor learning of locomotion in older adults? A randomized controlled trial. Clin. Interv. Aging **13** (2018). https://doi.org/10.2147/CIA.S164401
48. S. Kumar, A. Sharma, K. Mamun, T. Tsunoda, A deep learning approach for motor imagery EEG signal classification, in *2016 3rd Asia-Pacific World Congress on Computer Science and Engineering (APWC on CSE)*, 2016. https://doi.org/10.1109/APWC-on-CSE.2016.017
49. S.R. Sreeja, J. Rabha, K.Y. Nagarjuna, D. Samanta, P. Mitra, M. Sarma, Motor imagery EEG signal processing and classification using machine learning approach, in *2017 International Conference on New Trends in Computing Sciences (ICTCS)*, 2017. https://doi.org/10.1109/ICTCS.2017.15
50. M. Erdogan, B. Ozkan, A. Karasan, I. Kaya, Selecting the best strategy for industry 4.0 applications with a case study, in *Industrial Engineering in the Industry 4.0 Era*, (Springer, Cham, 2018), pp. 109–119
51. V. Terziyan, S. Gryshko, M. Golovianko, Patented intelligence: Cloning human decision models for industry 4.0. J. Manuf. Syst. **48**, 204–217 (2018). https://doi.org/10.1016/j.jmsy.2018.04.019
52. J. Mayor, S. Botero, J.D. González-Ruiz, Modelo de decisión multicriterio difuso para la selección de contratistas en proyectos de infraestructura: caso Colombia. Obras y Proy. **20**, 56–74 (2016). Accessed 13 Mar 2022. [Online]. Available: https://scielo.conicyt.cl/pdf/oyp/n20/art05.pdf
53. A. Jahan, K.L. Edwards, M. Bahraminasab, 4 – multi-criteria decision-making for materials selection, in *Multi-Criteria Decision Analysis for Supporting the Selection of Engineering Materials in Product Design*, ed. by A. Jahan, K.L. Edwards, M. Bahraminasab, 2nd edn., (Butterworth-Heinemann, Oxford, 2016), pp. 63–80
54. A. Kolios, V. Mytilinou, E. Lozano-Minguez, K. Salonitis, A comparative study of multiple-criteria decision-making methods under stochastic inputs. Energies **9**(7) (2016). https://doi.org/10.3390/en9070566
55. H. Karunathilake, E. Bakhtavar, G. Chhipi-Shrestha, H.R. Mian, K. Hewage, R. Sadiq, Chapter seven – decision making for risk management: A multi-criteria perspective, in *Methods in Chemical Process Safety*, ed. by F.I. Khan, P.R. Amyotte, vol. 4, (Elsevier, Cambridge, 2020), pp. 239–287
56. B. Ceballos, M.T. Lamata, D. Pelta, J.M. Sanchez, EL MÉTODO TOPSIS RELATIVO VS. ABSOLUTO. Rev. Electrónica Comun. y Trab. ASEPUMA **14**, 181–192 (2013)

57. E. Triantaphyllou, Multi-criteria decision making methods, in *Multi-Criteria Decision Making Methods: A Comparative Study*, ed. by E. Triantaphyllou, (Springer, Boston, 2000), pp. 5–21
58. W. Sałabun, A. Piegat, Comparative analysis of MCDM methods for the assessment of mortality in patients with acute coronary syndrome. Artif. Intell. Rev. **48**(4), 557–571 (2017). https://doi.org/10.1007/s10462-016-9511-9
59. J. Damidavičius, M. Burinskiene, J. Antuchevičiene, Assessing sustainable mobility measures applying multicriteria decision making methods. Sustainability **12**(15) (2020). https://doi.org/10.3390/su12156067
60. I. Siksnelyte, E. Zavadskas, D. Streimikiene, D. Sharma, An overview of multi-criteria decision-making methods in dealing with sustainable energy development issues. Energies **11**(10), 2754 (2018). https://doi.org/10.3390/en11102754
61. E.A. Adalı, A. Tuş, Hospital site selection with distance-based multi-criteria decision-making methods. Int. J. Healthc. Manag. **14**(2), 534–544 (2021). https://doi.org/10.1080/20479700.2019.1674005
62. X. Zhongyou, Study on the application of TOPSIS method to the introduction of foreign players in CBA games. Phys. Procedia **33**, 2034–2039 (2012). https://doi.org/10.1016/j.phpro.2012.05.320
63. V. Yadav, S. Karmakar, P.P. Kalbar, A.K. Dikshit, PyTOPS: A Python based tool for TOPSIS. SoftwareX **9** (2019). https://doi.org/10.1016/j.softx.2019.02.004
64. M. Miguel Ángel Quiroz Martínez, S. Ginnette Andreina Granda Villon, S. Davis Israel Maldonado Cevallos, M. Yelandi Leyva Vázquez, Comparative analysis to select an emotion recognition tool applying fuzzy decision maps and TOPSIS. Dilemas Contemp. Educ. Política y Valores **6**(6), 453–463 (2020)
65. L. Pérez-Domínguez, J. Luis Macías-García, K. Yohana Sánchez-Mojica, D. Luviano-Cruz, Comparación Método multi-criterio TOPSIS y MOORA para la optimización de un proceso de inyección de plástico. Mundo Fesc (14), 98–105 (2017). [Online]. Available: https://dialnet.unirioja.es/servlet/articulo?codigo=6559177
66. X. Zhou, Y. Hu, W. Liang, J. Ma, Q. Jin, Variational LSTM enhanced anomaly detection for industrial big data. IEEE Trans. Industr. Inform. **17**(5), 3469–3477 (2021). https://doi.org/10.1109/TII.2020.3022432
67. I.H. Sarker, Machine learning: Algorithms, real-world applications and research directions. SN Comput. Sci. **2**(3), 160 (2021). https://doi.org/10.1007/s42979-021-00592-x
68. B. Bajic, I. Cosic, M. Lazarevic, N. Sremčev, A. Rikalovic, Machine learning techniques for smart manufacturing: Applications and challenges in industry 4.0, in *9th International Scientific and Expert Conference TEAM 2018*, (2018), pp. 29–38. Available: https://www.researchgate.net/publication/328290180_Machine_Learning_Techniques_for_Smart_Manufacturing_Applications_and_Challenges_in_Industry_40
69. A.A.A. Elhag, High-dimensional learning, in *Towards Data Science* (2022). https://towardsdatascience.com/high-dimensional-learning-ea6131785802. Accessed 10 July 2022
70. S. Theodoridis, K. Koutroumbas, Chapter 1 – introduction, in *Pattern Recognition*, ed. by S. Theodoridis, K. Koutroumbas, 4th edn., (Academic, Boston, 2009), pp. 1–12
71. V. Rodríguez-Diez, J.F. Martínez-Trinidad, J.A. Carrasco-Ochoa, M. Lazo-Cortés, C. Feregrino-Uribe, R. Cumplido, A fast hardware software platform for computing irreducible testors. Expert Syst. Appl. **42**(24), 9612–9619 (2015). https://doi.org/10.1016/j.eswa.2015.07.037
72. B. Venkatesh, J. Anuradha, A review of feature selection and its methods. Cybern. Inf. Technol. **19**(1), 3–26 (2019). https://doi.org/10.2478/CAIT-2019-0001
73. N. Martínez, M. León, Z. García, Features selection through FS-testors in case-based systems of teaching-learning, in *MICAI 2007: Advances in Artificial Intelligence*, 2007, pp. 1206–1217
74. I. Mejía-Guevara, A. Kuri, *Evolutionary Feature and Parameter Selection in Support Vector Regression*, Lecture Notes in Artificial Intelligence (LNAI), vol. 4827 (Springer, Heidelberg, Berlin 2007)

75. J.F. Martínez-Trinidad, A. Guzman-Arenas, The logical combinatorial approach to pattern recognition, an overview through selected works. Pattern Recogn. **34**, 741–751 (2001). https://doi.org/10.1016/S0031-3203(00)00027-3
76. M. Lazo-Cortes, J. Ruiz-Shulcloper, E. Alba-Cabrera, An overview of the evolution of the concept of testor. Pattern Recogn. **34**(4), 753–762 (2001). https://doi.org/10.1016/S0031-3203(00)00028-5
77. V. Rodríguez-Diez, J.F. Martínez-Trinidad, J.A. Carrasco-Ochoa, M.S. Lazo-Cortés, The impact of basic matrix dimension on the performance of algorithms for computing typical Testors, in *10th Mexican Conference, MCPR 2018*, (México: Springer, 2018), pp. 41–50. https://doi.org/10.1007/978-3-319-92198-3_5
78. A. Lias-Rodríguez, A. Pons-Porrata, *Un nuevo Algoritmo de Escala Exterior para el Cálculo de los Testores Típicos* (Centro de Estudios de Reconocimiento de Patrones y Minería de Datos, Santiago de Cuba, 2005), p. 10. [Online]. Available: http://www.cerpamid.co.cu/sitio/files/publicaciones/1034921953BR_RECPAT09.pdf
79. J.P. Gómez, F.E.H. Montero, J.C. Sotelo, J.C.G. Mancilla, Y.V. Rey, RoPM: An algorithm for computing typical testors based on recursive reductions of the basic matrix. IEEE Access **9**, 128220–128232 (2021). https://doi.org/10.1109/ACCESS.2021.3112385
80. A.E. Gallegos Acosta, F.J. Álvarez Rodríguez, M.D. Torres Soto, A. Torres Soto, *Identificación de factores de riesgo en patologías médicas mediante métodos de selección de subconjuntos de características [recurso electrónico]* (Universidad Autónoma de Aguascalientes, Aguascalientes, 2018)
81. R.A. Vázquez, S. Godoy-Calderón, Using testor theory to reduce the dimension of neural network models. Res. Comput. Sci. **28**, 93–103 (2007)
82. G.R. Yang, X.-J. Wang, Artificial neural networks for neuroscientists: A primer. Neuron **107**(6), 1048–1070 (2020). https://doi.org/10.1016/j.neuron.2020.09.005
83. O.I. Abiodun, A. Jantan, A.E. Omolara, K.V. Dada, N.A. Mohamed, H. Arshad, State-of-the-art in artificial neural network applications: A survey. Heliyon **4**(11), e00938 (2018). https://doi.org/10.1016/j.heliyon.2018.e00938
84. J. Ribeiro, R. Lima, T. Eckhardt, S. Paiva, Robotic process automation and artificial intelligence in industry 4.0 – A literature review. Procedia Comput. Sci. **181**, 51–58 (2021). https://doi.org/10.1016/j.procs.2021.01.104
85. F. Rozo-García, Revisión de las tecnologías presentes en la industria 4.0. Rev. UIS Ing. **19**(2), 177–191 (2020). https://doi.org/10.18273/revuin.v19n2-2020019
86. C.A. Guaillazaca González, A. Valeria Hernandez, Clasificador de Productos Agrícolas para Control de Calidad basado en Machine Learning e Industria 4.0. Rev. Perspect. **2**(2), 21–28 (2020). https://doi.org/10.47187/perspectivas.vol2iss2.pp21-28.2020
87. J.M. Peña Lorenzo, *Aplicación de técnicas de aprendizaje profundo (deep learning) para la detección de objetos en industria 4.0* (Universidad de Valladolid, Valladolid, 2020)
88. N.V. Ramírez Pérez, M. Laguna Estrada, N.N. Rubín Ramírez, UN ACERCAMIENTO A LA INDUSTRIA 4.0 A TRAVÉS DE REDES NEURONALES PARA LA REDUCCIÓN DE SCRAP EN UNA EMPRESA AUTOMOTR´iZ. Pist. Educ. **41**(133), 551–568 (2019). Accessed 11 July 2022. [Online]. Available: http://www.itc.mx/ojs/index.php/pistas/article/view/2144
89. M. Medina Carmona, J.V. Sánchez Andrés, *La Electricidad del Cerebro: Los Secretos de la Actividad Cerebral*, 1st edn. (National Geographic, México, 2017)
90. W. Rivas-Asanza, B. Mazon-Olivo, F. Mejia, Capítulo 1: Generalidades de las redes neuronales artificiales, in *Redes Neuronales Artificiales Aplicadas al Reconocimiento de Patrones*, (Universidad Técnica de Machala, Washington, DC, 2018), pp. 11–35
91. O.I. Abiodun et al., Comprehensive review of artificial neural network applications to pattern recognition. IEEE Access **7**, 158820–158846 (2019). https://doi.org/10.1109/ACCESS.2019.2945545
92. I.A. Basheer, M. Hajmeer, Artificial neural networks: Fundamentals, computing, design, and application. J. Microbiol. Methods **43**(1), 3–31 (2000). https://doi.org/10.1016/S0167-7012(00)00201-3

93. Q. Zhang, H. Yu, M. Barbiero, B. Wang, M. Gu, Artificial neural networks enabled by nanophotonics. Light Sci. Appl. **8**(1), 42 (2019). https://doi.org/10.1038/s41377-019-0151-0
94. F. Amato, A. López, E.M. Peña-Méndez, P. Vaňhara, A. Hampl, J. Havel, Artificial neural networks in medical diagnosis. J. Appl. Biomed. **11**(2) (2013). https://doi.org/10.2478/v10136-012-0031-x
95. S. Parra, M. Torrens, *La Inteligencia Artificial: el camino hacia la ultrainteligencia* (RBA Coleccionables, Barcelona, 2017)
96. R. Sánchez-Reolid, A. García, M. Vicente-Querol, L. Fernández-Aguilar, M. López, A. González, Artificial neural networks to assess emotional states from brain-computer interface. Electronics **7**(12), 384 (2018). https://doi.org/10.3390/electronics7120384
97. A.E. Gallegos Acosta, M.D. Torres Soto, A. Torres Soto, E.E. Ponce de León Sentí, Contrastación de algoritmos de aprendizaje automático para la clasificación de señales EEG. Res. Comput. Sci. **148**(8), 515–525 2020. Accessed 15 Apr 2021. [Online]. Available: https://www.rcs.cic.ipn.mx/2020_149_8/Contrastacion de algoritmos de aprendizaje automatico para la clasificacion de senales EEG.pdf

Index

A
Agent-based simulation, 215, 216, 221–223
Algorithms, 5, 22, 45, 72, 103, 179, 229, 244, 276, 288, 308
Analytic hierarchy process (AHP), 127, 129, 130, 134, 138, 141, 245, 247, 314
Architecture, 24, 44, 48, 52–53, 59–61, 67, 77, 94, 124, 177, 181, 204, 211, 275–285, 320
Artificial intelligence (AI), vii, 7, 10, 11, 13, 21–23, 26, 45, 53, 71, 72, 102, 103, 118, 124–125, 180–181, 190, 203, 216, 223, 238, 245, 309–311, 318
Artificial neural network (ANN), 8–10, 23, 47, 181, 253, 269, 308, 315, 319–321, 323, 329
Assistive devices, 71–95
Augmented reality (AR), vii, 5, 21–24, 29, 203, 309–311
Automation, 4–9, 14, 73, 124, 125, 148, 150, 157–160, 163, 164, 171, 189, 203, 309
Azure Cloud, 27–29, 31, 35

B
Battery management system (BMS), 175–192
Biometrics, 244–246, 249, 250, 256, 258, 262, 264, 265
Blockchain, 14, 275–285, 311

C
Computer vision, 4, 7, 8, 10–11, 15, 23, 24, 26–33, 35, 36, 38, 48, 71–95
Convolutional, 8, 24, 44, 48, 52, 76, 77, 118, 179

D
Data analytics for applications in industry 4.0, 102
Deep learning, 8–11, 23, 44, 45, 48, 59, 67, 76, 77, 92, 95, 104, 118, 181, 253, 254, 269, 290, 311, 320
Deep neural networks, 8, 10, 11, 44, 58, 117, 245, 249
Density-Based Spatial Clustering of Applications with Noise (DBSCAN), 103, 104, 106, 110–113, 116
Digitalization, 176, 178, 191, 244, 307
Distribution route optimization, 287–303

E
Edge-computing, 3–16, 73
EEG signals, 307, 308, 311, 312, 322, 324, 326, 329, 330
Electrochemical impedance spectroscopy, 175–192
Entropy, 11, 50, 127, 129, 134, 138, 141
Erbium-doped fiber amplifier (EDFA), 200–202, 205–211
Experimental design, 146, 147, 162

F
Failure detection, 21–40, 61, 105, 106, 111, 113, 116, 118
Fourth industrial revolution, 4, 123, 124, 145, 309

L. C. Méndez-González et al. (eds.), *Innovation and Competitiveness in Industry 4.0 Based on Intelligent Systems*, EAI/Springer Innovations in Communication and Computing, https://doi.org/10.1007/978-3-031-29775-5

G

Genetic algorithm (GA), 13, 177, 180–184, 186, 189–192, 289
Geo-simulation, 215–239

H

HealthCare, 57, 244, 278, 307, 310, 311, 330
HoloLens, 22, 24, 26–30, 34–36, 38–40

I

Industrial Internet of things (IIoT), 4–7, 14, 125, 275–285
Industry 4.0 (I4.0), vii, viii, 3–16, 21, 26, 39, 43–68, 71, 72, 102, 115, 118, 119, 123–141, 145–171, 175–192, 197–211, 243–270, 277, 287–303, 307–311, 313, 317, 320, 330
Initial states, 230, 275–285
Injection moulding, 15, 101–119
Intelligent traffic systems, 230–232
Internet of things (IoT), vii, 4, 5, 10, 11, 13, 14, 21, 24, 39, 72, 73, 102, 123, 125, 176–178, 180, 189–191, 202, 203, 231, 238, 275, 277–278, 287–292, 302, 309–311, 313, 318

M

Machine learning, vii, 3–16, 21, 23, 45–52, 72, 92, 101–119, 176, 249, 250, 318–320
Manufacturing processes, 4, 5, 7, 10, 14, 22–24, 26, 101, 102, 104, 105, 109, 113, 116, 118, 309
Melanoma, 43–68
Mixed reality, vii, 21–40, 61
Motor imagery (MI), 307, 311, 312, 322, 324, 327
Multicriteria decision making (MCDM), 124, 127, 313–315

N

Neural networks, vii, 8, 9, 13, 14, 23, 24, 44, 46–48, 50, 52, 58, 59, 72, 74, 76, 77, 95, 117, 179, 245, 249, 253, 269, 290, 308, 321, 323, 324, 326, 327

O

Opensource, 137
Optical Amplifier, 197–211
Optimization, 39, 60, 77, 93, 94, 145–171, 179–181, 183, 203, 216, 231, 245, 252–254, 287–303, 320

P

Parameters optimisation, 105, 117, 119, 146

R

Raman amplifier (RA), 200–201, 205, 207–211

S

SHERPA, 123–141
Small-and medium-sized enterprises (SMEs), 123–141
SOA, 202, 205–211
Speech-to-text, 83, 88
Steepest ascent, 145–171
Support vector machine (SVM), 68, 72, 74–78, 80–83, 86, 87, 90–95

T

Technique for Order Preference by Similarity to Ideal
Solution (TOPSIS), 245, 256, 257, 308, 314–317, 321, 323, 327, 329, 330
Text-to-speech, 83, 88
Traffic accidents, 216, 218, 220–221, 226, 227, 231, 237
Traffic congestion, 215–220, 230, 231
Traffic planning, 218
Typical testors theory, 308, 315, 319, 321, 323, 325–327, 329, 330

U

Unsupervised learning, 46, 103–104, 106, 117–119

V

Value added network, 7
Virtual reality (VR), 5, 21, 22, 26
Vision, 4, 7, 8, 10–11, 15, 22–24, 26–33, 35, 36, 38, 48, 71–95, 229, 320
Visually impaired, 71–95

W

WDM, 197–211

Y

YOLO algorithm, 39, 78, 94

MIX
Papier aus verantwortungsvollen Quellen
Paper from responsible sources
FSC® C105338

If you have any concerns about our products,
you can contact us on
ProductSafety@springernature.com

In case Publisher is established outside the EU,
the EU authorized representative is:
Springer Nature Customer Service Center GmbH
Europaplatz 3, 69115 Heidelberg, Germany

Printed by Libri Plureos GmbH
in Hamburg, Germany